Ciliopathies

Ciliopathies
A reference for clinicians

Edited by

Thomas D. Kenny

National Institute for Health Research Evaluations, Trials and Studies Coordinating Centre, University of Southampton, UK

Philip L. Beales

Professor of Medical and Molecular Genetics, UCL Institute of Child Health, London, UK

OXFORD
UNIVERSITY PRESS

OXFORD
UNIVERSITY PRESS

Great Clarendon Street, Oxford, OX2 6DP,
United Kingdom

Oxford University Press is a department of the University of Oxford.
It furthers the University's objective of excellence in research, scholarship, and education by publishing worldwide. Oxford is a registered trade mark of Oxford University Press in the UK and in certain other countries

First Edition published in 2014
Impression: 1

Published in the United States of America by Oxford University Press
198 Madison Avenue, New York, NY 10016, United States of America

British Library Cataloguing in Publication Data
Data available

Library of Congress Control Number: 2013937929

ISBN 978–0–19–965876–3

Printed and bound by
CPI Group (UK) Ltd, Croydon, CR0 4YY

Preface

Thomas D. Kenny

Until recently the diseases described in this book would have been considered nothing more than a rag-bag of mostly rare, often serious, invariably complex clinical conditions. Each was long recognised as a discrete clinical entity in its own right, but together they have overlapping pathological elements. It is apparent that, far from the beliefs of the past, cilia are much more than vestigial cellular organelles, but we are some way short of comprehending the full extent of the implications of this.

It was my privilege to be the medical adviser responsible for three NHS services that managed these conditions (Alström syndrome, primary ciliary dyskinesia and Bardet–Biedl syndrome). These conditions had a profound affect on me. They are rare, complex and serious. They have no cure, and no specific treatment, but far from being hopeless. Well-organised multidisciplinary clinical services can produce benefits for the patients they see (Kenny et al. 2008) and advanced the understanding of the conditions.

As I gained understanding from the peculiarities of one of these conditions (Paisey et al. 2011) and applied that understanding to the organisation of the services for other conditions I began to have a respect for the underlying biological processes that cause the overlapping phenotypical elements of these diseases and the wider group of other ciliopathies.

Due to their rarity, outside of specific services, many patients with ciliopathies get a 'raw deal'. Many clinicians will spend their whole careers without caring for a patient with one of these diseases, I never saw one while I was in general practice and thinking back I do not know if I would have recognised one if I had. The patients and the support groups that represent them have repeatedly expressed a desire for a reference text that they could point their primary and secondary care clinicians to. They complain that their diagnoses seemed to take 'forever' and that other than the hand-full of clinicians for each condition no one seems to have heard of them or do the simple things they need to help them stay as healthy as they can.

This book aims to be just such a text. Firstly, I hope it will help anyone, prompted to open and read it, to understand and care for a patient with a specific ciliopathy. Secondly, if it could also spark an interest in the underlying biological processes, and maybe guide the identification of a patient with a ciliopathy who would otherwise be missed, that would be excellent.

And thirdly, given ubiquitous nature of cilia in our cells, there is a likelihood that there are ciliopathies out there that have not yet been recognised as such. And maybe, just maybe, this will fall into the hands of a bright inquisitive clinician and it will be the catalyst to the recognition of a disease as a new ciliopathy.

References

Kenny, T. D., Jessop, E. G. & Gutteridge, W. H. 2008. Monitoring clinical quality in rare disease services—experience in England. *Orphanet J Rare Dis* **3**, 23.

Paisey, R., Barrett, T., Carey, C., Hiwot, T., Cramb, R., Bower, L., et al. 2011. Rare disorders presenting in the diabetic clinic: an example using audit of the NSCT adult Alström clinics. *Practical Diabetes* **28**, 340–343.

Contents

List of abbreviations *ix*

Gene list *xi*

List of contributors *xvii*

1 Towards the diagnosis of a ciliopathy *1*
Philip L. Beales and Thomas D. Kenny

2 Alström syndrome *8*
Richard Paisey

3 Jeune syndrome and the ciliary chondrodysplasias *30*
Miriam Schmidts

4 Joubert syndrome and Joubert syndrome-related disorders *64*
Victoria Harrison and Andrea H. Németh

5 Bardet–Biedl syndrome *92*
Philip L. Beales and Elizabeth Forsythe

6 Leber congenital amaurosis and other non-syndromic retinal ciliopathies *108*
Thomas D. Kenny, Philip L. Beales and Ronald Roepman

7 Meckel–Gruber syndrome *132*
Gabrielle Wheway and Colin A. Johnson

8 Nephronophthisis *150*
Shalabh Srivastava and John A. Sayer

9 Oral–facial–digital type I syndrome *162*
Brunella Franco

10 Autosomal dominant polycystic kidney disease *174*
Richard Sandford

11 Autosomal recessive polycystic kidney disease *194*
Carsten Bergmann

12 Primary ciliary dyskinesia *218*
Claire Hogg

13 Usher syndrome *238*
Maria Bitner-Glindzicz and Zubin Saihan

14 Syndromes not yet proven to be ciliopathies *262*
Kate Baker

Index *275*

List of abbreviations

AAV	adeno-associated virus
ACE	angiotensin converting enzyme
ACLS	acrocallosal syndrome
ACS	acrocallosal syndrome
ADPKD	autosomal dominant polycystic kidney disease
AFP	α-fetoprotein
ALT	alanine aminotransferase
ARPKD	autosomal recessive polycystic kidney disease
AST	aspartate transaminase
ATD	asphyxating thoracic dystrophy
BOR	branchio-oto-renal (syndrome)
BUN	blood urea nitrogen
CBF	ciliary beat frequency
CBS	cystathionine beta synthase
CD	cone dystrophy
CED	cranial–ectodermal dysplasia
CEP	centrosomal protein
CF	cystic fibrosis
CFTR	cystic fibrosis transmembrane conductance regulator
CGH	comparative genomic hybridisation
CHF	congenital hepatic fibrosis
CKD	chronic kidney disease
CLKT	combined liver and kidney transplantation
CNS	central nervous system
COACH	**c**erebellar vermis hypo- / aplasia, **o**ligophrenia, congenital **a**taxia, ocular **c**oloboma, and **h**epatic fibrosis (syndrome)
CORS	**c**erebellar–**o**cular–**r**enal (syndrome)
CPAP	continuous positive airway pressure
CRD	cone–rod dystrophy
CRISP	Consortium for Radiologic Imaging Study of PKD
CSF	cerebrospinal fluid
CT	computed tomography
DKA	Dekaban–Arima (syndrome)
DPM	ductal plate malformation
DRC	dynein regulatory complex
ECG	electrocardiogram
EGF	epidermal growth factor
eGFR	estimated GFR
EM	electron microscopy
ERG	electro-retinogram
ESRF	end-stage renal failure
EVC	Ellis–van Creveld (syndrome)
GCKD	glomerulocystic kidney disease
GCPS	Grieg cephalopolysyndactyly syndrome
GFR	glomerular filtration rate
GGT	gamma glutamyl transpeptidase (or γ-glutamyl transpeptidase)
GI	gastrointestinal
HALT	Halt Progression of Polycystic Kidney Disease (a study group)
HDL	high-density lipoprotein
Hh	hedgehog
HLS	hydrolethalus syndrome
ICA	intracerebral aneurysm, or intracranial aneurysm
IFT	intraflagellar transport
IPT	immunoglobulin-like, plexin, transcription factor
JATD	Jeune asphyxating thoracic dystrophy
JBTS	Joubert–Bolthauser syndrome
JS	Joubert syndrome
JSRD	Joubert syndrome-related disorders
LCA	Leber congenital amaurosis
MET	mechanotransduction
MKS	Meckel–Gruber syndrome
MOPD I	microcephalic osteodysplastic primordial dwarfism type I
MORM	**m**ental retardation, **o**besity, congenital **r**etinal dystrophy and **m**icropenis in males (syndrome)
MRI	magnetic resonance imaging
MSS	Mainzer–Saldino syndrome

mTOR	mammalian target of rapamycin
MTS	molar tooth sign
OFD	oral–facial–digital
OMA	oculomotor apraxia
OMIM	Online Mendelian Inheritance in Man database (www.OMIM.org)
ORF	open reading frame
NAD	nicotinamide adenine dinucleotide
NGS	Next Generation sequencing
NPHP	nephronophthisis
NRDT	Danish National Registry on Regular Dialysis and Transplantation
PbH1	parallel beta-helix 1
PCP	planar cell polarity
PHS	Pallister–Hall syndrome
PKD	polycystic kidney disease
PLD	polycystic liver disease
PML	premyelocytic leukaemia (gene product)
RAAS	renin–angiotensin–aldosterone system
RID	RPGR interacting domain
RPE	retinal pigment epithelium
RPGR	RP GTPase regulator
RPGRIP1	RP GTPase regulator interacting protein 1
RRT	renal replacement therapy
Shh	sonic hedgehog
SLS	Senior–Løken syndrome
SLO	Smith–Lemli–Opitz syndrome
TCTN	tectonic
TEM	transmission electron microscopy
TIPS	trans-jugular intrahepatic stent
TKV	total kidney volume
TMEM	transmembrane
TRP	transient receptor potential
TSC	tuberous sclerosis
TSH	thyroid stimulating hormone
VHL	von Hippel–Lindau (disease)
VLDL	very low density lipoprotein
V_2R	vasopressin V_2 receptor

Gene list

Some causative genes in ciliopathies and ciliopathy-related disorders. Other genes are mentioned elsewhere in the book.

Gene ID	Syndrome	Type of protein encoded	Function in cilia biology
AHI1	Joubert syndrome	Jouberin contains 7 WD repeats	Component of the tectonic-like complex. Localises to transition zone of cilium
AIPL1	Leber congenital amaurosis	Contains 3 TPR repeats	May be important in protein trafficking and/or protein folding and stabilization
ARL6 (*BBS3*)	Bardet–Biedl syndrome	Ras superfamily, small GTPase	Localises to ciliated cells in *Caenorhabditis elegans* and undergoes IFT; localises to basal body and ciliary gate in mammalian cell lines; component of BBSome
ARL13B	Joubert syndrome (classical form)	Ras superfamily, small GTPase	Localises to ciliary axoneme in mouse and human tissues
ATXN10	Nephronophthisis		Interacts with NPHP5
BBS4	Bardet–Biedl syndrome	Contains 10 TPR repeats	Targets cargo to pericentriolar material; component of BBSome
BBS5	Bardet–Biedl syndrome	Uncharacterised	Localises to ciliated cells in *C. elegans* and mammalian tissues; knockdown causes loss of flagella in *Chlamydomonas*; component of BBSome
BBS7	Bardet–Biedl syndrome	Six-bladed β-propeller structure; sequence similarity to BBS1 and BBS2	Component of BBSome
BBS10	Bardet–Biedl syndrome	Type II chaperonin	Probable molecular chaperone. As part of the BBS/CCT complex may play a role in the assembly of BBSome.
BBS12	Bardet–Biedl syndrome	Type II chaperonin	—
CCD2A	Joubert syndrome	Coiled-coil and C2 domain-containing	Tagged/over-expressed form localises to basal body in mammalian cell lines; cilia absent in patient fibroblasts

Gene ID	Syndrome	Type of protein encoded	Function in cilia biology
CCD2A	Joubert-spectrum disorders	Coiled-coil and C2 domain-containing	Tagged/over-expressed form localises to basal body in mammalian cell lines; cilia absent in patient fibroblasts
CCD2A	Mental retardation + retinitis pigementosa	Coiled-coil and C2 domain-containing	Tagged/over-expressed form localises to basal body in mammalian cell lines; cilia absent in patient fibroblasts
C2ORF71	Retinitis pigementosa	Uncharacterised; proline-rich domain	Tagged/over-expressed form localises pericentrosomal location and ciliary axoneme; downregulated in Bbs4–/– mouse retinas
CEP290	Joubert syndrome-related disorders	Uncharacterised	Tagged/over-expressed form localises to centrosomes and ciliary base in mammalian cell lines
CEP290	Leber congenital amaurosis	Uncharacterised	Tagged/over-expressed form localises to centrosomes and ciliary base in mammalian cell lines
CEP290	Meckel syndrome	Uncharacterised	Tagged/over-expressed form localises to centrosomes and ciliary base in mammalian cell lines
CRB1	Leber congenital amaurosis	Belongs to the Crumbs protein family	Plays a role in photoreceptor morphogenesis in the retina. May maintain cell polarization and adhesion.
CRX	Leber congenital amaurosis	Contains 1 homeobox DNA-binding domain	Essential for the maintenance of mammalian photoreceptors
DYNC2H1	Jeune syndrome	Dynein-heavy chain	*Dyn2hc1–/–* MEFs have abnormal cilia with swollen tip; abnormal cilia in patient chondrocytes
DYNC2H1	Short-rib polydactyly type III	Dynein-heavy chain	*Dyn2hc1–/–* MEFs have abnormal cilia with swollen tip; abnormal cilia in patient chondrocytes
DNAI1	Primary ciliary dyskinesia	intermediate chain dynein, belonging to the large family of motor proteins	Force-generating proteins responsible for the sliding movement in axonemes
DNAI2	Primary ciliary dyskinesia	Dynein intermediate chain family, and is part of the dynein complex of respiratory cilia and sperm flagella	Force-generating proteins responsible for the sliding movement in axonemes
DNAH5	Primary ciliary dyskinesia	An axonemal heavy chain dynein	force-generating protein with ATPase activity

Gene ID	Syndrome	Type of protein encoded	Function in cilia biology
DNAH11`	Primary ciliary dyskinesia	Member of the dynein heavy chain family	Force-generating proteins responsible for the sliding movement in axonemes
EVC	Ellis–van Creveld syndrome	Uncharacterised; contains leucine zipper, nuclear localisation signals, putative transmembrane domains	Endogeneous Evc localises to the basal body in mouse chondrocytes
EVC2	Ellis–van Creveld syndrome	Uncharacterised	—
EVC2	Weyer acrodental dysostosis	Uncharacterised	—
GLIS2	Nephronophthisis	Kruppel-like zinc finger transcription factor	Endogeneous Glis2 localises to the ciliary axoneme in MDCK cells
GUC2YD	Leber congenital amaurosis		
HYLS1	Hydrolethalus	Uncharacterised; putative transcription factor with nuclear localisation signal	HYSL1 is only found in organisms with centrioles, it localises to centrioles in worms and frogs, it is required for cilia formation in *C. elegans* ciliated neurons and *Xenopus laevis* mucociliary epithelium
IFT80	Jeune asphyxiating thoracic dystrophy	Intraflagellar transport protein	Endogeneous Ift80 localises to basal bodies and ciliary axoneme in ATDC5 cells; required for normal ciliogenesis in *Tetrahymena thermophila*
IFT122	Sensenbrenner syndrome	Intraflagellar transport protein	Reduced cilia length in patient fibroblasts and zebrafish Ift122 morphants
IFT43	Sensenbrenner syndrome	Intraflagellar transport protein	Ift88 and Ift57 accumulate at the ciliary tip of patient fibroblasts
INPP5E	MORM – mental retardation, truncal obesity, retinal dystrophy, micropenis	Inositol phosphatase	*Inpp5e–/–* mice have reduced number and shorter cilia in cystic renal epithelia, and *Inpp5e–/–* MEFs have fewer cilia when tyrosine kinase receptor signalling is activated; endogeneous Inpp5e localises to ciliary axonemes in MEFs
INPP5E	Joubert syndrome	Inositol phosphatase	*Inpp5e–/–* mice have reduced number and shorter cilia in cystic renal epithelia, and *Inpp5e–/–* MEFs have fewer cilia when tyrosine kinase receptor signalling is activated; endogeneous Inpp5e localises to ciliary axonemes in MEFs

Gene ID	Syndrome	Type of protein encoded	Function in cilia biology
INPP5E	Cerebello-oculo-renal syndrome	Inositol phosphatase	*Inpp5e–/–* mice have reduced number and shorter cilia in cystic renal epithelia, and Inpp5e–/– MEFs have fewer cilia when tyrosine kinase receptor signalling is activated; endogeneous Inpp5e localises to ciliary axonemes in MEFs
INVS	Nephronophthisis type II	contains multiple ankyrin domains and two IQ calmodulin-binding domains	Endogeneous Invs is present in a punctate pattern along the ciliary axoneme of MDCK cells
KIF7	Acrocallosal syndrome	Kinesin motor protein	Kif7 localises to the base of cilia in the absence of Shh, and to the tip in presence of Shh; cilia are longer in patient fibroblasts
KIF7	Hydrolethalus	Kinesin motor protein	Kif7 localises to the base of cilia in the absence of Shh, and to the tip in presence of Shh; cilia are longer in patient fibroblasts
KTU	Primary ciliary dyskinesia	highly conserved cytoplasmic protein	Required for cytoplasmic pre-assembly of axonemal dyneins
LCA5/ CORF152	Leber congenital amaurosis	Lebercilin; uncharacter ised; coiled-coil domain containing	Endogeneous lebercilin localises to the ciliary axoneme in RPE and IMCD3 cells
LRAT	Leber congenital amaurosis	Phosphatidylcholine—Retinol O-Acyltransferase	Transfers the acyl group from the sn-1 position of phosphatidylcholine to all-trans retinol, producing all-trans retinyl esters. Plays a critical role in vision
LRRC50/ DNAAF1	Primary ciliary dyskinesia	Leucine Rich Repeat Containing	cilium-specific and is required for the stability of the ciliary architecture
MKKS	McKusick-Kaufmann/ Bardet-Biedl syndrome (BBS6)	sequence similarity with other members of the chaperonin family	As part of the BBS/CCT complex may play a role in the assembly of BBSome
NEK1	Short-rib polydactyly Majewski type (type II)	Kinase domain, nuclear localisation and export signals, coiled-coil domains	Endogeneous NEK1 localises to the basal body, and Nek1 is required for ciliogenesis
NEK8	Nephronophthisis	Rcc1 domain containing; related to never in mitosis A (NIMA)	GFP-NEK8 localises to centrosomes and cilia; cilia are longer in mice with Nek8 mutations (jck mouse)
NPHP1	Joubert syndrome	Belongs to the nephrocystin-1 family. Contains 1 SH3 domain.	may play a role in the control of epithelial cell polarity

Gene ID	Syndrome	Type of protein encoded	Function in cilia biology
NPHP4	Nephronophthisis	Uncharacterised	Involved in the organization of apical junctions in kidney cells together with NPHP1 and RPGRIP1L/NPHP8
NPHP5 (IQCB5)	Senior–Løken syndrome	IQ domain protein	Endogeneous NPHP5 localised to the ciliary axoneme.
NPHP6 (CEP290)	Joubert syndrome	13 coiled-coil domains, SMC ('structural maintenance of chromosomes' domain), nuclear localisation signal, 6 KID motifs, homology with tropomyosin, ATP/GTP binding motif (P-loop)	Endogeneous CEP290 localises to the centrosome
RDH12	Leber congenital amaurosis	Belongs to the short-chain dehydrogenases/reductases (SDR) family.	Might be the key enzyme in the formation of 11-cis-retinal from 11-cis-retinol during regeneration of the cone visual pigments.
RPE65	Leber congenital amaurosis	Belongs to the carotenoid oxygenase family	Plays important roles in the production of 11-cis retinal and in visual pigment regeneration.
RPGRIP1	Leber congenital amaurosis	Belongs to the RPGRIP1 family. Contains 1 C2 domain.	Essential for RPGR function and is also required for normal disk morphogenesis
RPGRIP1L	Cerebello-oculo-renal syndrome	3 N-terminal coiled-coil domains, C-terminal RPGR-interacting domain, 2 central C2 motifs	
RPGRIP1L	Meckel syndrome	3 N-terminal coiled-coil domains, C-terminal RPGR-interacting domain, 2 central C2 motifs	
RSPH9	Primary ciliary dyskinesia	Belongs to the flagellar radial spoke RSP9 family	component of the axonemal radial spoke head
RSPH4A	Primary ciliary dyskinesia	Belongs to the flagellar radial spoke RSP4/6 family	component of the axonemal radial spoke head
SDCCAG8	Senior–Løken syndrome	N-terminal globular domain, nuclear localisation signal, 8 coiled-coil domains	Endogeneous SDCCAG8 is localised adjacent to the centrosome, in centrosomal appendages
SDCCAG8	Bardet–Biedl syndrome	N-terminal globular domain, nuclear localisation signal, 8 coiled-coil domains	Endogeneous SDCCAG8 is localised adjacent to the centrosome, in centrosomal appendages

Gene ID	Syndrome	Type of protein encoded	Function in cilia biology
TCTN2	Joubert syndrome	Belongs to the tectonic family	Abnormal ciliogenesis in *Tctn2–/–* mouse embryonic fibroblasts and neural tubes; interacts with MKS1
TMEM67	Meckel syndrome (MKS3) Joubert Syndrome 6 Nephronophthisis 11	Transmembrane protein (predicted)	Part of the tectonic-like complex which is required for tissue-specific ciliogenesis and may regulate ciliary membrane composition
TMEM216	Joubert-related syndromes	Transmembrane protein (predicted)	Endogeneous TMEM216 present in/adjacent to the basal body/ciliary base; knockdown of Tmem216 resulted fewer cilia and disrupted apical docking of centrosomes2
TMEM216	Meckel syndrome	Transmembrane protein (predicted)	
TTC21B	Nephronophthisis	Intraflagellar transport protein 139	*Ttc21b–/–* mice have abnormal nodal cilia; retrograde IFT is impaired, and cilia are shorted, following knockdown of Ttc21b in IMCD3 cells
TULP1	Leber congenital amaurosis	Belongs to the TUB family	Required for normal photoreceptor function and for long-term survival of photoreceptor cells. Interacts with cytoskeleton proteins and may play a role in protein transport in photoreceptor cells
WDR35	Sensenbrenner syndrome	AKA IFT121; Contains 5 WD repeats.	Component of the IFT complex A (IFT-A), required for retrograde ciliary transport. Required for ciliogenesis.

List of contributors

Philip L. Beales
Institute of Child Health,
University College London, UK

Thomas D. Kenny
National Institute for Health Research
Evaluations, Trials and Studies
Coordinating Centre,
University of Southampton, UK

Richard Paisey
Torbay Hospital, South Devon Healthcare
NHS Foundation Trust, UK

Miriam Schmidts
Institute of Child Health,
University College London, UK

Victoria Harrison
Department of Clinical Genetics,
Southampton Region Clinical Genetics
Service, UK

Andrea H. Németh
Nuffield Department of Clinical
Neurosciences,
University of Oxford, UK

Elizabeth Forsythe
Institute of Child Health,
University College London, UK

Ronald Roepman
Nijmegen Medical Centre,
St. Radboud University, The Netherlands

Gabrielle Wheway
Leeds Institute of Molecular Medicine,
St James's University Hospital, UK

Colin A. Johnson
Leeds Institute of Molecular Medicine,
St James's University Hospital, UK

Shalabh Srivastava
Institute of Genetic Medicine,
Newcastle University, UK

John A. Sayer
Institute of Genetic Medicine,
Newcastle University, UK

Brunella Franco
Department of Translational Medicine,
Frederico II University,
and Telethon Institute for Genetics
and Medicine, Naples, Italy

Richard Sandford
Academic Laboratory of Medical Genetics,
University of Cambridge, UK

Carsten Bergmann
Center for Human Genetics, Bioscientia,
Ingelheim,
and Department of Nephrology & Center
for Clinical Research,
University Hospital Freiburg, Germany

Claire Hogg
Royal Brompton and Harefield NHS
Foundation Trust, London, UK

Maria Bitner-Glindzicz
Institute of Child Health,
University College London, UK

Zubin Saihan
Institute of Ophthalmology,
University College London, UK

Kate Baker
Department of Medical Genetics,
Addenbrooke's Hospital,
Cambridge, UK

Chapter 1

Towards the diagnosis of a ciliopathy

Philip L. Beales and Thomas D. Kenny

The structure and function of cilia

Motile cilia and flagella (which share structural identity) are important for cell motility and for organising fluid flow across the cell surface. For example, cells lining the respiratory tract oviducts, epididymis and ependymal surface of the brain bear large clusters of motile cilia, which beat in concert to generate a wave-like motion. By contrast, the ubiquitous immotile or 'primary' cilium is present as a solitary cellular extension. It is sessile in nature and was long regarded a vestigial remnant of its motile cousin. This view has now been superseded following several studies indicating that primary cilia serve an essential sensory purpose in transducing extracellular information to the cell interior within multiple tissue types both during development and adult life (see Singla & Reiter, 2006 for a review). More recently, studies have indicated that the cilium is a central structure for regulation of key signal transduction pathways including the sonic hedgehog, platelet-derived growth factor, Wnt canonical and non-canonical (e.g. Ca^{2+} and planar cell polarity) pathways (reviewed by Michaud & Yoder, 2006).

With the exception of higher plants and fungi, most eukaryotic cells bear apical protrusions. In vertebrates, cilia are present throughout each tissue; however, amongst invertebrates they are confined to the sensory neurons, where they serve to sense changes in the environment including chemical stimuli and even vibration.

To appreciate the role primary cilia play in disease pathogenesis it is important to understand their structure and function. Motile cilia are long thin protrusions, extending up to 20μm from the cell surface, that tend to be concentrated in large numbers on the apical surface of cells and beat in coordinated waves to clear mucus from the respiratory epithelium, drive sperm along the Fallopian tubes, and move cerebrospinal fluid in the brain ventricles and spinal cord. In cross section, these cilia are constructed from a '9 + 2' arrangement of microtubules, in which nine microtubule doublets surround a central inner pair. The outer and inner doublets are connected by radial spokes, which are used to bend the outer doublets relative to the inner, producing a sheer force necessary to bend the cilium. Small dynein arms facilitate movement (Figure 1.1).

Primary cilia usually lack the central microtubule doublet, have a '9 + 0' arrangement and are generally immotile, with the unusual exception of cilia covering the node (or organiser)

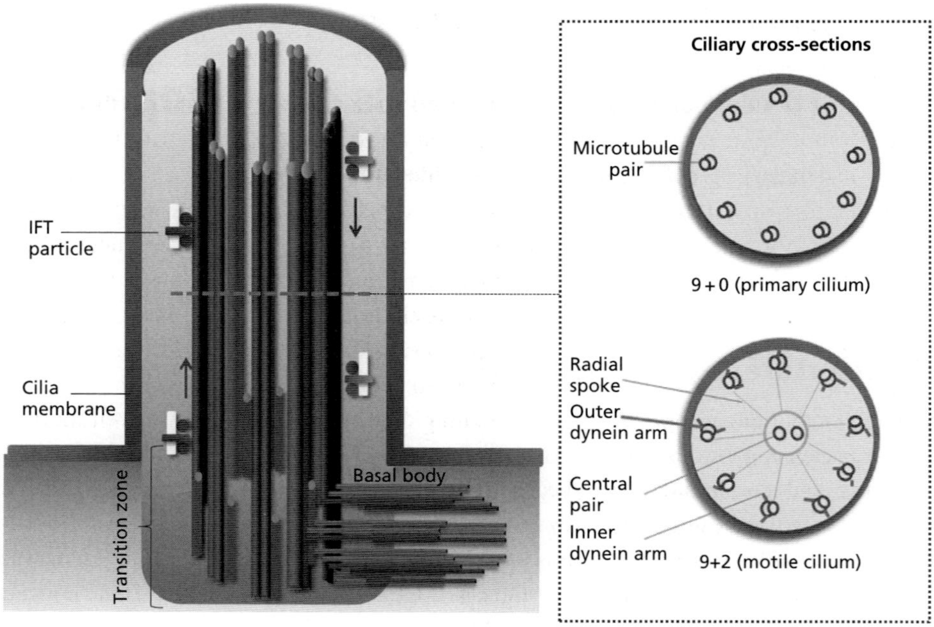

Figure 1.1 The typical ultrastructure of the cilium revealing the microtubular axoneme comprising nine outer doublets surrounding a central microtubule pair (note: the inner pair are typically absent in non-motile cilia), the dynein arms (in motile cilia) that attach to the doublets through the spokes. Image courtesy of Dr Miriam Schmidts.

of the vertebrate embryo. The primary cilium exists as a single apical appendage (compared with motile cilia). The function of primary cilia, once regarded as vestigial (Webber & Lee, 1975), is complex and variable.

At the base of the cilium lies the basal body, a cylindrical structure perpendicular to the cell membrane, anchoring the cilium in the cytoplasm. The basal body also acts as a nucleation point from which the cilium extends out from the cell. Microtubule fibres project from the basal body to the proximal region or transition zone, from which a nascent cilium grows.

Intraflagellar transport

Cilia lack the host machinery to manufacture proteins, and so all associated proteins required for ciliary biogenesis or function must be transported from the cytoplasm. An evolutionarily conserved system for ferrying proteins within the cilium either toward the tip (anterograde) and from the tip (retrograde) is employed, termed intraflagellar transport (IFT). Anterograde transport involves the loading of cargo proteins bound for the cilium onto an IFT particle, which in turn is attached to a kinesin motor protein complex (Rosenbaum & Witman, 2002). Retrograde transport is facilitated by the dynein–dynactin motor complex (Rosenbaum & Witman, 2002). Loading of cargo is regulated at the basal body and transition zone.

Diseases can arise from cilia dysfunction

The primary ciliary dyskinesias (including Kartagener syndrome), characterised by bronchiectasis, infertility and occasionally situs inversus/isomerism, can be directly ascribed to ciliary motile dysfunction most often secondary to structural deficits within the ciliary axoneme (spoke or dynein). The spectrum of ciliopathies has widened to include a much larger number of syndromal entitities (the primary ciliopathies) in which either ciliary structure or function has been shown to be abnormal, or a causative gene product has been localised to the ciliary apparatus or to associated complexes and pathways. Most of the disease-related proteins are not expressed in the cilium itself, but rather at its base, in the basal body, transition zone or centrosome. A growing body of evidence indicates that they likely serve other functions within the cytoplasm. Nevertheless, the common phenotype likely arises out of sub-optimal ciliary signalling. The assertion that these syndromes are caused exclusively by cilia dysfunction is probably overly inclusive but will suffice until such time as the underlying disease-causing mechanisms have been elucidated.

The best-studied mouse model of defective IFT is the Oak Ridge polycystic kidney mouse, originally described as a model for human recessive polycystic kidney disease (Lehman et al. 2008). The Oak Ridge polycystic kidney mouse arose through integration of a transgene into an intron of *Ift88* resulting in a hypomorphic allele (Tg737) and disrupting the expression and function of the polaris. The phenotype includes dishevelled fur, polydactyly, hepatic and pancreatic ductal cysts, retinal degeneration, skeletal defects, cerebellar hypoplasia, hydrocephalus, growth retardation and late-onset obesity (Lehman et al. 2008). This was the first mammalian model to establish a connection between cystic kidney disease and ciliary dysfunction.

Classifying ciliopathies

Many of the ciliopathies have long been recognised as discrete clinical entities but only in the last few years has this disparate collection of rare and clinically perplexing disorders been reclassified as a group in terms of overlapping phenotype and pathophysiology (Badano et al. 2006; Baker & Beales, 2009). These revelations have come about through, albeit, an incomplete understanding of the underlying biology in which multiple factors interact to determine phenotype, perhaps by influencing quantitative and qualitative aspects of ciliary function in different tissues. In support of this view, it has been demonstrated that mutations in the same ciliary genes can give rise to quite different syndromes (Baala et al. 2007; Bergmann et al. 2008; Hoefele et al. 2007; Karmous-Benailly et al. 2005; Leitch et al. 2008).

Hence prudent classification and diagnosis on phenotypic, genotypic and ultimately physiological grounds is challenging but nonetheless important at the clinical level for accurate prognosis, counselling, antenatal screening and management.

Classification of the ciliopathies is based on either cilia type involvement (motile or primary) or on phenotypic similarities (e.g. skeletal involvement, obesity). The list of confirmed ciliopathies continues to grow rapidly. At a clinical level, the primary ciliopathies

Table 1.1 Common association of clinical features in ciliopathies

Disease	Bardet–Biedl	Oral–facial–digital type I	Meckel	Joubert	Jeune
Retinitis pigmentosa	✓		✓	✓	✓
Renal cystic disease	✓	✓	✓	✓	✓
Polydactyly	✓	✓	✓	✓	✓
Situs inversus	✓		✓	✓	
Cognitive impairment	✓	✓	✓	✓	
Hepatic disease	✓	✓	✓	✓	
Skeletal defects		✓	✓	✓	✓
Posterior brain defect			✓	✓	

share a number of overlapping features which, although they can occur in isolation or in association with other unrelated anomalies, may in combination be predictive of other potential disorders of cilia function (Badano et al. 2006; Baker & Beales, 2009) (Table 1.1). By identifying core features of primary ciliopathies such as retinitis pigmentosa, renal cystic disease, polydactyly, situs inversus and brain anomalies, Baker and Beales (2009) attempted to predict further ciliopathies from sources such as the London Dysmorphology Database and the Online Mendelian Inheritance in Man (NLM). Their screening strategy yielded 127 putative ciliopathy conditions. On the basis of clinical features and known genes/protein functions, these 127 disorders, which shared phenotypic overlap of at least two features of the core nine, were judged to be either *known* ciliopathies ($n = 14$), *likely* ciliopathies ($n = 16$), *possible* ciliopathies ($n = 72$) or *unlikely* ciliopathies ($n = 25$) (Baker & Beales, 2009). This list is necessarily and by design inclusive of all conditions with phenotypic overlap, not all of which will be caused by direct disruption to cilia or indirect disruption to ciliary signalling processes. However, many if not most of the genes implicated in the aetiology of these conditions do in fact code for proteins with putative ciliary functions or upstream/downstream interactions.

Since the discovery of mutations in IFT genes in human disorders, the ciliopathies can also be categorised according to the presence or absence of skeletal involvement (Badano et al. 2006). To date the total number of proven ciliopathies is 20 caused by mutations in at least 95 genes (Table 1.2).

Diagnosing a ciliopathy

Clinical diagnosis

Based on the presence of a combination of the core clinical features associated with known ciliopathies, it is possible to diagnose a ciliopathic disorder even if it has not previously been described. We propose using the algorithm (Figure 1.2) in clinical practice to assist in diagnosis where there is a high index of suspicion.

Table 1.2 List of known ciliopathies and their underlying genes

Syndrome	Gene ID
Acrocallosal syndrome	*KIF7*
Bardet–Biedl syndrome	*BBS1, BBS2, ARL6 (BBS3), BBS4, BBS5, MKKS, BBS7, TTC8, PTHB1, BBS10, TRIM32, BBS12, MKS1, CEP290,* C2orf86
Cerebello-oculo-renal syndrome	*INPP5E, RPGRIP1L*
Ellis–van Creveld syndrome	*EVC, EVC2*
Hydrolethalus	*HYLS1, KIF7*
Jeune asphyxiating thoracic dystrophy	*IFT80, TTC21B, WDR19, DYNC2H1*
Joubert syndrome	*AHI1, INPP5E, MKS3, NPHP1, NPHP6* (*CEP290*), *TCTN3, ARL13B, CEP290, TMEM216, TTC21B, CXORF5, KIF7, TMEM138, C5ORF42, CEP41, TMEM237, TCTN1, TMEM231, RPGRIP1L, ZNF423, CC2D2A*
Leber congenital amaurosis	*AIPL1, CEP290, CRB1, CRX, GUC2YD, LCA5/CORF152, LRAT, RDH12, RPF65, RPGRIP1, TULP1*
Meckel syndrome	*CEP290, MKS3, RPGRIP1L, TMEM67, TMEM216*
Mental retardation + retinitis pigementosa	*CC2D2A*
MORM—mental retardation, truncal obesity, retinal dystrophy, micropenis	*INPP5E*
Nephronophthisis	*ATXN10, GLIS2, NEK8, NPHP4, TTC21B, NPHP1, NPHP5 (IQCB5), NPHP6 (CEP290), NPHP3, RPGRIP1L, INVS*
Autosomal dominant polycystic kidney disease	*PKD1, PKD2, PKHD1*
Primary ciliary dyskinesia	*DNAI1, DNAI2, DNAH5, DNAH11', KTU, LRRC50, RSPH9, RSPH4A, TXNDC3*
Senior–Løken syndrome	*NPHP5* (*IQCB5*), *SDCCAG8*
Sensenbrenner syndrome	*IFT122, IFT43, WDR35*
Short-rib polydactyly Majewski type (type II)	*NEK1*
Short-rib polydactyly type III	*DYNC2H1*
Weyer acrodental dysostosis	*EVC2*

Molecular diagnosis

Given the broad range of phenotypes that can arise from mutations in the same ciliopathy gene (e.g. *CEP290* in Meckel, NPHP, Bardet–Biedl and Joubert syndromes), the option to utilise high throughput molecular assays such as next generation sequencing is of mounting importance in the clinical setting (Coppieters et al. 2010). With over 1000 known ciliary proteins, it is vital to develop disease-specific panels for accurate diagnosis of ciliopathies or even new gene discovery as reported by the Hildebrandt laboratory (Otto et al.

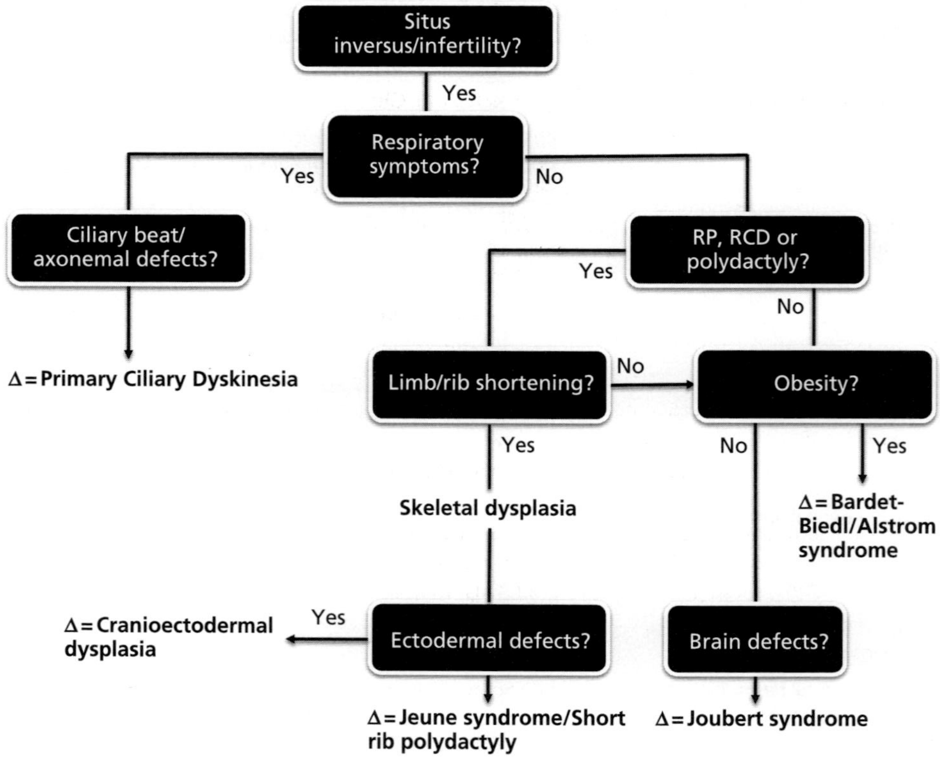

Figure 1.2 Clinical diagnostic algorithm.

2010). Using the NimbleGen™ 385k platform the authors designed a ciliopathy candidate exon capture array, which contains oligonucleotides from 828 nephronophthisis-related ciliopathy (NPHP-RC) candidate genes. Using next-generation sequencing, they detected 12 different truncating mutations in a novel gene, *SDCCAG8*, in 10 NPHP-RC families. This study demonstrated loss of function in *SDCCAG8* as a novel cause of a retinal–renal ciliopathy and validates exome capture analysis for broadly heterogeneous single-gene disorders.

The adoption of clinical exomes or the design of targeted panels (Haloplex, Truseq) is currently under way in many diagnostic genetics laboratories and is a cost-effective approach particularly suited to this pleiotropic group of conditions.

References

Baala, L., Audollent, S., Martinovic, J., Ozilou, C., Babron, M. C., Sivanandamoorthy, S., et al. 2007. Pleiotropic effects of *CEP290* (NPHP6) mutations extend to Meckel syndrome. *Am J Hum Genet* **81**, 170–179.

Badano, J. L., Mitsuma, N., Beales, P. L. & Katsanis, N. 2006. The ciliopathies: An emerging class of human genetic disorders. *Annu Rev Genomics Hum Genet* **7**, 125–148.

Baker, K. & Beales, P. L. 2009. Making sense of cilia in disease: The human ciliopathies. *Am J Med Genet, C, Semin Med Genet* **151C**, 281–295.

Bergmann. C., Fliegauf, M., Brüchle, N. O., Frank, V., Olbrich, H., Kirschner, J., et al. 2008. Loss of nephrocystin-3 function can cause embryonic lethality, Meckel–Gruber-like syndrome, situs inversus, and renal–hepatic–pancreatic dysplasia. *Am J Hum Genet* **82**, 959–970.

Coppieters, F., Lefever, S., Leroy, B. P. & De Baere, E. 2010. CEP290, a gene with many faces: Mutation overview and presentation of CEP290base. *Hum Mutat* **31**, 1097–1108.

Hoefele, J., Wolf, M. T., O'Toole, J. F., Otto, E. A., Schultheiss, U., Dêschenes, G., et al. 2007. Evidence of oligogenic inheritance in nephronophthisis. *J Am Soc Nephrol* **18**, 2789–2795.

Karmous-Benailly, H., Martinovic, J., Gubler, M. C., Sirot, Y., Clech, L., Ozilou, C., et al. 2005. Antenatal presentation of Bardet–Biedl syndrome may mimic Meckel syndrome. *Am J Hum Genet* **76**, 493–504.

Lehman, J. M., Michaud, E. J., Schoeb, T. R., Aydin-Son, Y., Miller, M. & Yoder, B. K. 2008. The Oak Ridge polycystic kidney mouse: Modeling ciliopathies of mice and men. *Dev Dyn* **237**, 1960–1971.

Leitch, C. C., Zaghloul, N. A., Davis, E. E., Stoetzel, C., Diaz-Font, A., Rix, S., et al. 2008. Hypomorphic mutations in syndromic encephalocele genes are associated with Bardet–Biedl syndrome. *Nat Genet* **40**, 443–448.

Michaud, E. J. & Yoder, B. K. 2006. The primary cilium in cell signaling and cancer. *Cancer Res* **66**, 6463–6467.

Otto, E. A., Hurd, T. W., Airik, R., Chaki, M., Zhou, W. & Stoetzel, C. 2010. Candidate exome capture identifies mutation of SDCCAG8 as the cause of a retinal–renal ciliopathy. *Nat Genet* **42**, 840–850.

Rosenbaum, J. L. & Witman, G. B. 2002. Intraflagellar transport. *Nat Rev Mol Cell Biol* **3**, 813–825.

Singla, V. & Reiter, J. F. 2006. The primary cilium as the cell's antenna: Signaling at a sensory organelle. *Science* **313**, 629–633.

Webber, W. A. & Lee, J. 1975. Fine structure of mammalian renal cilia. *Anat Rec* **182**, 339–343.

Chapter 2

Alström syndrome

Richard Paisey

History of the eponym

Carl Henry Alström was born in Vasteras in Sweden in 1907 and was 52 years old when he published case studies made over 13 years of first one case and then two second cousins with progressive retinal degeneration, obesity, neuronal hearing loss and insulin resistance (Alström et al. 1959). He recognised that the early hearing loss and preserved intelligence distinguished the condition from Laurence–Moon–Bardet–Biedl syndrome, and that the familial nature of the disorder made autosomal recessive inheritance likely. His insight was made in the context of his previous training in psychiatry and distinguished work on the links between tuberculosis and the health consequences of schizophrenia (Alström et al. 1949). He had also led treatments of epilepsy and studies into its inheritance as physician at the neurological clinic in the Serafimerlasarettet Hospital in Stockholm. In 1950 he published significant work on the social effects of epilepsy (Alström, 1950). He was admirably qualified and experienced in mental health, social medicine, neurology and metabolic medicine. His career spanned the distinction between insulin resistant and insulin sensitive diabetes (Himsworth 1939), the discovery of DNA in 1953 by Watson and Crick (Watson & Crick, 1953) and the first insulin assay in 1959 by Yallow and Berson (Yallow & Berson, 1960). The patience and thoroughness with which he used all these skills and insights to describe the syndrome makes it appropriate that it should be named after him.

Epidemiology

It was recognised from the initial description of Alström syndrome in 1959 that its occurrence conformed to an autosomal recessive pattern. Childhood onset retinal dystrophy, commonly with sensori-neural hearing loss sets it in the clinical context of a large number of deaf blind syndromes ranging from acquired prenatal infections such as rubella, to other recessive conditions notably Wolfram syndrome, Bardet–Biedl syndrome, Leber's congenital amaurosis and Usher syndrome. However, the early onset of nystagmus and photophobia in most cases, coupled in many with a high rate of reversible infantile cardiomyopathy and obesity, are enough to make a presumptive diagnosis of Alström syndrome (Russell-Eggitt et al. 1998) As a result clinically diagnosed cases were reported over the next 30 years from Japan, Germany, the USA, India and, in particular, Canada where the eight cases described for the first time the occurrence of infantile cardiomyopathy (Garg et al.

1991; Ikeda et al. 1974; Marshall et al. 1997; Michaud et al. 1996). Collaboration between the French Canadian group and the Jackson Laboratory in Maine led to the discovery of a founder effect from genealogical study and localisation of the gene to chromosome 2p occurred in 1997 (Marshall et al. 1997). The frequency and diagnostic utility of infantile cardiomyopathy led to the identification of 22 cases in the UK in the following year, some of which had been diagnosed as Leber's congenital amaurosis (Russell-Eggitt et al. 1998). A further study from another group in the UK endorsed this finding and reported delayed male puberty due to hypogonadism and high incidence in a consanguineous group of Pakistani origin in the Leeds–Bradford area (Macari et al. 1998). More cases were reported from North Africa, China, Belgium and Turkey confirming the global occurrence of the condition (Liu et al. 2009; Macari et al. 1998; Satman et al. 2002; Van den Abeele et al. 2001). Nevertheless, fewer than 100 cases had been reported at this time. The *ALMS1* gene was discovered independently in 2001 by investigators in Southampton, UK, and the Jackson Laboratory, USA (Collin et al. 2002; Hearn et al. 2002). Mutation analysis to support clinical diagnosis was now possible and this has confirmed the worldwide distribution of the syndrome. In the last 10 years many more case reports, often genetically confirmed, have brought the numbers recognised to nearly 1000 worldwide. Cases related to consanguinity have been increasingly recognised, particularly from Saudi Arabia, Turkey and the Pakistani community in Pakistan and the UK (Aldahmesh et al. 2009; Bond et al. 2005; Satman et al. 2002). An overview of 183 cases has further defined the phenotype, demonstrating that major organ fibrosis is common, of varying intensity and age of onset (Marshall et al. 2005). Thus cardiac, renal or hepatic dysfunction may dominate evolution of the condition. In some cases useful vision is maintained into the third decade of life. This finding raises the possibility that onset of retinal dystrophy might be even later in some cases leading to an undiagnosed pool of phenotypically mild cases. In a recent report eight cases were described in whom the diagnosis was made as late as 20 to 36 years of age in patients attending adult diabetic clinics in the UK (Paisey et al. 2011). Subsequently, kyphoscoliosis, hypothyroidism, polcystic ovarian syndrome, coronary artery disease and dyslipidaemia have been linked with the syndrome (Jatti et al. 2012; Marshall et al. 2005; Paisey et al. 2004). Hypertriglyceridaemia may be severe enough to cause pancreatitis. Very severe insulin resistance is virtually always present and may increase with age despite lessening obesity (Minton et al. 2006). Acanthosis nigricans is common. Type 2 diabetes develops in 70% by the third decade, but varies in incidence in Europe and Canada (Bettini et al. 2012; Mokashi & Cummings, 2011).

Clinical features

These will be discussed in chronological order of presentation during life, commencing with an overview of the complications attributed to the syndrome. This clinical outline relies heavily on several sources: the multi-author collaboration describing the clinical characteristics of 182 cases in *Archives of Internal Medicine*, 2005 (Marshall et al. 2005), GeneReviews updated 2011 (Marshall et al. 2011), the experience of the National Specialist

Clinical Team review clinics at Torbay Hospital, UK, and Birmingham Children's Hospital, UK, the Alström Clinic in Padua, Italy, Alström UK, and Alström International (Bar Harbor, Maine, USA).

History

The earliest manifestations of Alström syndrome in the majority of known cases are nystagmus and pronounced photophobia in the first months of life, often when fixation develops at 7–8 weeks. In approximately 40% sudden collapse occurs between 6 and 12 weeks of age due to dilated cardiomyopathy. There may well be a family history of sudden infant death from cardiac problems in an older sibling, though more often than not genetic confirmation of the diagnosis in that sibling will not have been possible. An almost pathognomonic feature is the capacity for recovery from infant cardiomyopathy if the condition is recognised and treated intensively and promptly (Bond et al. 2005). Education of families with an infant at risk is essential as onset is often precipitous. Exercise tolerance, electrocardiogram (ECG) and echocardiogram may all return to normal during childhood though recurrence of cardiac dysfunction in adolescence is well described. In contrast to the infant cardiomyopathy, visual loss is progressive throughout the first three decades of life sometimes punctuated by sudden complete visual loss. Sensori-neural hearing loss is frequently noted during childhood with slow progression over a period of years. An increased frequency of suppurative middle ear disease has been noted in childhood which abates in adolescence. Many parents describe hyperphagia and an absence of satiety after meals. Obesity is often pronounced in childhood, but modifiable by calorie restriction and exercise. Adult body mass index ranges from 28 to 38 kg/m^2. Pronounced subcutaneous fat is present in contrast to insulin resistant syndromes associated with lipodystrophy (Paisey et al. 2008). Visceral fat accumulation also occurs as shown on abdominal computed tomography and magnetic resonance imaging (MRI) scans. Obesity can lessen in adult life though one small study has shown that this has been associated with an increase in insulin resistance (Minton et al. 2006). Allowing for the sensory impairment developmental milestones are usually normal, though severe hypoxia occurring during infant cardiomyopathy or other inherited or acquired conditions can result in educational challenge. It seems unlikely that there is a specific intellectual impairment of any kind in the syndrome, as noted by Alström himself. Amongst 40 subjects who have attended the Alström adult National Clinical Service Team (NSCT) clinics at Torbay Hospital many have examples of impressive achievements.

Recurrence of cardiomyopathy and development of diabetes, dyslipidaemia, hepatic or renal impairment often bring the young person with Alström syndrome to medical attention in adolescence. This, coupled with progressive loss of sight, can unsurprisingly give rise to severe emotional crises. Emotional, psychological and practical support for families at this time is crucial.

Although type 2 diabetes, renal and hepatic fibrosis, and recurrent cardiomyopathy can occur simultaneously during the second and third decades of life there is compelling evidence that these fibrotic processes are independent of diabetes. Death from severe cardiomyopathy and progression to end-stage renal failure can occur even when glucose tolerance has remained normal.

Deaths from infantile cardiomyopathy, adolescent cardiomyopathy, hepatic cirrhosis and pulmonary infections (see the section 'Respiratory system') have resulted in a median life expectancy of 21years. However, a substantial minority of patients known to the NHS rare disease clinics (NSCT) are alive and well in the fourth and fifth and even sixth decades of life (Paisey et al. 2011). Of these, eight have confirmed pathological mutations in both copies of the *ALMS1* gene though in two only one mutation has so far been discovered.

The kidney

Chronic kidney disease stages 1 to 3 are very common from the third decade, remaining stable in many over 10 years or more whilst in 10% progression to end-stage renal failure (ESRF) is rapid. This is not usually related to uropathy (see the section 'Urological problems') but sudden deterioration in end-stage renal failure (EGFR) may be precipitated by septicaemia. Renal histopathology shows interstitial tubular fibrosis and some glomerulosclerosis (Marshall et al. 2005). Cystic kidney disease is uncommon. Patients may remain asymptomatic until severe renal failure has developed. Hypertension is common with disturbed microvascular function (an increased augmentation index (Smith et al. 2007) and sometimes microalbuminuria or modest proteinuria).

Nephrogenic, and cranial diabetes insipidus might be anticipated in Alström syndrome because of the renal fibrosis and pituitary dysfunction, respectively. In fact these have rarely been reported, and are not described in genetically confirmed cases.

The heart

Sub-acute development or recurrence of clinically important cardiomyopathy presents most commonly in late adolescence or the early twenties. Clinical signs are muted as jugular venous pressure and peripheral oedema are obscured by subcutaneous fat. Shortness of breath and unexpectedly severe hypoxia in the context of respiratory tract infections may unmask deterioration in cardiac function. Non-specific ECG changes and subtle decline in echocardiographic measure of left ventricular ejection fraction occur. Cardiac MRI has shown fibrotic changes in all subjects with Alström syndrome so far, including right ventricular changes and a restrictive pattern in adults (Loudon et al. 2009). This can remain stable and not clinically significant for many years. Serum *brain natriuretic peptide* (NT BNP) levels may be helpful in monitoring progress, but can also be influenced by treatments for hypertension and heart failure and by changes in renal function.

Coronary artery disease

The combination of insulin resistance, type 2 diabetes, dyslipidaemia and renal failure has been shown to be a potent cause of atherosclerosis progressing to coronary artery disease in the general population. Acute coronary syndrome requiring coronary angioplasty has not been reported in the syndrome in this context (Jatti et al. 2012).

The liver

Non-alcoholic hepatic steatosis is very common. Fibrosis and cirrhosis with portal hypertension have also been described-the latter in up to 10% of Alström subjects. Jaundice and severe disorder of liver enzymes is uncommon and once again clinical examination baffled by the thick subcutaneous and visceral abdominal fat. Diagnosis therefore rests on hepatic ultrasound to discover severe fibrosis or portal hypertension and upper gastrointestinal (GI) endoscopy, computed tomography scan or transjugular hepatic venogram to visualise and treat varices. Fully developed cirrhosis of the liver with gastro-oesophageal varices have occurred in adolescence, but may present later (Awazu et al. 1997; Chang et al. 2000; Chou et al. 2000; Connolly et al. 1991; Quiros-Tejeira et al. 2001).

Smooth muscle dysfunction

Gastro-oesophageal reflux disease

Acid reflux, occasionally leading to vomiting is frequent, is usually controlled by proton pump inhibitors, though it occasionally requires gastric fundal plication. There is no evidence that this is the cause of pulmonary fibrosis found in the syndrome.

Caecal volvulus

This has been described in one sibling pair with Alström syndrome so far (Khoo et al. 2009) and may thus represent another as yet unknown recessive condition occurring in this family. However, in view of the increased prevalence of upper GI and bladder dysmotility this may represent another complication of the syndrome.

Urological problems

Delay in voiding urine is common in female subjects. Rarely, this symptom progresses to severe detrusor urethral dyskinesia with the need for intermittent self-catheterisation and even construction of an ileal conduit. Standard histological examination of smooth muscle in affected tissues has been normal.

Insulin resistance, glucose intolerance and dyslipidaemia

A post-receptor insulin resistance is manifest from infancy with accompanying hyperinsulinaemia, low serum high-density lipoprotein (HDL), high serum triglycerides and frequent progression to type 2 diabetes by the second or third decade of life (Bettini et al. 2012; Mokashi & Cummings, 2011; Paisey et al. 2004, 2008).

Diabetes presents and develops in a similar manner to non-syndromic type 2 diabetes, but at a much younger age and lower body mass index (typically 28–32 kg/m^2). Stimulated serum C-peptide levels are typically 2000 to 10,000 pmol/l. Healthy diet, aerobic exercise and metformin are often successful in achieving good glycaemic control at first (Lee et al. 2009). If not, incretin analogues have been found to be effective in 50%, though insulin may be required, as indicated in the section 'Metabolism' (Wu et al. 2003). Acanthosis nigricans is very common, whilst a pseudo acromegaloid appearance is attributed to insulin resistance in a minority of cases. Severe hypertriglyceridaemia (>8 mmol/l) develops in up to one third of patients with Alström syndrome. In some, levels >20 mmol/l have precipitated acute pancreatitis (Paisey, 2009; Wu et al. 2003). Low serum HDL cholesterol levels are also common and expected in the presence of severe insulin resistance (95% <1.0 mmol/l in adults)

Thyroid failure

Hypothyroidism is common from adolescence onwards with low levels of thyroid stimulating hormone (TSH), low levels of free T4 levels and absent thyroid autoimmunity. Goitre is uncommon clinically. Standard replacement therapy with L-thyroxine is indicated in approximately 30% of cases. Screening with free T4 as well as TSH levels is crucial as most hypothyroidism is secondary.

Male puberty

In contrast to the accelerated bone age and resultant short stature, puberty is often arrested. This can result from secondary hypogonadism with low serum gonadotrophin levels and testosterone, testicular fibrosis with primary hypogonadism or a combination of both. Secondary sexual characteristics may be normal, though penile development and testicular size are often reduced. Fertility has not yet been described in male patients with the syndrome and confirmed pathological mutations in both copies of the gene. Testosterone replacement therapy is indicated for completion of puberty, muscle strength, libido and bone protection. Screening with serum testosterone and gonadotrophin levels is necessary.

Female puberty

This can occur normally but may be affected by polycystic ovarian syndrome presumed to be a consequence of the insulin resistance and obesity. Fertility has not been conclusively described in females with Alström syndrome but two cousins have been reported with apparently mild Alström syndrome clinically and proven fertility (Iannello et al. 2004). Genetic confirmation of the diagnosis was not available at that time.

Growth hormone secretion

Published reports have suggested that there is a subtle defect in dynamics of Insulin like growth factor binding protein 1 (IGFBP-1), acid labile fraction and IGFBP-2 in 15 Alström subjects, whilst others have found partial growth hormone deficiency in two siblings, and severe growth hormone deficiency in three unrelated cases (Alter & Moshang, 1993; Maffei et al. 2007; Tai et al. 2003). MRI scans of the pituitary have shown small pituitary glands or, in one case, an empty sella.

Musculoskeletal problems

Functional, reversible cervical kyphosis is common, which progresses to fixed deformity in adult life in some. Thoracic kyphosis is also frequent, whilst pronounced thoracic kyphoscoliosis is relatively unusual (three of 40 adults in the adult NSCT clinic).

Most patients with Alström syndrome have flat small feet and small hands with short fingers.

Respiratory system

Chronic obstructive pulmonary disease is said to be common in the syndrome but without good evidence. The majority of subjects do not smoke. Respiratory function tests as expected tend to show a restrictive defect resulting from kyphosis and poor inspiration. Spirometry is difficult to perform with Alström patients because of problems with the seal around the mouthpiece and difficulties co-ordinating the breathing routine. Pulse oximetry during a 6-min walk is very often normal with oxygen saturations >95%. High-resolution pulmonary computed tomography scanning and lung histology at autopsy has shown that a non-inflammatory alveolar fibrosis can occur.

The combination of restrictive lung defect, myocardial fibrosis and diabetes contribute to a propensity to upper and lower respiratory tract infections. Prompt diagnosis and intensive therapy are essential in community acquired pneumonia or perioperative respiratory infection. Unexpectedly, severe hypoxia and circulatory collapse are described and prolonged intensive care is often necessary. Death from cerebral hypoxia is otherwise a real threat (Khoo et al. 2009; Tiwari et al. 2010).

Annual review

In the light of the many and varied medical problems which may be associated with the syndrome and successful palliative treatments for many of them an annual review process is strongly recommended. A plan which has developed from the NSCT clinics in the UK is outlined in Figure 2.1.

Range of the phenotype

Many autosomal recessive conditions exhibit wide variations in phenotype. Early studies in cystic fibrosis and phenylketonuria are examples of this. Variations in clinical course in these conditions could result from genotype–phenotype interaction, the effects of modifier genes, or environmental influences (Atar & Körperich, 2010; Bremer et al. 2008; Collaco et al. 2008; Dorfman et al. 2008; Gámez et al. 2000) It is likely that analogous influences will affect the presentation of Alström syndrome.

Sensory impairment

The cardinal features of blindness and deafness can vary widely in age of onset and even more in rate of progression. Retinal dystrophy presents as progressive visual impairment, severe photophobia, and nystagmus usually starting between 6 weeks and 2 years of age. Most lose all perception of light by the end of the second decade but some retain the ability to read large print into the third decade at least. Hearing loss is even more variable. Progressive high-tone loss is often documented in the first decade, but normal hearing and audiograms have been demonstrated in a minority. Hearing loss may progress to the

Consider General well-being:

- Psychology—adjustment to condition
- Mobility
- Eyesight and visual aids-Braille, IT
- Hearing and appropriate hearing aids
- Growth and development

Ask specifically about:

- Shortness of breath
- Tiredness
- Problem with passing water
- Heartburn
- What treatments are being taken?
- Any side effects?

Examination:

- Observe mobility, foot condition and spine
- Weight, height, blood pressure
- Acanthosis
- Listen to heart and lungs
- Audiogram-hearing aid check
- Visual acuity check

Investigation—blood tests:

- Urea and electrolytes, EGFR
- Liver function tests
- Haematology
- Haemoglobin A1c and blood glucose
- Serum triglycerides and cholesterol
- Thyroid function
- Oestrogen or testosterone, FSH and LH imaging
- Bladder scan post micturition if urinary symptoms

Physiological tests:

- ECG
- Echocardiogram
- Cardiac MR to stage at adolescence
- Six minute walk
- Hepatic ultrasound
- Respiratory function tests consistent with age
- Urodynamics if symptoms
- Foot examination
- Duplex arterial scan in renal failure

Figure 2.1 History from Alström person and family.

severe or moderately severe range (40–70 dB) by the end of the first to second decade, though in some subjects it can remain stable for many years. No genetic or environmental factors have been shown conclusively to influence sensory loss.

Major organ fibrosis

Cardiac MRI, hepatic imaging and chronic kidney disease staging all suggest that these organs exhibit some degree of fibrosis in virtually all cases. This is borne out by autopsy studies (Marshall et al. 2005). Nevertheless the rate of decline in function varies enormously. In the least severely affected survival to the fourth and fifth decades is well described with minimal renal impairment, fatty liver only and good cardiac function. In contrast, death from cardiomyopathy occurs in 25% in adolescence or the twenties; hepatic cirrhosis with portal hypertension and variceal bleeding has developed as early as 15 years of age; renal dysfunction can progress to end-stage renal failure before 20 years. The interplay of these changes may cause deterioration in organ function, for example when renal failure impinges on cardiac function. Sudden irreversible worsening of renal failure can occur with septicaemia. Although the source of infection may be the urinary tract there is no evidence that ascending urinary tract infection is the main cause of renal fibrosis in the majority of cases.

Uncommon anatomical problems

Kyphoscoliosis and smooth muscle disorders

These appear to be 'all or none' phenomena in a minority (<10%) of Alström patients. There is so far no clear genotype–phenotype interaction. Modifier genes or the co-inheritance of other recessive or even dominant gene defects could be involved.

Metabolic disorders

Moderate obesity, insulin resistance, hyperinsulinaemia, low HDL cholesterol and high serum triglyceride levels are almost always present. However, just as in the background population, progression to type 2 diabetes varies widely. There is a higher prevalence of obesity and greater rate of type 2 diabetes in Canadian compared to Italian children with Alström syndrome. This suggests an effect of environment on progression of glucose tolerance in the syndrome (Bettini et al. 2012; Mokashi & Cummings, 2011). Later presentation of type 2 diabetes with co-existing unexplained visual impairment has led to late diagnosis of the syndrome in a recent series of cases (Paisey et al. 2011).

Finally, the reliance on retinal dystrophy to anchor the diagnosis may have resulted in failure to detect mildly affected individuals. The phenotype may be more varied than we yet realise, although a search amongst young onset obese type 2 diabetic persons did not reveal any with *ALMS1* gene mutations (Patel et al. 2006).

Diagnosis of Alström syndrome

Carl Henry Alström's original description of the syndrome relied entirely on careful clinical observation and family history and this clearly remains the starting point for making a provisional diagnosis of the syndrome. More detailed investigation and genetic testing are then undertaken.

Family history

The diagnosis is much more likely if there is an autosomal recessive history in the family of infant cardiomyopathy or early onset retinal dystrophy. Where Alström syndrome has been genetically confirmed in affected siblings, cousins or ancestors then mutation analysis to confirm the diagnosis in the suspected new case is straightforward. Even then the clinical manifestations are important especially if there is consanguinity as other recessive conditions may co-exist within the family.

Clinical features

Two striking presentations make the diagnosis a strong possibility. The first is *abrupt onset infantile dilated cardiomyopathy*, especially if apparent complete recovery occurs. The rarity of this form of cardiomyopathy and its high frequency in Alström syndrome make it worthwhile to consider *ALMS1* mutation analysis in all such cases unless an alternative diagnosis such as one of the mitochondrial diseases or Pompe syndrome is known to be the cause.

The second is *early onset visual impairment due to retinal dystrophy accompanied by nystagmus and intense photophobia*. This has been comprehensively investigated in a series of 22 cases of apparent Leber's optic atrophy where infant cardiomyopathy suggested Alström syndrome to be the correct diagnosis (Russell-Eggitt et al. 1998). This was later confirmed in all cases by genetic testing. Early onset retinal dystrophy may be an isolated finding. In that case prominence of night blindness or macular degeneration may guide diagnosis to Best's syndrome or Usher's syndrome, whilst the presence of polydactyly indicates Bardet–Biedl syndrome.

With increased awareness of the syndrome and more readily available genetic testing it has become clear that some cases may present late. So far all genetically confirmed cases have developed retinal dystrophy by the second decade, though the possibility remains that the syndrome could be compatible with a delay in visual loss until later in life. Preservation of visual acuity sufficient to read large print into the third decade is well documented. In some circumstances isolated visual loss has been attributed to Leber's optic atrophy or retinitis pigmentosa with the correct diagnosis finally being made as a result of presentation of type 2 diabetes in early adult life (Paisey et al. 2011). This phenomenon of mis-diagnosis has been common in the past 20 years with understandable stress for the families concerned. Prompt diagnosis in childhood has crucial advantages for families, above all the resolution of uncertainty and clarity in offering prognosis and genetic counselling. The families-based charity Alström Syndrome UK support group has provided crucial support and guidance in this area. Once Alström syndrome is confirmed early recognition and treatment of hearing loss, organ dysfunction, hormone deficiencies, dyslipidaemia and diabetes can be undertaken. With this in mind it is appropriate to consider flow charts for *ALMS1* gene mutation testing in children and adults (see Figure 2.2).

Interesting features of Alström syndrome

Predictable consequences

Pathological mutations in the *ALMS1* gene are now known to result in variably progressive retinal degeneration, neuronal deafness, major organ fibrosis and severe post-receptor insulin resistance with associated obesity, type 2 diabetes and dyslipidaemia. These features

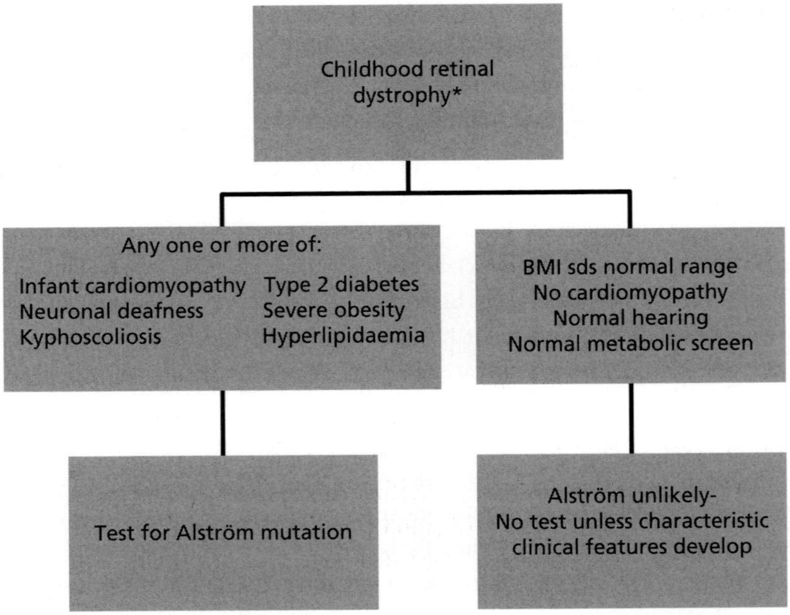

Figure 2.2 Suggested *ALMS1* gene testing in childhood onset retinal dystrophy.
*severe photophobia is usual.

predicate blindness and deafness as well as the potential for cardiac, renal and hepatic dysfunction and atherosclerosis. However, some aspects of the condition remain unexplained.

Variability

The variance in development of all the major complications discussed still remains to be linked with genotype, modifier genes or environment. Moreover the precise mechanism linking ALMS protein dysfunction with cardiac, hepatic and renal fibrosis is not understood.

Reversible infant cardiomyopathy

The devastating suddenness of onset and almost unique capacity for full recovery of cardiomyopathy in 40% of Alström infants is remarkable. There is a fundamental change in cardiomyocyte bio-energetics in the neonatal period with a switch from carbohydrate to fatty acid metabolism marked by decline in malonyl CoA expression and many other linked alterations in gene expression. This requires adaptation in cell trafficking perhaps baffled in the syndrome by impairment of microtubular function. The complexity of the physiology involved is well described in a wide ranging review article (Stanley et al. 2005). A greater understanding of cardiac substrate handling in the syndrome should inform mechanisms in non-ischaemic heart failure and its treatment in the wider population as well as in Alström adults.

Kyphoscoliosis

There are many acquired and genetic causes of kyphoscoliosis, amongst the latter Ehlers–Danlos syndrome and osteogenesis imperfecta (Baumann et al. 2012; Beighton et al. 1983; De Paepe & Malfait, 2012; Rohrbach et al. 2011). The precise sequence of events whereby a defect in ciliary protein found in every cell can cause the variety skeletal abnormalities found in Alström syndrome remains obscure.

Peripheral nervous system

There is a striking absence of peripheral neuropathy in Alström syndrome subjects despite the long duration of type 2 diabetes, dyslipidaemia and renal disease in many older patients. This is in contrast to an age- and gender-matched group of young onset type 2 diabetic persons (Paisey et al. 2009). It may be that impairment of microtubular function would reduce expression of glucose transporter 2 receptors and prevent excessive glucose flux in nervous tissue. Whether this or another mechanism is in operation the finding should open up a fruitful area of investigation exploring susceptibility to peripheral neuropathy in a range of conditions particularly diabetes.

Genetics

Discovery of the *ALMS1* gene

Carl Henry Alström correctly inferred from his initial family studies that the syndrome was likely to be inherited in a Mendelian autosomal recessive manner. A series of careful studies linking geneology, phenotype and linkage patterns in Acadian (Newfoundland French Canadians) enabled the gene to be localised to the short arm of chromosome 2 (Marshall et al. 1997). The gene was co-discovered by further analysis of a larger cohort of Alström subjects by Marshall and colleagues and by Wilson from a key translocation through the gene in one case (Collin et al. 2002; Hearn et al. 2002).

Variation in genotype

Pathological mutations in both copies of the *ALMS1* gene have been identified in the majority of families with typical Alström phenotype. Over 80% of mutations have been found in exons 8, 10 and 16 (Joy et al. 2007) and 80 different mutations have so far been identified. Even in populations with many generations of consanguinity there are many cases of compound heterozygosity as well as homozygosity for causative mutations (Aldahmesh et al. 2009). Larger studies will be required to conclusively establish whether there are genotype–phenotype correlations in the syndrome. So far, mutations in other genes have not been associated with the phenotype.

Future studies

With new more powerful techniques for exome sequencing it will be possible to identify many more mutations in the gene and link them to phenotype in Alström patients as well as common conditions including obesity, insulin resistance, cardiomyopathy and major organ fibrosis. The recent location of the promoter region for *ALMS1* (Hearn et al. 2005; Purvis et al. 2010) will eventually allow correlation of genetic heterogeneity in this region with variations in severity of the major manifestations of Alström syndrome. Modifier genes and epigenetic influences on transcription factors could also exert an influence on phenotype as is emerging in the case of cystic fibrosis (Lele, 2009). These studies are, however, more straightforward in cystic fibrosis where 70% of cases result from the same F508 deletion and calibration of severity of phenotype is readily determined by respiratory function tests.

Prenatal diagnosis

Pre-marriage, pre-conception, pre-implantation and prenatal diagnosis can be undertaken in Alström families where the precise gene causing mutations have been identified in the family. Where consanguinity is an issue autozygosity mapping is important as exemplified in the identification of causative mutations in kindreds with the related ciliopathy Bardet–Biedl syndrome (White et al. 2007). The extent of autozygosity in the wider population and its growing application in identifying recessive disease causing mutations has been well described (Wang et al. 2009). Finally, the phenomenon of autozygosity highlights the potential for the risk of more than one recessive disorder in a family.

Physiology of Alström syndrome

In the past 10 years there has been a watershed of information leading to the realisation of the pivotal role of the cilium in eukaryotic cells. Much of this new understanding has come from identification of pathological mutations causing a number of rare inherited conditions now collectively referred to as ciliopathies (Adams et al. 2007; Badano et al. 2006; Tobin & Beales, 2009; Waters & Beales, 2011). The primordial cilium has long been known to be a component of all eukaryotic cells but until recently considered inactive. The discovery of the genes associated with polycystic kidney disease, primary ciliary dyskinesia, Leber's optic atrophy, Joubert syndrome, Bardet–Biedl syndrome, Meckel–Gruber syndrome, Alström syndrome and Usher syndrome quickly led to the association of these conditions with dysfunction of ciliary proteins. The cilium, basal bodies/centrosomes and linked microtubules are now known to be comprised of 1000 proteins and to be crucial in cell physiology. This involves sensing external stimuli, transference of proteins between cell compartments, focussing dynamic changes in microtubular generation including cell division, differentiation of cells and genesis of organ shape and symmetry. This very basic role in cell physiology underlies the potential for multisystem complications in the ciliopathies. The ALMS protein localises to the centrosome in all cells but its precise function is not known. It also associates with actin of microtubules in the cochlea (Collin et al. 2012). The selective nature and variable severity of organ dysfunction in Alström patients remain unexplained. Visual and hearing impairment characteristic of the syndrome might be expected because of the role of the cilium in replenishment of photoreceptor cell complexes in the retina and transduction of vibration to the auditory nerve from the hair cells of the organ of Corti. However, there is no evidence of anosmia or primary ciliary dyskinesia. This is surprising in view of the crucial role of cilia in olfactory cells and the respiratory epithelium. The influence of the cilium in the planar cell polarity pathway might expected to give rise to structural congenital heart disease, situs inversus, mental retardation or scoliosis but only the latter has been shown to occur in Alström syndrome so far and that in a minority of families.

Apparent receptor abnormalities in Alström syndrome are likely to be linked to an effect of the mutated protein on intracellular microtubular function most notably resulting in insulin resistance. Associated obesity and dyslipidaemia confirm post-insulin receptor/intracellular insulin resistance. The most likely mechanism is reduced expression of glucose transporters in response to insulin binding. This would particularly affect muscle cells in which glucose transporter 4 is the port of entry for glucose into this tissue and of high affinity. A likely consequence would be diversion of glucose to the liver with excess triglyceride synthesis and very low density lipoprotein (VLDL) secretion leading to fatty liver, high serum triglyceride and low HDL cholesterol levels typical of both Alström syndrome and the metabolic syndrome. The variable degree of hypertriglyceridaemia and heterogeneity of type 2 diabetes in Alström patients is likely to result from activity of modifier genes and interaction with lifestyle. The precise relationship between ciliopathies, appetite control and obesity is unclear, though there is one report in a mouse model of morphological change in the hypothalamic cell (Heydet et al. 2012).

Other endocrine disorders such as male hypogonadism, partial growth hormone deficiency, hypothyroidism and polycystic ovarian syndrome are probably caused by a combination of endocrine gland fibrosis, insulin resistance and hypothalamic/pituitary under-activity. Major organ fibrosis commonly affects heart kidney and liver and is most likely related to the primary ciliary defect rather than a consequence of obesity, arteriosclerosis or diabetes. There are no reports of organ specific or non-organ specific autoantibodies in the syndrome. The relationship between the ALMS protein dysregulation and apoptosis is under investigation.

There is a significant clinical interaction between major complications in the syndrome. Most obviously suitability for renal transplantation requires a body mass index less than 35 kg/m^2, good control of hyperlipidaemia and diabetes and adequate cardiac function. The haemodynamic effects of variable renal, cardiac and hepatic dysfunction can lead to a wide range of effects on blood pressure and renal perfusion.

Treatment of Alström syndrome

Psychosocial needs

From diagnosis all active family members should be involved in full discussion of the implications of the syndrome as appropriate for the age of the child. It is important to clarify that not all those with the syndrome progress to major organ failure but that regular surveillance will identify problems early and allow best balance of treatments to be developed.

When school age approaches a clear perspective should be developed with each family to inform decisions about education. Whether mainstream or special schooling for the blind is chosen a care plan including nutrition, access to exercise and integration in the community is the ideal. The Alström Syndrome UK support group has been an effective agency in this work.

Visual impairment

No known treatments have been discovered which alter the progressive visual loss caused by the severe cone–rod dystrophy, so that at present monitoring and support in Braille and computer skills are key. Childhood onset visual impairment is the rule but some individuals maintain capacity for navigational vision and ability to read large print until the third decade. Interventions must therefore be individualised. The most distressing early symptom, often evident in the first 6 months of life, is photophobia. Great day-to-day distress can be avoided by the provision of dark glasses. Cataracts are common, unrelated to the presence or absence of diabetes but can theoretically develop early enough to affect residual vision. Cataract extraction has been performed both to maximise visual acuity and for cosmetic reasons.

Braille and computer speech synthesisers have been enormously helpful to many Alström young people and both should be discussed and introduced as promptly as is feasible. Mobility can be greatly improved by proper training in the use of a white stick and by a guide dog.

Deafness

As with visual loss there are no treatments available to reverse or prevent progression of loss of hearing. The onset of and extent sensori-neural deafness is much more variable than the blindness. It is crucial to treat any otitis media which may compound neural deafness with a conductive problem and to introduce hearing aids in good time to enhance learning and social interaction. A minority of Alström subjects progress to severe hearing loss sufficient to qualify for cochlear implant. This has been successful in one case, but any residual natural auditory function is abolished by the procedure.

Cardiomyopathy

In infancy prompt recognition of the cause of sudden collapse and intensive treatment with oxygen, diuretics and ventilation if necessary are life-saving. Subsequent recovery is usual and the heart structurally normal unless there is a superimposed congenital heart defect. However, the influence of the cilium on the planar cell polarity pathway could result in structural cardiac abnormalities rather than being the consequence of co-inheritance of another recessive disorder. Cardiomyopathy develops in 25% in adolescence sometimes in the context of previous infant cardiac failure but also *de novo*. It is recommended that treatment at this time is guided by serial electrocardiography, echocardiography and cardiac MRI scanning on at least one occasion. Both left and right ventricular fibrosis is usually present (Corbetti et al. 2012; Loudon et al. 2009). Renal impairment and type 2 diabetes are common and hepatic cirrhosis may also be present. Some subjects can, as a result, experience a significant fall in systemic blood pressure and EGFR if renin–angiotensin–aldosterone inhibition is too intense. In these circumstances hydralazine can be a safer hypotensive agent. This may not be unique to Alström syndrome as shown by the recent ALTITUDE trial in hypertensive type 2 diabetic subjects where addition of Aliskerin to angiotensin2 blockade in type 2 diabetic participants caused pre-renal failure and stroke (Angeli et al. 2012). There are no trials to guide best cardioprotective therapy, so empirical treatment with angiotensin-converting enzyme (ACE) inhibition, A2 blockade, beta blockers and diuretics is necessary with careful monitoring of blood pressure and renal function. Atrial fibrillation has been reported in the syndrome with successful control with digoxin and restoration of sinus rhythm by cardioversion. Two UK patients have had dual chamber pacemakers to synchronise ventricular contraction. Lastly, one teenage patient with Alström syndrome has had a successful cardiac transplant associated with valvular heart disease (Goerler et al. 2007). Two UK patients in their third decade have died post-operatively after cardiac transplantation for severe cardiomyopathy. More studies are required to define the key indications and optimum timing of cardiac transplantation in the syndrome.

Metabolism

Severe insulin resistance is always a feature from childhood and has been attributed to intracellular microtubular dysfunction. However, in all other respects the resultant dyslipidaemia and glucose intolerance follow the same heterogeneous pattern and range of responses to treatment as those seen in non-syndromic type 2 diabetes ('Intensive blood-glucose control with sulphonylureas or insulin compared with conventional treatment and risk of complications in patients with type 2 diabetes (UKPDS 33). UK Prospective Diabetes Study (UKPDS) Group,' 1998). Reduced calorie intake and exercise are often sufficient to minimise glucose intolerance in the first two decades. Thereafter metformin is effective, most usefully followed by the addition of incretin analogues if required (Paisey, 2009). A minority require the addition of insulin to control osmotic symptoms in the third decade but earlier if obesity is severe (Bettini et al. 2012; Marshall et al. 1997; Minton et al. 2006; Mokashi & Cummings, 2011; Paisey et al. 2011; Pirgon et al. 2009). Mild hypertriglyceridaemia (2 to 5 mmol/l) is an almost invariable finding, though in some, levels >20 mmol/l require nicotinic acid derivatives to prevent pancreatitis (Paisey, 2009; Paisey et al. 2004, 2008; Wu et al. 2003). A 48-h fast has been successful in re-setting the triglyceride levels below 5 mmol/l when pancreatitis has occurred or levels are >50 mmol/l.

A few patients have required such high doses of insulin (>1000 units daily) to control glycaemia that U500 insulin has been employed to good effect. Whole pancreas transplant has been successful in one phenotypically type 1 diabetic Alström person and several worldwide have benefitted from insulin pump therapy.

Atherosclerosis

In view of the postulated impairment of intermediary metabolism through slowed intracellular trafficking and the combination of diabetes, renal and hepatic fibrosis a high degree of caution is required in prescription of statins. One case of acute coronary syndrome due to coronary atheroma has been reported. There is hope for longer survival too with renal transplantation, effective treatment of diabetes and hyperlipidaemia and prompt treatment of intercurrent infections. It is notable also that the risk factors for atheroma will have been present for over 20 years by the time an Alström person has reached 25 years of age! In practice it has been possible to introduce statins at low or moderate doses (20 mg simvastatin daily or 10 mg rosuvastatin twice weekly) in six subjects >25 years with serum cholesterol/HDL cholesterol ratio >5. Monitoring for muscle pain and raised serum creatine kinase levels is important.

Hepatic steatosis, fibrosis and cirrhosis

No treatment specific to the syndrome has been discovered. If variceal bleeding occurs then beta blocker therapy, banding, trans-jugular intrahepatic stent (TIPS) procedure and accepted treatments for hepatic encephalopathy have all been effective.

Renal failure

A growing number of Alström patients have been offered haemodialysis for up to 3 years successfully and a number have had successful live related donor or cadaveric renal transplantation.

Urethral–detrusor dys-synergia

A minority of female patients present in teenage years with painless delayed bladder emptying. In some this progresses to painful erratic bladder spasms with disordered urethral tightening. Most cases respond to alpha blockers in conventional doses, whilst some have also learned intermittent bladder self-catheterisation. Two UK patients have progressed to urinary diversion procedures involving cystectomy and ileal bladders.

Gastro-oesophageal disorders

Reflux oesophagitis is frequent and usually responds to protein pump inhibitors in conventional doses. Gastric plication has been successful in a severe case with proven acid reflux and oesophageal spasm with normal bowel transit.

Infection

The relationship between the ALMS protein dysfunction and susceptibility to infection remains unclear. A number of anecdotal reports have implicated bacterial lower respiratory tract infection—community acquired or post-operative—in severe hypoxia and death. As discussed previously the combination of restrictive lung capacity, cardiac fibrosis, diabetes and possibly muted response to infection sets the scene for this catastrophe.

Early intensive antibiotic treatment is indicated for respiratory tract infections. Planned post-operative ventilation has proven to be successful in pre-empting surgery associated problem. Morbidity and mortality from respiratory infections are much increased in osteogenesis imperfecta due to kyphoscoliosis (McAllion & Paterson, 1996) and this factor may be under-estimated in Alström patients.

Musculoskeletal system

Chronic pain can result from kyphoscoliosis with impingement of the lower rib cage on the iliac crest, muscle spasm due to flat everted feet and resulting biomechanical problems, and spondylitis. Simple measures such as insoles, exercises and analgaesics are often effective. Major surgery for scoliosis and very severe spondylitis (DISH) has been successfully undertaken.

Gene therapy and stem cell therapy

The place of these new fundamental treatment options in Alström syndrome is unclear. The possibility of a 'global cure' for all the manifestations of the syndrome seems unlikely in view of the wide range of organs involved some of which like the kidney are comprised of several different tissues. The *ALMS1* gene and ALMS protein are very large. This makes it unlikely that whole gene therapy or protein repletion will be feasible. It is to be hoped that once the actions of the ALMS protein have been elucidated that treatments may emerge based on its downstream effects in the photoreceptor, organ of Corti, heart, liver and kidney (Mockel et al. 2012; Redin et al. 2012).

References

Adams, N. A., Awadein, A., Toma, H. S. 2007. The retinal ciliopathies. *Ophthalmic Genet* **28**, 113–125.

Aldahmesh, M. A., Abu-Safieh, L., Khan, A. O., Al-Hassnan, Z. N., Shaheen, R., Rajab, M., et al. 2009. Allelic heterogeneity in inbred populations: The Saudi experience with Alström syndrome as an illustrative example. *Am J Med Genet A* **149A**, 662–665.

Alström, C. H. 1950. A study of epilepsy in its clinical, social and genetic aspects. *Acta Psychiatr Neurol Suppl* **63**, 1–284.

Alström, C. H., Gentz, C., Lindblom, K. 1949. (Not available). *Poumon* **5**, 97–108.

Alström, C. H., Hallgren, B., Nilsson, L. B., Asander, H. 1959. Retinal degeneration combined with obesity, diabetes mellitus and neurogenous deafness: A specific syndrome (not hitherto described) distinct from the Laurence-Moon-Bardet-Biedl syndrome: a clinical, endocrinological and genetic examination based on a large pedigree. *Acta Psychiatr Neurol Scand Suppl* **129**, 1–35.

Alter, C. A., Moshang, T. Jr, 1993. Growth hormone deficiency in two siblings with Alström syndrome. *Am J Dis Child* **147**, 97–99.

Angeli, F., Reboldi, G., Mazzotta, G., Poltronieri, C., Garofoli, M., Ramundo, E., et al. 2012. Safety and efficacy of aliskiren in the treatment of hypertension and associated clinical conditions. *Curr Drug Saf* **7**, 76–85.

Atar, M., Körperich, E. J. 2010. Systemic disorders and their influence on the development of dental hard tissues: A literature review. *J Dent* **38**, 296–306.

Awazu, M., Tanaka, T., Sato, S., Anzo, M., Higuchi, M., Yamazaki, K., et al. 1997. Hepatic dysfunction in two sibs with Alström syndrome: Case report and review of the literature. *Am J Med Genet* **69**, 13–16.

Badano, J.L., Mitsuma, N., Beales, P.L., Katsanis, N. 2006. The ciliopathies: An emerging class of human genetic disorders. *Annu Rev Genomics Hum Genet* **7**, 125–148.

Baumann, M., Giunta, C., Krabichler, B., Rüschendorf, F., Zoppi, N., Colombi, M., et al. 2012. Mutations in FKBP14 cause a variant of Ehlers-Danlos syndrome with progressive kyphoscoliosis, myopathy, and hearing loss. *Am J Hum Genet* **90**, 201–216.

Beighton, P., Spranger, J., Versveld, G. 1983. Skeletal complications in osteogenesis imperfecta. A review of 153 South African patients. *S. Afr. Med J.* **64**, 565–568.

Bettini, V., Maffei, P., Pagano, C., Romano, S., Milan, G., Favaretto, F., et al. 2012. The progression from obesity to type 2 diabetes in Alström syndrome. *Pediatr Diabetes* **13**, 59–67.

Bond, J., Flintoff, K., Higgins, J., Scott, S., Bennet, C., Parsons, J., et al. 2005. The importance of seeking ALMS1 mutations in infants with dilated cardiomyopathy. *J Med Genet* **42**, e10.

Bremer, L. A., Blackman, S. M., Vanscoy, L. L., McDougal, K. E., Bowers, A., Naughton, K. M., et al. 2008. Interaction between a novel TGFB1 haplotype and CFTR genotype is associated with improved lung function in cystic fibrosis. *Hum Mol Genet* **17**, 2228–2237.

Chang, K. W., Hou, J. W., Lin, S. J., Kong, M. S. 2000. Alstrom syndrome with hepatic dysfunction: Report of one case. *Acta Paediatr Taiwan* **41**, 270–272.

Chou, P. I., Chen, C. H., Chen, J. T., Wen, L. Y., Wu, D. A., Feldon, S. E. 2000. Alström syndrome with subclinical insulin-resistant diabetes and hepatic dysfunction: A family report. *J Pediatr Ophthalmol Strabismus* **37**, 179–182.

Collaco, J.M., Vanscoy, L., Bremer, L., McDougal, K., Blackman, S.M., Bowers, A., et al. 2008. Interactions between secondhand smoke and genes that affect cystic fibrosis lung disease. *JAMA* **299**, 417–424.

Collin, G. B., Marshall, J. D., Ikeda, A., So, W. V., Russell-Eggitt, I., Maffei, P., et al. 2002. Mutations in ALMS1 cause obesity, type 2 diabetes and neurosensory degeneration in Alström syndrome. *Nat Genet* **31**, 74–78.

Collin, G. B., Marshall, J. D., King, B. L., Milan, G., Maffei, P., Jagger, D. J., Naggert, J. K. 2012. The Alström syndrome protein, ALMS1, interacts with α-actinin and components of the endosome recycling pathway. *PLoS One* **7**, e37925.

Connolly, M. B., Jan, J. E., Couch, R. M., Wong, L. T., Dimmick, J. E., Rigg, J. M. 1991. Hepatic dysfunction in Alström disease. *Am J Med Genet* **40**, 421–424.

Corbetti, F., Razzolini, R., Bettini, V., Marshall, J. D., Naggert, J., Tona, F., et al. 2012. Alström Syndrome: Cardiac Magnetic Resonance findings. *Int J Cardiol.* 10 April 2012, [Epub ahead of print].

De Paepe, A., Malfait, F. 2012. The Ehlers-Danlos syndrome, a disorder with many faces. *Clin Genet* **82**, 1–11.

Dorfman, R., Sandford, A., Taylor, C., Huang, B., Frangolias, D., Wang, Y., et al. 2008. Complex two-gene modulation of lung disease severity in children with cystic fibrosis. *J Clin Invest* **118**, 1040–1049.

Gámez, A., Pérez, B., Ugarte, M., Desviat, L. R. 2000. Expression analysis of phenylketonuria mutations. Effect on folding and stability of the phenylalanine hydroxylase protein. *J. Biol. Chem.* **275**, 29737–29742.

Garg, R. A., Singh, J., Mathur, B. B. 1991. Alstrom syndrome. *Indian Pediatr* **28**, 799–801.

Goerler, H., Warnecke, G., Winterhalter, M., Müller, C., Ballmann, M., Wessel, A., et al. 2007. Heart-lung transplantation in a 14-year-old boy with Alström syndrome. *J Heart Lung Transplant* **26**, 1217–1218.

Hearn, T., Renforth, G. L., Spalluto, C., Hanley, N. A., Piper, K., Brickwood, S., et al. 2002. Mutation of ALMS1, a large gene with a tandem repeat encoding 47 amino acids, causes Alström syndrome. *Nat Genet* **31**, 79–83.

Hearn, T., Spalluto, C., Phillips, V. J., Renforth, G. L., Copin, N., Hanley, N. A., et al. 2005. Subcellular localization of ALMS1 supports involvement of centrosome and basal body dysfunction in the pathogenesis of obesity, insulin resistance, and type 2 diabetes. *Diabetes* **54**, 1581–1587.

Heydet, D., Chen, L. X., Larter, C. Z., Inglis, C., Silverman, M. A., Farrell, G. C., et al. 2012. A truncating mutation of Alms1 reduces the number of hypothalamic neuronal cilia in obese mice. *Dev Neurobiol* 73, 1–13.

Himsworth, H. P. and Kerr, R. B. 1939. Insulin-sensitive and insulin-insensitive types of diabetes mellitus. *Clin Sci* **4**, 119.

Iannello, S., Bosco, P., Camuto, M., Cavaleri, A., Milazzo, P., Belfiore, F. 2004. A mild form of Alström disease associated with metabolic syndrome and very high fasting serum free fatty acids: two cases diagnosed in adult age. *Am J Med Sci* **327**, 284–288.

Ikeda, Y., Morita, Y., Matsuo, Y., Akanuma, Y., Itakura, H. 1974. [A case of Alström syndrome associated with situs inversus totalis and characteristic liver cirrhosis (author's transl)]. *Nippon Naika Gakkai Zasshi* **63**, 1303–1311.

Jatti, K., Paisey, R., More, R. 2012. Coronary artery disease in Alström syndrome. *Eur J Hum Genet* **20**, 117–118.

Joy, T., Cao, H., Black, G., Malik, R., Charlton-Menys, V., Hegele, R. A., et al. 2007. Alstrom syndrome (OMIM 203800): A case report and literature review. *Orphanet J Rare Dis* **2**, 49.

Khoo, E. Y. H., Risley, J., Zaitoun, A. M., El-Sheikh, M., Paisey, R. B., Acheson, A. G., et al. 2009. Alström syndrome and cecal volvulus in 2 siblings. *Am J Med Sci.* **337**, 383–385.

Lee, N.-C., Marshall, J. D., Collin, G. B., Naggert, J. K., Chien, Y.-H., Tsai, W.-Y., et al. 2009. Caloric restriction in Alström syndrome prevents hyperinsulinemia. *Am J Med Genet A* **149A**, 666–668.

Lele, R. D. 2009. Epigenetics—gene silencing. *J Assoc Physicians India* **57**, 60–66.

Liu, L., Dong, B., Chen, X., Li, J., Li, Y. 2009. Identification of a novel ALMS1 mutation in a Chinese family with Alström syndrome. *Eye (London)* **23**, 1210–1212.

Loudon, M. A., Bellenger, N. G., Carey, C. M., Paisey, R. B. 2009. Cardiac magnetic resonance imaging in Alström syndrome. *Orphanet J Rare Dis* **4**, 14.

Macari, F., Lautier, C., Girardet, A., Dadoun, F., Darmon, P., Dutour, A., et al. 1998. Refinement of genetic localization of the Alström syndrome on chromosome 2p12-13 by linkage analysis in a North African family. *Hum Genet* **103**, 658–661.

Maffei, P., Boschetti, M., Marshall, J. D., Paisey, R. B., Beck, S., Resmini, E., et al. 2007. Characterization of the IGF system in 15 patients with Alström syndrome. *Clin Endocrinol. (Oxford)* **66**, 269–275.

Marshall, J. D., Bronson, R. T., Collin, G. B., Nordstrom, A. D., Maffei, P., Paisey, R. B., et al. 2005. New Alström syndrome phenotypes based on the evaluation of 182 cases. *Arch Intern Med* **165**, 675–683.

Marshall, J. D., Ludman, M. D., Shea, S. E., Salisbury, S. R., Willi, S. M., LaRoche, R. G., et al. 1997. Genealogy, natural history, and phenotype of Alström syndrome in a large Acadian kindred and three additional families. *Am J Med Genet* 73, 150–161.

Marshall, J. D., Maffei, P., Beck, S., Barrett, T. G., Paisey, R. B. 2013. Clinical utility gene card for: Alström syndrome. *Eur J Hum Genet* 24 April 2013. doi: 10.1038/ejhg.2013.61. [Epub ahead of print].

McAllion, S. J. & Paterson, C. R. 1996. Causes of death in osteogenesis imperfecta. *J Clin Pathol* **49**, 627–630.

Michaud, J. L., Héon, E., Guilbert, F., Weill, J., Puech, B., Benson, L., et al. 1996. Natural history of Alström syndrome in early childhood: Onset with dilated cardiomyopathy. *J Pediatr* **128**, 225–229.

Minton, J. A. L., Owen, K. R., Ricketts, C. J., Crabtree, N., Shaikh, G., Ehtisham, S., et al. 2006. Syndromic obesity and diabetes: Changes in body composition with age and mutation analysis of ALMS1 in 12 United Kingdom kindreds with Alstrom syndrome. *J Clin Endocrinol Metab* **91**, 3110–3116.

Mockel, A., Obringer, C., Hakvoort, T. B. M., Seeliger, M., Lamers, W. H., Stoetzel, C., et al. 2012. Pharmacological modulation of the retinal unfolded protein response in Bardet–Biedl syndrome reduces apoptosis and preserves light detection ability. *J Biol Chem* **287**, 37483–37494.

Mokashi, A. & Cummings, E. A. 2011. Presentation and course of diabetes in children and adolescents with Alstrom syndrome. *Pediatr Diabetes* **12**, 270–275.

Paisey, R. B. 2009. New insights and therapies for the metabolic consequences of Alström syndrome. *Curr Opin Lipidol* **20**, 315–320.

Paisey, R., Barrett, T., Carey, C., Hiwot, T., Cramb, R., White, A., et al. 2011. Rare disorders presenting in the diabetic clinic: An example using audit of the NSCT adult Alström clinics. *Pract Diabetes* **28**, 340–343.

Paisey, R. B., Carey, C. M., Bower, L., Marshall, J., Taylor, P., Maffei, P., Mansell, P. 2004. Hypertriglyceridaemia in Alström's syndrome: Causes and associations in 37 cases. *Clin Endocrinol (Oxford)* **60**, 228–231.

Paisey, R. B., Hodge, D. & Williams, K. 2008. Body fat distribution, serum glucose, lipid and insulin response to meals in Alström syndrome. *J Hum Nutr Diet* **21**, 268–274.

Paisey, R. B., Paisey, R. M., Thomson, M. P., Bower, L., Maffei, P., Shield, J. P. H., et al. 2009. Protection from clinical peripheral sensory neuropathy in Alström syndrome in contrast to early-onset type 2 diabetes. *Diabetes Care* **32**, 462–464.

Patel, S., Minton, J. A. L., Weedon, M. N., Frayling, T. M., Ricketts, C., Hitman, G. A., et al. 2006. Common variations in the *ALMS1* gene do not contribute to susceptibility to type 2 diabetes in a large white UK population. *Diabetologia* **49**, 1209–1213.

Pirgon, Ö., Atabek, M. E. & Tanju, I. A. 2009. Metabolic syndrome features presenting in early childhood in Alström syndrome: A case report. *J Clin Res Pediatr Endocrinol* **1**, 278–280.

Purvis, T. L., Hearn, T., Spalluto, C., Knorz, V. J., Hanley, K. P., Sanchez-Elsner, T., et al. 2010. Transcriptional regulation of the Alström syndrome gene ALMS1 by members of the RFX family and Sp1. *Gene* **460**, 20–29.

Quiros-Tejeira, R. E., Vargas, J. & Ament, M. E. 2001. Early-onset liver disease complicated with acute liver failure in Alstrom syndrome. *Am J Med Genet* **101**, 9–11.

Redin, C., Le Gras, S., Mhamdi, O., Geoffroy, V., Stoetzel, C., Vincent, M.-C., et al. 2012. Targeted high-throughput sequencing for diagnosis of genetically heterogeneous diseases: Efficient mutation detection in Bardet–Biedl and Alstrom Syndromes. *J Med Genet* **49**, 501–512.

Rohrbach, M., Vandersteen, A., Yiş, U., Serdaroglu, G., Ataman, E., Chopra, M., et al. 2011. Phenotypic variability of the kyphoscoliotic type of Ehlers–Danlos syndrome (EDS VIA): Clinical, molecular and biochemical delineation. *Orphanet J Rare Dis* **6**, 46.

Russell-Eggitt, I. M., Clayton, P. T., Coffey, R., Kriss, A., Taylor, D. S., Taylor, J. F. 1998. Alström syndrome. Report of 22 cases and literature review. *Ophthalmology* **105**, 1274–1280.

Satman, I., Yilmaz, M. T., Gürsoy, N., Karşidağ, K., Dinççağ, N., Ovali, T., et al. 2002. Evaluation of insulin resistant diabetes mellitus in Alström syndrome: A long-term prospective follow-up of three siblings. *Diabetes Res Clin Pract* **56**, 189–196.

Smith, J. C., McDonnell, B., Retallick, C., McEniery, C., Carey, C., Davies, J. S., et al. 2007. Is arterial stiffening in Alström syndrome linked to the development of cardiomyopathy? *Eur J Clin Invest* **37**, 99–105.

Stanley, W. C., Recchia, F. A. & Lopaschuk, G. D. 2005. Myocardial substrate metabolism in the normal and failing heart. *Physiol Rev* **85**, 1093–1129.

Tai, T.-S., Lin, S.-Y. & Sheu, W.H.-H. 2003. Metabolic effects of growth hormone therapy in an Alström syndrome patient. *Horm Res* **60**, 297–301.

Tiwari, A., Awasthi, D., Tayal, S. & Ganguly, S. 2010. Alstrom syndrome: A rare genetic disorder and its anaesthetic significance. *Indian J Anaesth* **54**, 154–156.

Tobin, J. L. & Beales, P. L. 2009. The nonmotile ciliopathies. *Genet Med* **11**, 386–402.

UK Prospective Diabetes Study (UKPDS) Group, 1998. Intensive blood-glucose control with sulphonylureas or insulin compared with conventional treatment and risk of complications in patients with type 2 diabetes (UKPDS 33). *Lancet* **352**, 837–853.

Van den Abeele, K., Craen, M., Schuil, J. & Meire, F. M. 2001. Ophthalmologic and systemic features of the Alström syndrome: Report of 9 cases. *Bull Soc Belge Ophtalmol* 67–72.

Wang, S., Haynes, C., Barany, F. & Ott, J. 2009. Genome-wide autozygosity mapping in human populations. *Genet Epidemiol* **33**, 172–180.

Waters, A. M. & Beales, P. L. 2011. Ciliopathies: An expanding disease spectrum. *Pediatr Nephrol* **26**, 1039–1056.

Watson, J. D. & Crick, F. H. 1953. Molecular structure of nucleic acids; a structure for deoxyribose nucleic acid. *Nature* **171**, 737–738.

White, D. R. A., Ganesh, A., Nishimura, D., Rattenberry, E., Ahmed, S., Smith, U. M., et al. 2007. Autozygosity mapping of Bardet–Biedl syndrome to 12q21.2 and confirmation of FLJ23560 as BBS10. *Eur J Hum Genet* **15**, 173–178.

Wu, W.-C., Chen, S.-C., Dia, C.-Y., Yu, M.-L., Hsieh, M.-Y., Lin, Z.-Y., et al. 2003. Alström syndrome with acute pancreatitis: A case report. *Kaohsiung J Med Sci* **19**, 358–361.

Yallow, R. S. & Berson, S. A. 1960. Immunoassay of endogenous plasma insulin in man. *J Clin Invest* **39**, 1157–1175.

Chapter 3

Jeune syndrome and the ciliary chondrodysplasias

Miriam Schmidts

History of the eponym

Ciliary chondrodysplasias represent a group of rare inherited diseases impairing skeletal development, collectively caused by recessive mutations in ciliary genes. They can be subdivided into different groups of severity, clinical phenotype and underlying genetic defects.

Short-rib–polydactyly syndromes

This group of perinatally lethal skeletal dysplasias is characterised by severe narrowing of the thorax leading to pulmonary hypoplasia, short limbs (micromelia) and polydactyly. Five subtypes are phenotypically distinguished:

- SRPS I (Saldino–Noonan type, OMIM #263530, Saldino & Noonan, 1972)
- SRPS II (Majewski type, OMIM #263520, Majewski et al. 1971)
- SRPS III (Verma–Naumoff type, OMIM #263510, Verma et al., 1975; Naumoff et al. 1977),
- SRPS IV (Beemer–Langer type, OMIM #269860, Beemer et al. 1983)
- SRPS V (OMIM #614091, Mill et al. 2011)

Oral–facial–digital syndrome 4

Oral–facial–digital syndrome IV (OFD IV, OFD syndrome with tibia defects, Mohr–Majewski syndrome, Baraitser–Burn syndrome, OMIM #258860) was first described by Burn et al. 1984 and Baraitser, 1986. The latter noticed a phenotypical overlap with SRPS Majewski type with pre- and post-axial polydactyly of the hands and feet, tibia hypoplasia and oral and facial defects.

Asphyxating thoracic dystrophy

Asphyxating thoracic dystrophy (ATD, Jeune syndrome, Jeune asphyxating thoracic dystrophy, JATD, OMIM #208500) was first described in 1955 by Jeune et al. and followed by Maroteaux and Savart in 1964. JATD patients present with variable degree of rib shortening, typical pelvis configuration with trident acetabular roof and acetabular spurs and rarely exhibit polydactyly.

Mainzer–Saldino syndrome

Mainzer–Saldino syndrome (MSS, cono-renal syndrome, OMIM #266920) was recognised in 1970 by Mainzer et al. who described two patients with cone-shaped epiphyses of the hand, retinal disease and deterioration of renal function. A narrow ribcage, craniosynostosis and liver involvement may be present as well and the disease was first associated with ciliary malfunction by Perrault et al. 2012.

Cranial–ectodermal dysplasia

Cranial–ectodermal dysplasia (CED, Sensenbrenner syndrome, Levin syndrome, OMIM #218330), a combination of dolichocephalus due to craniosynostosis of the sutura sagittalis, epicanthus, very thin, sparse and slow growing hair, teeth abnormalities, brachydactyly and short ribs, was described by Levin et al. in 1977. CED is counted towards the ciliary chondrodysplasias since Walczak–Sztulpa et al. identified biallelic mutations in *IFT122* as the underlying cause in an affected sib pair in 2010.

Ellis–van Creveld syndrome

Ellis–van Creveld syndrome (EVC, chondroectodermal dysplasia, mesoectodermal dysplasia, OMIM #225500) was first observed by Richard Ellis and Simon van Creveld in the 1940 but McKusick et al. described a large consanguineous Amish family in Pennsylvania in 1964 who exhibited acromelic dwarfism, polydactyly of the hands dysplastic nails, teeth abnormalities and cardiac defects. As Langer described 1968, radiological features in EVC are very difficult to distinguish from those observed in JATD; however, cardiac defects are common in EVC while JATD patients may rather present with additional renal, liver and retinal disease.

Weyers acrofacial dysostosis

Weyers acrofacial dysostosis (Curry–Hall syndrome, OMIM #193530), first described by Weyers, 1952, is allelic to EVC but inherited in an autosomal dominant pattern. Patients present with a milder phenotype of polydactyly, dentition anomalies and dystrophic nails.

Epidemiology

All the ciliary chondrodysplasias described in this chapter are very rarely observed in non-consanguineous populations with frequencies below 1 in 200,000 or even below 1 in 1,000,000, but might occur more frequently in isolated and consanguineous populations due to their autosomal recessive inheritance pattern. As larger patient studies have not been published to date, the exact frequency of this disease group remains unclear.

For JATD (Jeune syndrome), a prevalence of approximately 1 in 200,000 has been observed in Western countries (Oberklaid et al. 1977) which may also apply to other SRPS forms but exact numbers are unknown to date. The frequency of CED (Sensenbrenner syndrome) and MSS in the population also remains unknown; however, these diseases seem to occur even more rarely than JATD so their expected frequency is well below 1 in 200,000; however, multiple SRPS IV cases have been described among Hungarian Roma families by Kovacs et al. in 2006. OFD IV occurs presumably with a similar frequency of less than 1 in 200,000. Ellis–van Creveld syndrome (EVC) is thought to affect less than 1 in 1,000,000 people but many of the cases described in the literature to date have been observed within the Amish Community in Pennsylvania. The prevalence of Weyers acrofacial dysostosis, which often co-occurs in EVC pedigrees, remains unknown.

Clinical features

All ciliary chondrodysplasias are characterised by developmental skeletal defects, mainly affecting limbs, ribs and sometimes also the craniofacial skeleton. While all SRPS subtypes are lethal perinatally due to cardiorespiratory failure, JATD, MSS, CED, OFD IV and EVC patients often survive longer due to less severely narrowed ribcage (Baujat et al. 2013; Huber and Cormier-Daire, 2012; Schmidts et al. 2013). Most ciliary chondrodysplasias also exhibit additional extraskeletal features such as renal, liver, heart or eye disease. In the following section, disorders are grouped according to phenotypic as well as genotypic similarities (see Figure 3.1 and Table 3.1)

SRPS spectrum group 1: SRPS II, SRPS IV and OFD IV

SRPS II (Majewski syndrome)

SRPS II patients present with hydropic appearance at birth exhibiting a very narrow thorax with horizontally oriented ribs, protuberant abdomen and shortened tubular bones with smooth ends. In contrast to some of the other SRPS subtypes, SRPS II is defined by

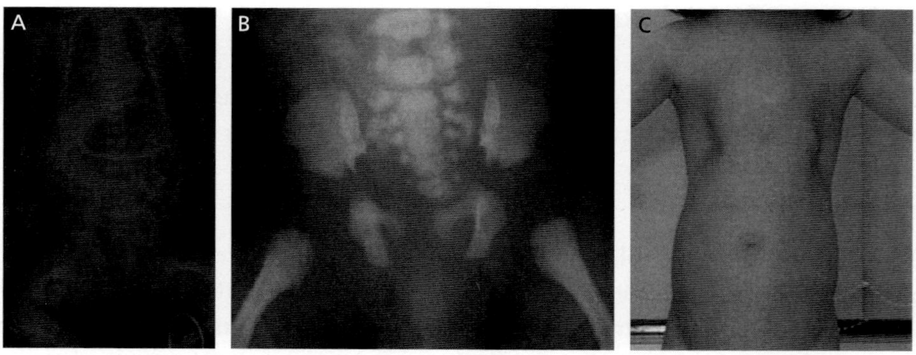

Figure 3.1 Typical radiological and clinical findings in ciliary chondrodysplasias. Example shown for JATD: (A) Newborn, short femora, narrow thorax with horizontal ribs and protuberating abdomen. (B) Pelvis: Acetabular spurs. (C) thorax configuration in a JATD patient aged 3 years.
Images courtesy of Hulya Kayserili, Istanbul, Turkey and Bernhard Zabel, Freiburg, Germany.

Table 3.1 Summary of the clinical, radiological and genetic findings in ciliary chondrodysplasias

Group	Name	Rib shortening	Micromelia	Polydactyly	Other defining radiological features	Additional features	Prognosis	Causative gene
1	SRPSII (Majewski)	+++	+	Consistent (pre and post axial)	Disproportionally small and oval tibia. Normal pelvis.	Cleft lip and palate. Cystic renal disease. Genital abnormalities.	Perinatal lethality	*NEK1 DYNC2H1*
1	SRPSIV (Beemer–Lange)	+++	+	Rare	Similar to SRPSII, tibia not small	Cleft lip	Perinatal lethality	-
1	OFD4 (Mohr–Majewski)	+ to +++	+	Common (pre- and post-axial)	Disproportionally small and oval tibia. -	Cleft/pseudo cleft of the lip and palate. Lobulated tongue.	Dependent on degree of thoracic constriction	*TCTN3*
2	SRPSI (Saldino–Noonan)	+++	++	Common	Pointed ends of long bones, defective ossification, hypoplastic ilia, and acetabular spurs.	Cystic kidneys, Great vessel, gastrointestinal - and genitourinary malformations	Perinatal lethality	-
2	SRPSIII (Verma–Naumoff)	+++	+	Common	Rounded ends of long bones. Longitudinal spurs. Hypoplastic ilia, acetabular spurs. Short cranial base, bulging forehead depressed nasal bridge.		Perinatal lethality	*DYNC2H1, IFT80*
3	JATD (Jeune)	+ to ++	+	Rare	Pelvis with trident acetabular roof, lateral spikes and short iliac bones with Irregular metaphyses. Bowing of the femur. Small middle phalanges.	Renal and hepatic disease. Retinal degeneration (rare).	Dependent on degree of thoracic constriction/Severity of Renal Involvement	*DYNC2H1, IFT80, WDR19, TTC21B IFT140*

Table 3.1 (continued) Summary of the clinical, radiological and genetic findings in ciliary chondrodysplasias

Group	Name	Rib shortening	Micromelia	Polydactyly	Other defining radiological features	Additional features	Prognosis	Causative gene
3	Mainzer–Saldino (conorenal syndrome)	+ to +++	(+)	-	Cone shaped epiphyses of the phalanges.	Early onset retinopathy and renal disease. Hepatic fibrosis.	Dependent on degree of thoracic constriction/severity of renal involvement	*IFT140*
3	SRPSV	+++	+	Common	Similar to SRPSIII but with acromesomelic hypomineralisation and campomelia.	Renal dysplasia, gut malrotation	Perinatal lethality	*WDR35*
3	Cranioectodermal dysplasia (CED, Sensenbrenner)	+ to +++	(+)	-	Craniosynostosis. *Dolichocephalus*	Dysplastic teeth. Sparse, fine slow growing hair. Renal disease.	Variable. Probably mainly due to renal failure in the past	*WDR35, IFT43, WDR19, IFT122*
4	Ellis–van Creveld (EVC)	+ to +++	+	Consistent (post-axial)	Pelvis: Short ilium, acetabular spurs, irregular metaphyses. Bowing of the femur. Small middle phalanges.	Dysplastic teeth and fingernails. Cardiac malformations especially atrial septation defects	Defined by degree of thoracic constriction/cardiac defects	*EVC1, EVC2*
4	Weyers acrofacial dysostosis	-	+	Consistent	No thoracic phenotype	Dysplastic teeth and nails	No significant lethality	*EVC1, EVC2*

a disproportionately short and oval-shaped tibia (shorter than the fibula), normal pelvic development and the presence of median cleft lip and palate and shortened tubular bones with smooth ends. Patients consistently have pre- and post-axial polydactyly. Larynx and epiglottis malformations, renal cysts, genital abnormalities and cerebral malformation are also observed in this variant. Distinguishing SRPS II from OFD IV can be difficult as both disorders present with dysplastic tibia (Chen et al. 1980; Majewski et al. 1971; Spranger et al. 1974, 2002).

SRPS IV (Beemer–Langer syndrome)

SRPS IV was first described in a case report detailing the findings in two unrelated individuals presenting with median cleft of the upper lip, narrow chest with short horizontal ribs and short, bowed limbs (Beemer et al., 1983). Clinical presentation is strikingly similar to SRPSII with hydropic appearance at birth, short long bones, small thorax and protuberant abdomen but polydactyly is often absent. If present polydactyly can be pre- or post-axial. The main phenotypic difference to SRPSII is the lack of a dysplastic tibia as the tibia in SRPS IV is relatively well tubulated and longer than the fibula. Further, tubular bones have smooth metaphyseal margins. Cases with extraskeletal malformations such as renal involvement, brain malformations or congenital heart defects have been described in the literature (Beemer et al. 1983; Beighton et al. 1992; Kovacs et al. 2006; Lin et al. 1991; Passarge, 1983; Spranger et al. 2002; Yang et al. 1991).

Oral–facial–digital syndrome 4

OFD IV (orofacial syndrome IV, Mohr–Majewski syndrome) combines features of Majewski type II short-rib–polydactyly syndrome and Mohr syndrome (OFD type II), such as severe tibial dysplasia which is also seen in SRPS II and likewise may present with lobulated tongue. Further, cleft and pseudo-cleft of the lip and or palate, have been described as well as pre- and post-axial polydactyly of the hands and feet. The thoracic constriction is of variable severity. Other features include coloboma of the eye, intrahepatic cysts and hepatic fibrosis, ambiguous genitalia, cystic dysplastic kidneys, anal atresia, brain malformations severe talipes equinovarus and severe bilateral deafness (Ades et al. 1994; Baraitser, 1986; Nevin & Thomas, 1989)

SRPS spectrum group 2: SRPS I, SRPS III

SRPS I (Saldino–Noonan syndrome)

As all SRPS subtypes, SRPS I is lethal in the new-born period. Patients have a hydropic appearance, small thorax with short horizontal ribs and marked micromelia resulting in 'flipper-like' limbs with marked metaphyseal irregularities in X-rays. Pre or post-axial polydactyly is consistently observed and ossification defects of vertebrae, pelvis as well as of the bones of the hands and feet occur frequently. The pelvis shows anomalies including small ilia with flattened acetabular roofs with ossified spurs, resembling the pelvis configuration in EVC and JATD. Extraskeletal manifestations range from polycystic kidneys, transposition of great vessels to atretic lesions of the gastrointestinal and genitourinary systems. Extraskeletal findings reported include disturbance of the sexual development (Marec et al. 1973; Richardson et al. 1977; Saldino & Noonan, 1972; Spranger et al. 2002).

SRPS III (Verma–Naumoff syndrome)

Apart from SRPS-typical features such as hydropic appearance at birth, short long bones, short horizontal ribs, narrow thorax and protuberant abdomen, SRPS III is defined by a short cranial base, bulging forehead, depressed nasal bridge and flat occiput. Radiologically, the long tubular bones show a corticomedullary demarcation, slightly widened metaphyses and marked longitudinal spurs. Patients usually present with pelvic abnormalities similar to those seen in SRPS I. Polydactyly is commonly observed though not completely penetrant. Cleft lip and palate have been observed as well as malformation of the larynx and epiglottis. Extraskeletal findings reported include disturbance of the sexual development (Bernstein et al. 1985; Naumoff, 1980; Naumoff et al. 1977; Spranger et al. 2002; Verma et al. 1975).

Group 3: JATD, SRPS V, MSS and CED

Jeune asphyxating thoracic dystrophy

The skeletal phenotype is mainly characterised by a narrow, sometimes bell-shaped thorax due to shortening of the ribs which restricts pulmonary development and may cause severe respiratory distress throughout the first two years of life. Respiratory problems account for most of the mortality in JATD (mortality rates have been estimated as high as 60% in the literature (Oberklaid et al. 1977) while we have observed a mortality of approx. 20% of life born children in a cohort of 130 families and recorded termination of pregnancy in another 20% of cases (Schmidts et al. 2013, and own unpublished data). However, rib growth seems to catch up after the first one or two years of life so that patients 'grow out' of the respiratory phenotype (Baujat et al. 2013; de Vries et al. 2010; Schmidts et al. 2013).

While renal involvement has been initially reported in up to 30% of JATD patients (Herdman & Langer 1968), our own observations (Schmidts et al. 2013) and a study from Baujat et al. (2013) agree with de Vries et al. (2010), suggesting it may occur in less than 20% of all cases but is frequently observed in association with mutations in the *IFT140* gene, especially if associated with additional retinal disease (Schmidts et al. 2013). Retinal disease in JATD is described as 'rarely observed' in the literature (Allen et al. 1979; Bard et al. 1978) and based on fundoscopy and patient history we noted retinopathy in less than 5% of cases in our cohort (Schmidts et al. 2013). However, this might underestimate the true epidemiology as some patients still might develop retinal disease later in life. Also fundoscopy can appear normal in cases without pigmentary deposits and electroretinogram (ERG) examination as early detector of retinal disease was not routinely performed in all JATD patients. In line with this, Baujat et al. observed ERG abnormalities in up to 50% of JATD patients in 2013. It is not clear to date how many of these patients will develop clinically manifest retinal disease at a later timepoint.

Other extraskeletal features include pancreatic lesions and (usually) mild liver disease (Hopper et al. 1979). One patient with JATD has been reported in the literature to have recieved liver transplantation (Yerian et al. 2003) and one out of 130 cases in our own cohort has passed away from liver failure (own unpublished data).

MSS

Identified as ciliary chondrodysplasia due to biallelic mutations in *IFT140* by Perrault et al. (2012), MSS represents an allelic disorder to JATD (Schmidts et al. 2013; Spranger et al. 2002). As in JATD, cone-shaped epiphyses are observed radiologically after the first year of life in frequent association with impaired renal function due to both (poly) cystic and nephronophthisis-like renal phenotypes and often early onset retinopathy (Perrault et al. 2012). Short distal phalanges and flattening of the femur epiphyses have been described by Mainzer et al. (1970), Popovic-Rolovic et al. (1976) and Robins et al. (1976). While most patients seem to manifest with renal symptoms during childhood and progress into end-stage renal disease before adulthood, a marked phenotypic variability concerning type of kidney disease, age of onset and speed of progression which has been noted both between and even within families (Perrault et al. 2012; Schmidts et al. 2013). A precise prognosis is therefore difficult to make. The average onset of retinal symptoms is rather difficult to define as not all patients published so far have been investigated using ERG and therefore the diagnosis sometimes relies on visual impairment and/or abnormal fundoscopy results; however, childhood onset seems common (Perrault et al. 2012). Retinal phenotypes range from pigmentary retinopathies to atypical non-pigmentary forms (Perrault et al. 2012; Schmidts et al. 2013). Further, few patients with MSS are found to suffer from hepatic involvement with cholestasis and fibrosis (Perrault et al. 2012) while both Mainzer et al. (1970) and Giedion (1979) found up to 25% of MSS patients to present with ataxia which was not confirmed by Perrault et al. (2012).

SRPS V

SRPS V has only been described by Kannu et al. in 2007 and Mill et al. in 2011 in two consecutively affected pregnancies of a mother from Maori descent from New Zealand. The foetuses presented very early in pregnancy with hydrops, narrow chest and severely shortened humeri and femora, bowed lower limbs, a lack of ossification in radii, ulnae, tibiae and fibulae as well as both hands and feet, hypoplasia of the scapulae and peritoneal calcifications. Post-axial polydactyly and syndactyly of both hands and feet was noted as well as short broad fingers and thumbs with rounded bulbous tips. Normal facial appearance was noted but bilateral cystic hygroma was described in both foetuses while only the second foetus presented with a posterior cleft palate. While the skeletal phenotype partially overlaps with SRPS I, II and III, the clinical picture in these foetuses was complicated by additional acromesomelic hypomineralisation and campomelia, therefore this seems to represent a different entity of SRPS. Visceral features included mild hypospadias, juxtamedullary and mainly glomerular cysts in the kidneys in the first foetus and malrotation of the gut in both foetuses. Ectopic spleen tissue was identified in the pancreas of the first foetus. No structural heart defects were observed. Both pregnancies were terminated before week 16 due to the severity of the findings on ultrasound examination (Mill et al. 2011)

CED

Patients with CED (or Sensenbrenner syndrome) present with characteristic craniofacial malformations including dolichocephaly as a result of craniosynostosis of the sutura sagittalis, high forehead, telecanthus and epicanthus, broad nasal bridge, low-set prominent ears, hypodontia and/or microdontia. Rhizomelic micromelia and brachydactyly are common features as well as variable narrowing of the thorax. Compared to SRPS subtypes and JATD, the thoracic restriction is usually milder. Many patients exhibit very fine, sparse and slow-growing hair and dysplastic fingernails have been reported (Eke et al. 1996; Lang & Young, 1991; Levin et al. 1977). Skin laxity was observed in several CED patients, potentially increasing the risk of hernias (Fry et al. 2009; Walcz ak et al., 2010). Visceral manifestations include progressive renal failure most likely due to a nephronophthisis-like renal phenotype with small hyperechogenic kidneys and the histological picture of tubular–interstitial nephritis and microscopic glomerular and tubular cysts, hepatic cysts and fibrosis/hepatic ductal plate malformation, inconsistent heart defects and retinal dystrophy (Eke et al. 1996; Konstantinidou, 2009; Lang & Young, 1991; Levin et al. 1977; Zaffanello et al. 2006). Several CED patients in the literature are developmentally delayed with one patient exhibiting microcephaly with a hypoplastic corpus callosum (Amar et al. 1997).

Group 4: EVC and Weyers acrofacial dysostosis

EVC

Referred to as 'six finger dwarfism' by McKusick et al. in 1964, the main features of EVC include short stature and acromelic shortening of the limbs, narrow thorax of variable degree, consistent post-axial polydactyly of both hands but not the feet and dysplastic nails and teeth (prenatal eruption of teeth, hypodontia and malformed teeth). Further, fusion of the hamate and capitate hand bones has been noted. Congenital heart defects are common, mainly as primary atrial septation defects affecting approximately half of all patients (Blackburn & Belliveau, 1971; McKusick et al. 1964). Hydrocephalus due to a Dandy–Walker malformation has also been described (Zangwill et al. 1988).

Weyers acrofacial dysostosis

Patients with Weyers acrofacial dysostosis present with polydactyly of the hands combined with dental abnormalities such as abnormal shape and number of teeth, nail dysplasia as well as short stature with short extremities. In contrast to EVC, thoracic constriction and/or visceral involvement is usually not observed (Curry & Hall, 1979; Roubicek & Spranger, 1984; Weyers, 1952).

Range of the phenotype

Skeletal manifestation

Thorax phenotype

Most ciliary chondrodysplasia patients present with a narrow or small thorax *in utero* and at birth. This is due to short horizontal ribs impairing pulmonary development and expansion in utero which in turn may lead to respiratory difficulties of variable degree after birth and represents the major cause of lethality of these disorders. While SRPS types I–V are inevitably lethal perinatally (Huber & Cormier-Daire, 2012), the severity of the thoracic phenotype is more variable in JATD (Schmidts et al. 2013) and EVC while MSS and Sensenbrenner syndrome patients seem to represent the mild end of the spectrum regarding thorax constriction (Levin et al. 1977; Perrault et al. 2012). OFD IV patients can be severely affected but have been reported to survive beyond the neonatal period (Baraitser, 1986; Burn et al. 1984). No thorax phenotype is observed in Weyers acrofacial dysostosis (Weyers, 1952). It is of note that patients with Schwachman–Diamond syndrome (SDS, OMIM#260400) may present with a thorax phenotype similar to what is observed in ciliary chondrodysplasias, especially JATD potentially leading to misdiagnosis as reported by Keogh et al. 2012 and Schaballie et al. 2013.

Polydactyly

While SRPS type II, V (only two cases described to date), EVC and OFD IV usually and SRPS type I and III often present with polydactyly, this is rarely observed in SRPS IV and JATD (Baujat et al. 2013; Elcioglu & Hall, 2002; Huber & Cormier-Daire, 2012; Schmidts et al. 2013; Yang et al. 1991). Polydactyly is usually not seen in MSS or CED patients (Levin et al. 1977; Perrault et al. 2012).

Pelvis

In JATD and EVC, the pelvis has a typical radiographic configuration with small ilia and trident acetabular roof with spur-like projections during the first year of life (Langer, 1968; Maroteaux & Savart, 1964). This is not observed in MSS (Perrault et al. 2012) or Sensenbrenner syndrome and can be a useful diagnostic tool to distinguish JATD cases with renal disease from MSS and CED. However, this appearance becomes less pronounced after the first year of life and therefore a lack of this sign in a patient beyond infancy does not exclude the diagnosis of JATD or EVC. The pelvis in SRPS I and III resembles the picture seen in JATD and EVC while a normal configuration is observed in SRPS II (Naumoff, 1980; Saldino & Noonan, 1972; Spranger et al. 2002).

Epiphyses

In both JATD and MSS cone-shaped epiphyses of the phalanges can often be observed radiologically from the age of 12 months onwards (Perrault et al. 2012; Schmidts et al. 2013). While pointed ends of the long bones are observed in SRPS I (Saldino & Noonan, 1972), rounded ends are found in SRPS III and V (Mill et al. 2011; Naumoff, 1980).

Tibia dysplasia

Small and ovoid tibiae are observed in SRPS I and OFD IV but not in SRPS IV (Elcioglu & Hall, 2002).

Clefting

Cleft of the lips and/or palate is observed in SRPS II, SRPS IV and OFD IV (Ades et al. 1994; Beemer et al. 1983; Majewski et al. 1971)

Other craniofacial features

Craniosynostosis leading to dolichocephalus, telecanthus, epicanthus, broad nasal bridge and large deep set ears are typical features of CED (Levin et al. 1977) and distinguish these patients from JATD and MSS patients. SRPS III patients exhibit a short cranial base, bulging forehead and a depressed nasal bridge (Naumoff, 1980) not observed in the other SRPS subtypes.

Extraskeletal manifestation

Eye

Clinically manifest retinal disease is frequently observed in MSS (Perrault et al. 2012) but rarely described in JATD (de Vries et al. 2010; Schmidts et al. 2013) although up to 50% of JATD patients present with an abnormal ERG (Baujat et al. 2013). This phenotype has been described in CED (Eke et al. 1996) but is not associated with EVC. It is unclear if SRPS forms are associated with retinal defects, due to the short lifespan of affected newborns no retinal investigations have been reported in the literature. Retinal disease has not been described in OFD IV in the literature but instead Ades et al. reprted colobomata in a foetus with OFD IV in 1994.

Kidney

Similar to the manifestation of retinal disease, all MSS and many CED patients develop renal symptoms (Levin et al. 1977; Perrault et al. 2012) whereas the majority of JATD patients described to date does not seem to exhibit clinically manifest renal disease (Baujat et al. 2013; de Vries et al. 2010; Schmidts et al. 2013). Renal cysts have also been found in one of the two foetuses affected with SRPS V (Kannu et al. 2007) and in OFD IV (Ades et al. 1994) as well as in SRPS I (Saldino & Noonan, 1972). Renal disease is not a feature of EVC.

Liver and pancreas

Hepatic and pancreatic lesions and fibrosis occur in some patients with ciliary chondrodysplasias, especially SRPS I, SRPS V, JATD, CED, MSS and OFD IV (Ades et al. 1994; Amar et al. 1997; Mill et al. 2011; Perrault et al. 2012; Saldino & Noonan, 1972; Yerian et al. 2003) but are not a feature of EVC or Weyers acrofacial dysostosis. Long-term data about the outcome of a larger patient cohort is missing but most patients with elevated liver enzymes seem not to progress into liver failure and symptomatic pancreatic disease appears similarly rare.

Heart

Structural heart defects are a frequent feature in EVC (McKusick et al. 1964) where they mainly occur as atrial septum defects but can also be associated with JATD (Schmidts et al. 2013) and CED (Amar et al. 1997) while MSS patients are not frequently affected (Perrault et al. 2012). Congenital heart defects have also been described in SRPS I (transposition of the great vessels; Saldino & Noonan, 1972) and SRPS IV (Kovacs et al. 2006).

Brain

Ciliary chondrodysplasias may be associated with malformations of the central nervous system although many surviving patients (particularly JATD patients) tend to show a normal psycho-motor development (Baujat et al. 2013; Schmidts et al. 2013). However, a subset of JATD patients with additional features of Joubert syndrome has been described by Lehman et al. 2010 and in the initial description of MSS, several patients were noted to suffer from ataxia (Mainzer et al. 1970, brain imaging was not performed). This was not confirmed by Perrault et al. (2012) and magnetic resonance imaging (MRI) examinations performed in this study returned normal results, despite the fact that several of the patients investigated were developmentally delayed. The authors suggested this could have occurred independently as those patients came from consanguineous families. Developmental delay was also found in several CED cases by Amar et al. (1997). Brain malformations occur frequently in SRPS IV (Lurie et al., 1994) as well as in other SRPS subtypes and OFD IV (Ades et al. 1994; Digilio et al. 1995; Thomas et al. 2012).

Ectoderm (nails, hair and teeth)

Ectodermal defects such as teeth and nail anomalies occur in EVC, Weyers craniofacial dysostosis and CED but not in JATD and MSS (Baujat et al. 2013; Levin et al. 1977; McKusick et al. 1964; Perrault et al. 2012; Roubicek & Spranger, 1984; Schmidts et al. 2013).

Tongue

A lobulated tongue is observed in OFD IV patients (Nevin & Thomas, 1989).

Laterality defects

Laterality defects do not seem to be a common feature of ciliary chondrodysplasia syndrome, so far they have only been described in SRPS type V in form of malrotation of the gut, respectively the caecum (Mill et al. 2011) and in a single JATD patient (Baujat et al. 2013).

Hearing impairment

In contrast to other ciliopathies such as Usher syndrome, hearing impairment is not a feature of ciliary chondrodysplasias except maybe for OFD IV where Nevin and Thomas described bilateral deafness in one case in 1989.

Diagnosis

The diagnosis of ciliary chondrodysplasias is mainly based on clinical observation, radiological studies and ultrasound (especially for antenatal cases).

Clinical observation and patient history

Common clinical features are short stature at birth or in early childhood, short extremities, short fingers and narrow thorax (in infants also often protruding abdomen). If polydactyly is present, this is a strong indicator for a ciliary chondrodysplasia; however, this is infrequently observed in JATD patients and usually absent in MSS and CED. Signs of ectodermal defects such as dysplastic nails and teeth indicate a case from the EVC/CED spectrum. CED patients additionally often have a history of unusual hair and may exhibit craniofacial abnormalities such a dolichocephalus, telecanthus and epicanthus not seen in EVC while EVC patients exhibit polydactyly not observed in CED. Reported renal and/or retinal involvement is suggestive of a case of MSS, JATD or CED as well as liver disease. A lobulated tongue distinguishes OFD IV from other ciliary chondrodysplasias described here but can also be observed in other OFD subtypes. Atrial septation defects often occur in EVC but may also be observed in other syndromes of the spectrum. SRPS I to V are perinatally lethal due to respiratory insufficiency as a consequence of the lung hypoplasia resulting from the constricted ribcage, these disorders can therefore largely be excluded in patients who survived beyond this age range. CED and especially MSS patients seem to have a rather mild respiratory course while there is a wide range of severity observed in JATD. As outlined above, special care should be taken to correctly diagnose Schwachman–Diamond syndrome amongst patients with narrow thorax; however, this can be diffcult in young children where typical symptoms such as exocrine pancreatic insufficiency and bone marrow failure might not be evident yet.

Radiology

Radiological studies are an important diagnostic tool in the diagnosis of ciliary chondrodysplasias. Some radiological features also help to distinguish different subtypes within the disease spectrum although this might not always be achievable with certainty due to phenotypic overlap (Elcioglu & Hall, 2002). While most patients (with the exception of Weyers acrofacial dysostosis) exhibit short and horizontal ribs and shortened long bones on skeletal surveys, a typical pelvis configuration with small ilia, trident acetabular roof and acetabular spurs is typically only observed in JATD, EVC, SRPS I and SRPS III. Metaphyseal irregularities are mainly observed in SRPS I and SRPS III. SRPS II patients tend to have larger ilia without acetabular spurring while they share the tibia dysplasia with OFD IV patients but ribs tend to be longer in OFD IV. While both JATD and MSS patients may exhibit cone-shaped epiphyses of the phalanges, the typical pelvis configuration noted in JATD is usually not found in CED and cone-shaped epiphyses can help to distinguish JATD from EVC cases. (Elcioglu & Hall, 2002; Huber & Cormier-Daire, 2012; Spranger et al. 2002). However, while the pelvis configuration is best examined before the age of 12 months, cone-shaped epiphyses usually only present later in life.

Evaluation of pulmonary function

Patients with a narrow thorax and/or symptoms of respiratory disease require evaluation of their pulmonary function and capacity including spirometry and polysomnography.

Sonography/echocardiography

Prenatal ultrasound plays a crucial role in the diagnostic of ciliary chondrodysplasias as in most cases shortening of the femora, a small ribcage and polydactyly can already be observed in utero. At which time-point in pregnancy these findings become apparent depends of the severity; in very severe cases this might be possible as early as 12 weeks into the pregnancy (Thomas et al. 2012) while in milder cases shortening of the femora might only be evident from 18 weeks onwards (Baujat & Merrer, 2007). Further, hyperechogenic kidneys, renal cysts, cleft palate, cardiac defects and liver fibrosis may also be diagnosed antenatally.

Post-natal ultrasound of the abdomen and the brain is indicated in all cases with a suspected ciliary chondrodysplasia. Special attention should be paid to the presence of hyperechogenic (small) kidneys, renal cysts, liver cysts and to the liver echogenicity as sign of hepatic fibrosis, to pancreatic malformations (if possible) and brain malformations. Also laterality defects should be excluded (Baujat et al. 2013; Thomas et al. 2012, Fliegauf et al. 2007). Renal abnormalities are suggestive of MSS, JATD or CED while they are uncommon in EVC (Perrault et al. 2012; Schmidts et al. 2013; Levin et al. 1977; Baujat & Merrer, 2007).

Similarily, an echocardiogram should be performed to rule out structural heart defects, especially if cardiac murmurs are present. Cardiac defects are often found in EVC and have been described in CED but are rarely observed in JATD or MSS (Amar et al. 1997; Baujat & Merrer, 2007; Baujat et al. 2013; Perrault et al. 2012; Schmidts et al. 2013).

Fundoscopy and electro-retinogram

Patients with ciliary chondrodysplasias should undergo ophthalmology consultation. As retinal changes are earliest detected by ERG preceding abnormal fundus findings and visual impairment and this examination can be performed non-invasively and painlessly via the skin, ERG seems to be the first-choice technique. It is of note, however, that in infants the retina might not yet be fully differentiated (Camuglia et al. 2011). Fundoscopy is indicated if the ERG shows abnormal findings or if it is not possible to perform ERG examinations. As not all cases of retinopathy associated with ciliary chondrodysplasias go along with fundus abnormalities (Perrault et al. 2012, and own unpublished data), a normal fundoscopy does not exclude retinal disease in those patients.

Laboratory findings

If a ciliary chondrodysplasia is suspected, laboratory parameters indicative of renal, hepatic or pancreatic involvement should be analysed. These parameters include serum electrolytes, creatinine, blood urea nitrogen (BUN), liver transaminases [aspartate transaminase (AST) and alanine aminotransferase (ALT)], alkaline phosphatase (APT), total and conjugated bilirubin and gamma glutamyl transpeptidase (GGT) (de Vries et al. 2010). To differentiate JATD from the non-ciliary chondrodysplasia condition Schwachman–Diamond syndrome exclusion of exocrine pancreatic insufficiency may be performed via investiagtion of elastase levels in the stool and neutropenia, thrombocytopenia and anemia should be exluded in cases with clinical signs such as frequent infections. Proteinuria and haematuria should be excluded from the urine. In case of respiratory symptoms, blood gas analysis is indicated (Baujat et al. 2013).

Genetic investigations

The proportion of cases caused by mutations in currently known causative genes depends on the individual condition: while the vast majority of EVC cases is caused by mutations in the *EVC1* and *EVC2* gene (D'Asdia et al. 2012) and causative mutations in *NEK1* and *DYNC2H1* are found in over two-thirds of all SRPS II cases (El Hokayem et al. 2012), only about 50% of all MSS are caused by IFT140 mutations (Perrault et al. 2012) and 50% of *OFD4* cases seem to result from mutations in *TCTN3* (Thomas et al. 2012). JATD is genetically more heterogeneous with about 50% of all patients harbouring mutations in known genes such as *DYNC2H1* (Baujat et al. 2013; Schmidts et al. 2013, and own unpublished data); similarly, multiple genes have been found to cause CED.

The fairly large number of different disease subtypes with overlapping phenotypes and genotypes as well as the substantial size of some of the genes known to be causative (e.g. *DYNC2H1* with 90 exons) made it difficult if not impossible in the past to establish a molecular diagnosis in the clinical setting. Hopefully, the introduction of Next Generation sequencing (NGS) will significantly facilitate genetic diagnostics for ciliary chondrodysplasias, either in form of specific gene panels to be analysed for ciliary chondrodysplasia patients or in form of clinical whole exome sequencing. Single-gene analysis, however, will have a fairly high success rate in EVC (*EVC1*, *EVC2*), SRPII (*NEK1*), OFD IV (*TCTN3*), SRPS V (*WDR35*) as well as MSS and JATD cases with renal and retinal involvement (*IFT140*), while in other cases gene panel analysis seems more feasible. In cases where targeted sequencing of the major causative gene does not reveal any mutations, progression a wider NGS-based technique seems advisable. As up to 10% of cases might be caused by deletions of coding areas of causative genes and smaller deletions are likely to be missed in comparative genomic hybridisation (CGH) arrays, copy number variant analysis of NGS data is highly recommended. Genetic analysis is best undertaken in a centre familiar with ciliary chondrodysplasias. Karyogram and fluorescence *in situ* hybridisation analysis seem only indicated in cases where the clinical phenotype indicates a chromosomal rearrangement.

Translational diagnostics: Immunofluorescence of patient fibroblasts, and electron microscopy (EM)

As outlined later in this chapter, mutations in genes causative in ciliary chondrodysplasias can lead to defects in ciliogenesis and/or intraflagellar transport. These defects can sometimes be observed in patient fibroblasts using immunofluorescence and/or electron microscopy techniques. Although these investigations are fairly labour intensive and do not seem feasible as a standard analysis technique, they might be helpful in specific cases to define the functional consequences of genetic defects identified (e.g. ultrastructural ciliary defects or accumulation of intraflagellar transport (IFT) particles such as IFT57 or IFT88 or hedgehog pathway components at the ciliary tip (Arts et al. 2011; Bredrup et al. 2011; Schmidts et al. 2013; Thiel et al. 2011) (see Figure 3.2)

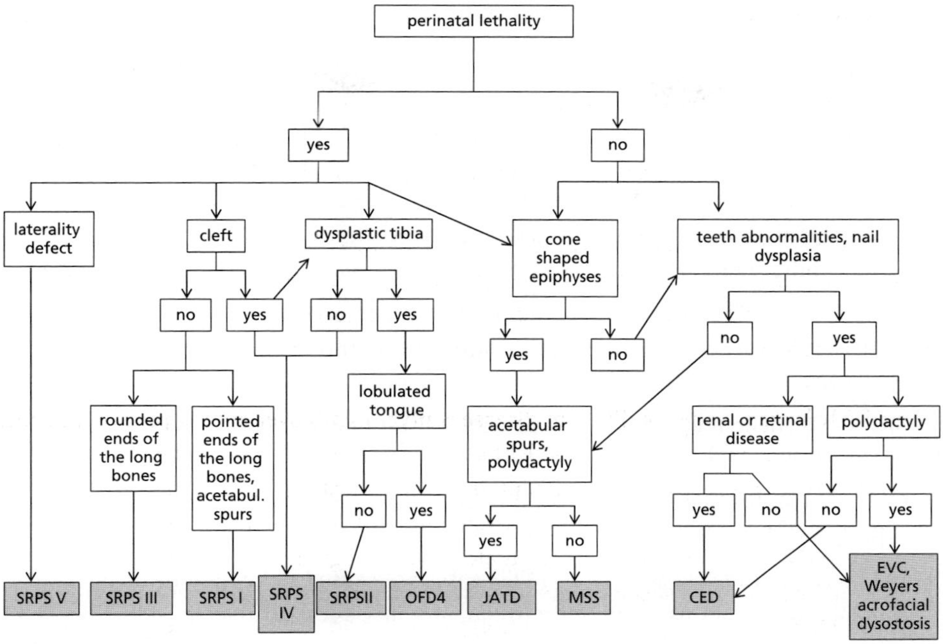

Figure 3.2 Simplified pathway of differential diagnosis for ciliary chondrodysplasias.

Causative genes

All ciliary chondrodysplasias (except Weyers acrofacial dysostosis which is inherited autosomal dominantly) are inherited in an autosomal recessive manner. Genetic overlap is observed between different SRPS types as well as between JATD and certain forms of SRPS and between JATD and MSS, the latter seem to be true allelic diseases. As for other ciliopathy diseases, for example Bardet–Biedl syndrome, Joubert syndrome and Meckel–Gruber syndrome where mutations in the gene *CEP290* have been shown to be able to cause all three phenotypes, mutations in several genes have not only been identified in patients with different types of chondrodysplasias (e.g. *DYNC2H1* in SRPS II, SRPS III and JATD (Dagoneau et al. 2009; El Hokayem et al. 2012; Merrill et al. 2009; Thiel et al. 2011), IFT80 mutations in JATD and SRPS III (Beales et al. 2007; Cavalcanti et al. 2011) and *IFT140* mutations in JATD and MSS (Perrault et al. 2012; Schmidts et al. 2013) but also mutations in *TTC21B* have been identified in one JATD patient as well as patients with Joubert syndrome and nephronophthisis (Davis et al. 2011) and mutations in *WDR35* can cause SRPS V and CED (Gilissen et al. 2010; Mill et al. 2011). Digenic inheritance of heterozygous mutations in two different genes has been reported in a single case of SRPS II (Thiel et al. 2011). If so-called 'triallelic inheritance', described in other ciliopathies such as Bardet–Biedl syndrome (Katsanis et al. 2011) occurs in ciliary dysplasias is currently unknown. However, as for other ciliopathies, there are indications that 'mutational load', that is the total of mutations in genes known to cause the disease, may also play a role in ciliary dysplasias (Schmidts et al. 2013). Similar to other genetically heterogeneous ciliopathies, ciliary chondrodysplasias have hugely benefited from the introduction of NGS which is currently revolutionising patient genotyping and gene discovery in the research setting and will soon also be available for clinical genetic diagnosis (see Table 3.2):

- **SRPS II**: Causative mutations in *DYNC2H1* and *NEK1* have been identified in SRPS II by Thiel et al. (2011) and seem to cause the majority of cases (El Hokayem et al. 2012). Digenic inheritance of a heterozygous NEK1 and a heterozygous DYNC2H1 mutation has been reported in a single case (Thiel et al. 2011).
- **SRPS III**: Mutations causing SRPS III have been identified in *DYNC2H1* (Dagoneau et al. 2009; Merrill et al. 2009) and *IFT80* (Beales et al. 2007).
- **SRPSV**: Biallelic mutations in *WDR35* have been found to cause SRPS V (Mill et al. 2011).
- **OFD IV (Mohr–Majewski)**: Mutations in *TCTN3* may cause up to 50% of all cases (Thomas et al. 2012).
- **JATD**: Approximately 50% of all JATD cases are caused by mutations in *DYNC2H1* (potentially mutations occur more frequently in Caucasian than in non-Caucasian cases) while mutations in *IFT80*, the first gene found to cause JATD (Beales et al. 2007), occur rarely (Baujat et al. 2013, own unpublished data). Further, mutations in WDR19 (Bredrup et al. 2011) and TTC21B (Davis et al. 2011) have been reported in single

Table 3.2 Genetic findings in ciliary chondrodysplasias: Genes indicated represent genes with identified causative mutations

	DYNC2H1	NEK1	IFT140	IFT80	TTC21B	WDR35	WDR19	IFT122	IFT43	EVC1, EVC2	TCTN3
SRPS II	■	■									
SRPS IV											
OFD IV											■
SRP I											
SRP III	■			■							
JATD	■		■	■	■						
MSS			■								
SRPV						■					
CED						■	■	■	■		
EVC										■	
Weyers acrofacial dysostosis										■	

JATD cases, while mutations in IFT140 seem to rarely be the cause of JATD in general but might account for a significant proportion of JATD cases with renal disease and retinal dystrophy (Schmidts et al. 2013).

- **Mainzer–Saldino**: Causative mutations in *IFT140* have been identified in approximately 50% of the cases investigated (Perrault et al. 2012; Schmidts et al. 2013).
- **CED**: CED is genetically heterogeneous with causative mutations found in *IFT122, IFT43, WDR19* and *WDR35* and likely more causative genes will be identified in the future (Arts et al. 2011; Bredrup et al. 2011; Gilissen et al. 2010; Walzcak et al, 2010).
- **EVC**: In most cases of EVC, biallelic causative mutations in *EVC1* and *EVC2* have been identified with mutations in *EVC1* accounting for 75% and mutations in *EVC2* accounting for 25% of the cases (D'Asdia et al. 2012; Ruiz-Perez et al. 2000, 2003).
- **Weyers acrofacial dysostosis**: Dominant mutations in *EVC1* and *EVC2* have been found to cause Weyers acrofacial dysostosis (d'Asdia et al., 2012; Ruiz-Perez et al. 2000).
- **SRPS I and SRPS IV**: The genetic basis of SRPS I and SRPS IV is still unknown.

Underlying ciliary defect

All genes implicated in ciliary chondrodysplasias to date except NEK1, EVC1 and EVC2 encode for proteins involved in IFT (Arts et al. 2011; Beales et al. 2007; Bredrup et al. 2011; Dagoneau et al. 2009; Gilissen et al. 2010; Merrill et al. 2009; Mill et al. 2011; Perrault et al. 2012; Ruiz-Perez et al. 2000, 2003, 2007; Thiel et al. 2011). EVC1 and EVC2 are found at the ciliary base (Ruiz-Perez et al. 2007) but do not seem to be involved in IFT while *NEK1* encodes a serine/threonine kinase involved in the control of cell-cycle dependent ciliogenesis (Thiel et al. 2011).

IFT ensures that structural ciliary components and cell signalling proteins reach their destination along the ciliary axoneme up to the ciliary tip (anterograde IFT, complex B). As the cilium does not have a protein synthesis machinery, proteins such as microtubule components, dynein arms and receptor proteins are produced in the cytosol of the cell and transported along microtubules to the base of the cilium where they are subsequently loaded onto IFT particles which help transport them along the ciliary axoneme. IFT is also required to translocate molecules back from the ciliary tip and the axoneme to the ciliary base, the cytosol of the cell or the cellular nucleus (retrograde IFT, complex A). IFT molecules hereby act as cargo adapters between the IFT motor components (kinesin for anterograde IFT, dynein-2 for retrograde IFT) and the proteins that require transport. (Cole & Snell, 2009; Hao et al. 2011; Rosenbaum & Witman, 2002). Dysfunctional IFT due to mutations in genes encoding for IFT components therefore leads to transport defects along the ciliary axoneme. In extreme cases, this can be so severe that the cilium cannot be built up or maintained which is often referred to as 'ciliogenesis defect' in the literature (Fliegauf et al. 2007). In milder cases, accumulation of ciliary components can be observed using immunofluorescence techniques on cultured cells (Arts et al. 2011; Schmidts et al. 2013) (see Figure 3.3).

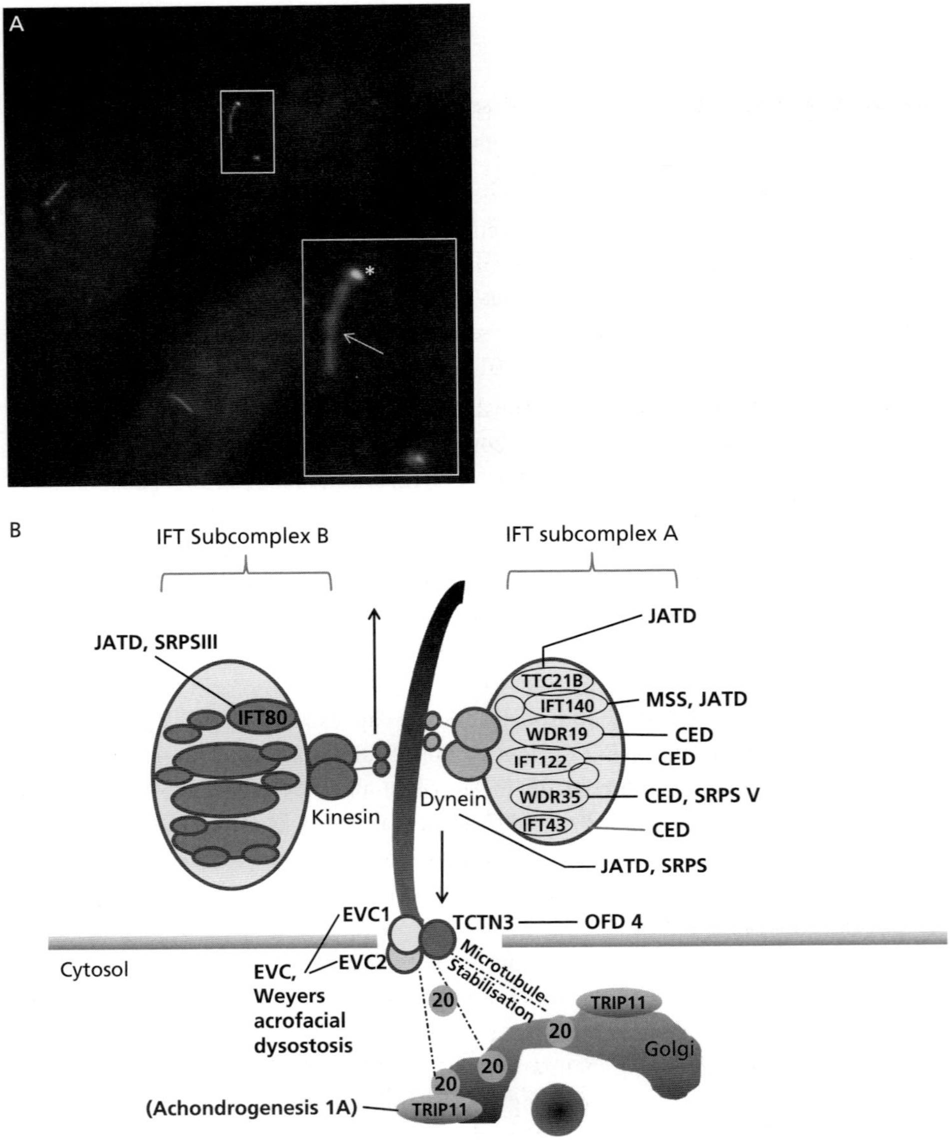

Figure 3.3 Visualisation of chondrocyte cilia and localisation of proteins affected by genetic defects in ciliary chondrodysplasias. (A). Chondrocyte cilia: Immunofluorescence microscopy after staining with anti-acetylated tubulin antibody, marking the ciliary axoneme (arrow) and anti-gamma-tubulin antibody marking basal body and centrioles (.indicated by a star). (B) Simplified schematic of IFT with genes indicated that have been shown to carry causative mutations in ciliary chondrodysplasias. Most genes implicated in ciliary chondrodysplasias encode for proteins of IFT complex A (retrograde IFT, motor protein: dyenin-2) while mutations in genes encoding for IFT-B complex proteins (anterograde IFT, motor protein: kinesin-2) have only been detected in JATD and SRPS III to date (DYNC2H1 is a component of the motor-protein dynein-2). EVC1, EVC2 and TCTN3 proteins are not part of IFT complexes but localise to the base of the cilium. Mutations in TRIP11 cause Achondrogenesis 1A (Smits et al. 2010), a chondrodysplasia not currently discussed as a ciliopathy although TRIP11 anchors IFT20 to the Golgi membrane and IFT20 in turn is involved in microtubule stabilisation and may be involved in transport of other IFT components to the cilium (Follit et al. 2008, 2009).

Physiological effect of the ciliary defect

As IFT is required to build and maintain the physical structure of a cilium, shortened or completely absent cilia on patient's fibroblasts have been observed in chondrodysplasias caused by mutations in IFT genes (Bredrup et al. 2011; Merrill et al. 2009) as well as in fibroblasts of patients with *NEK1* mutations (Thiel et al. 2011). Cilia length usually is assessed using immunofluorescence and/or electron microscopy techniques (Arts et al. 2011; Schmidts et al. 2013). As the cilium is thought to both transmit signals from outside to the cell into the cell (serving as an 'antenna') as well as signals from the cells being transmitted to the cilium (Ishikawa & Marshall, 2011), one can easily imagine that a lack of primary cilia critically interrupts the flow of cellular signalling information (Baker & Beales, 2009; Fliegauf et al. 2007; Jenkins & Beales, 2012).

However, ciliogenesis defects are not observed in all patients with mutations encoding for IFT genes or mutations in the gene encoding for the IFT motor protein DYNC2H1: in some patients with mutations in genes encoding for retrograde IFT proteins, cilia seem to be present on skin fibroblasts in normal numbers and seem to reach normal length but a ciliary accumulation of IFT components at the ciliary tip can be observed. This indicates that the more subtle dysfunction of IFT is sufficient to interrupt ciliary cell signalling pathways and cause developmental defects (Arts et al. 2011; Schmidts et al. 2013) This difference in severity of the ciliary phenotype in human patients compared to the ciliary phenotype in the corresponding knockout mouse models where strong ciliogenesis defects are observed (e.g. Dync2h1 knockout mouse, Ocbina et al. 2011) could result from the mice being true 'nulls' whereas the patients usually do not harbour two nonsense alleles and therefore probably are only hypomorphic for the disease (Jonassen et al. 2012; Rix et al. 2011; Schmidts et al. 2013).

Skeleton

The so-called hedgehog signalling pathway has been identified as crucially influencing chondrogenic (and subsequently osteogenic) proliferation and differentiation, an essential process in skeletal development (Kronenberg, 2003). Animal models for ciliary chondrodysplasias such as knockout mice for EVC (mutations in the *EVC1* and *EVC2* gene cause Ellis–van Creveld syndrome in humans), Dync2h1 and Ift80 (mutations in *DYNC2H1* and *IFT80* genes cause JATD and SRPS III in humans) indicate that IFT defects lead to imbalances in the hedgehog signalling pathway. This in turn seems to induce a premature stop of chondrogenic proliferation and induction of chondrogenic differentiation at the growth plates of the developing skeleton. With cells at the growth plate failing to divide, bone growth is severely slowed down or even arrested (Ocbina et al. 2011; Rix et al. 2011; Ruiz-Perez et al. 2007) (see Figure 3.4)

Kidney and eye

Renal and retinal disease occurs frequently in ciliopathies (Baker & Beales, 2009; Fliegauf et al. 2007; Jenkins & Beales, 2012) and can also be associated with ciliary chondrodysplasias,

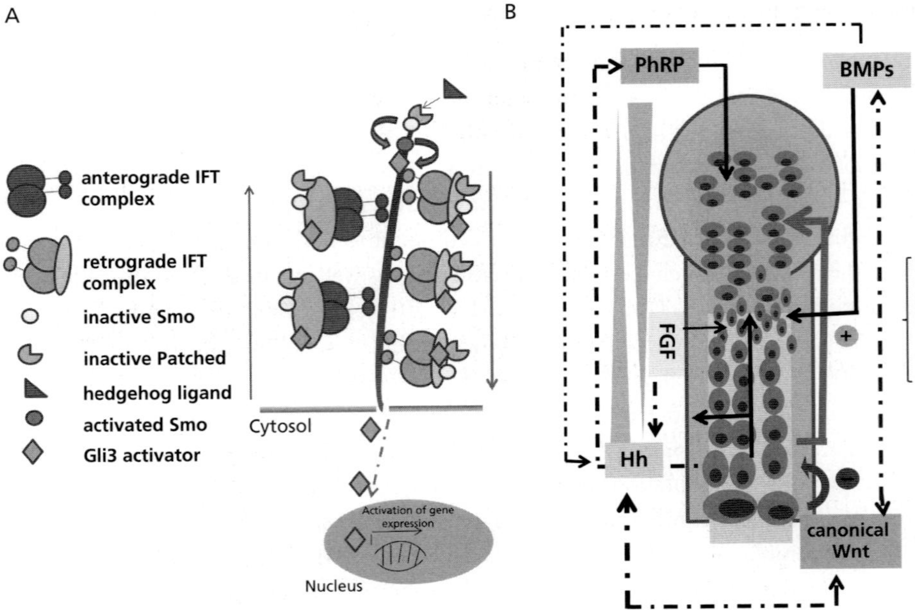

Figure 3.4 Simplified schematic of ciliary hedgehog signalling and other pathways and their influence on skeletal development. (A) Hedgehog signalling is dependent on an intact cilium and functioning IFT (Huangfu et al. 2003; Goetz et al. 2009) as signalling pathway components such as smoothened (smo) and patched localise to the cilium and require anterograde IFT to localise to the ciliary tip where binding of the hedgehog ligand activates the smoothened inhibitor patched which in turns releases smoothened. Activated smoothened releases GLI3 activator from its inhibitor SUFU (not shown). Gli3 activator is transported back to the cell where it activates genes involved in chondrogenic and osteogenic differentiation in the nucleus (adapted from Quinlan et al. 2008). (B) Simplified schematic to illustrate the process of bone elongation during development and growth influenced by different signalling pathways. Hedgehog (Hh), bone morphogenic protein (BMP), fibroblast growth factor (FGF) and PhRP (parathyroid hormone) are major regulators of proliferation and differentiation in the skeleton and the first two act via opposing gradients while FGF signalling inhibits chondrocyte proliferation (Long & Ornitz, 2013). While proliferation of chondrocyte precursor cells is repressed by canonical Wnt signalling in the metaphysis, proliferation and differentiation is induced by canonical Wnt signalling in the growth plate (Liu et al. 2008). The different signalling pathways are interlinked and imbalances between the pathways can lead to premature cellular differentiation and therefore premature stop of proliferation as observed in ciliary chondrodysplasias primarily due to aberrant hedgehog signalling arrested (Ocbina et al. 2011, Rix et al. 2011, Ruiz-Perez et al. 2007).

especially with MSS (Mainzer et al. 1970; Perrault et al. 2012), a subtype of JATD (Bernstein et al. 1974; Schmidts et al. 2013) and CED (Levin et al. 1977). Cells in the kidney tubules exhibit primary cilia in the same way as chondrocytes and loss of the physical structure of the cilium as well as loss of ciliary proteins has been shown to lead to cyst formation in mice (Davenport et al. 2007; Hossain et al. 2007; Lin et al. 2003; Moyer et al. 1994). Primary cilia therefore seem to play an essential role during renal tubular development, possibly via induction of orientated cell division and establishment of planar cell polarity (Fischer et al. 2006; Patel et al. 2008) to finally organise tubular elongation in the three dimensional space (Lienkamp et al. 2012). While hypomorphic Ift80 knockout mice do not exhibit any renal phenotype (and neither do human patients with IFT80 mutations known to date (Beales et al. 2007; Rix et al. 2011, own unpublished data), conditional knock out of Ift140 in the mouse kidney produces an early onset cystic phenotype (Jonassen et al. 2012). This phenotype resembles the renal phenotype observed in human JATD and MSS patients with IFT140 mutations (Perrault et al. 2012; Schmidts et al. 2013). Ift140 knockout mice show normally orientated mitotic spindle axis but hyper-proliferation of renal tubulus cells was noted in the pre-cystic epithelium while increased canonical Wnt and hedgehog signalling was only observed in cystic tissue (Jonassen et al. 2012), therefore neither disturbances in hedgehog nor Wnt signalling seem to initiate cyst formation in these mice but might contribute to cyst progression. How this relates to the human pathophysiology is not clear to date.

Retinal photoreceptors consist of two parts that are crucial for light processing: the inner and outer segment. These two segments are connected via a narrow bridge, the so-called 'connecting cilium' which shares similarities with the classical cilium in composition. The connecting cilium is essential for rhodopsin transport between the segments and defects in this transport mechanism lead to accumulation of rhodopsin inducing photoreceptor apoptosis and long-term retinal degeneration (Krock et al. 2009; Marszalek et al. 2000). Therefore, mutations in genes encoding for ciliary proteins may not only lead to developmental problems in tissues and organs which exhibit primary ciliary but also frequently cause retinal disease (Fliegauf et al. 2007; Pazour et al. 2002), occuring more in form of a degenerative process.

Clinical management

Ciliary chondrodysplasias are very rare diseases and no large long-term follow-up studies have been performed to date for most of the non-lethal forms. For SRPS I–V, only short-term palliative care is possible while OFD IV, JATD, MSS, CED and EVC patients require complex multidisciplinary clinical management approaches as outlined by Baujat and Merrer in 2007 for EVC and de Vries et al. in 2010 and Baujat et al. in 2013 for JATD.

Skeleton

Thorax

The main cause of lethality in all ciliary chondrodysplasias (except Weyers craniofacial dysostosis which does not affect the ribcage) is cardio-respiratory failure resulting from lung hypoplasia secondary to the constricted ribcage. The majority of fatalities occur perinatally and during the first years of life, the latter often due to respiratory decompensation caused by airway infections (Baujat et al. 2013; Huber & Cormier-Daire, 2012; Schmidts et al. 2013). While there are currently no therapeutic options for the perinatally lethal SRPS forms and, therefore, interruption of pregnancy is often offered to affected families, some therapies have been successfully developed for less severe cases of JATD. Care is usually provided in a multidisciplinary team built with paediatric respiratory physician and paediatric thorax surgeons. Most MSS and CED patients experience a rather mild respiratory disease cause.

Acute care after birth includes intubation and mechanical ventilation or continuous positive airway pressure (CPAP) ventilation through a face mask or nasal ventilation tube and or oxygen supply if the respiratory situation is unsatisfying. If patients cannot be weaned from ventilation or CPAP in a reasonable time frame, tracheotomy may be performed. In cases where the patient requires ventilation support only during the night, home care under the supervision of an appropriately trained nurse might be possible. Respiratory infections should be prevented wherever possible. As many patients seem to 'grow out' of the thoracic phenotype after the first years of life, invasive therapy is rarely required later in life (Baujat & Merrer 2007; de Vries et al. 2010; Schmidts et al. 2013). However, lung hypoplasia can be of life-threatening severity in some patients where sufficient gas exchange cannot be achieved with the conservative measures outlined above. In these cases, thoracic expansion surgery may be offered by few highly specialised centres worldwide. As this surgery imposes a significant risk, it is only undertaken in cases with no other therapeutic option (Davis et al. 1995, 2001, 2004; de Vries et al. 2010; Sharoni et al. 1998; Todd et al. 1986).

Pulmonary function should be assessed on a regular basis performing spirometry and volume measurements as well as polygraphic sleep studies (Baujat et al. 2013).

Polydactyly

Super-numerous fingers and/or toes are often removed shortly after birth depending on their impact on the functionality of hand and feet.

Hips, joints and spine

Unfortunately, no large clinical follow-up studies about the long-term impacts on the skeleton have been published at time of publishing this book although it seems well possible that patients with ciliary chondrodysplasias might experience a higher rate of secondary skeletal problems such as hip dysplasia, scoliosis, cervical spine compression, lumbar stenosis and osteoarthritis later in adulthood. Thorough screening for hip dysplasia is indicates for all new-borns and infants with ciliary chondrodysplasia as the anatomical pelvis configuration could predispose affected children. Early orthopaedic specialist consultation is recommended in case of clinical signs of the above (Baujat & Merrer, 2007; Baujat et al. 2013; Schmidts et al. 2013)

Stature and growth hormone treatment

Adult height is difficult to predict for ciliary chondrodysplasia patients as not many patients have been documented. In EVC, body heights ranging from 119 cm to 167 cm have been reported (Baujat & Merrer, 2007; Oliveira da Siva et al. 1980; Renier et al. 1975). For JATD, although many patients are of short stature at birth, at infancy and in early childhood, catch-up growth seems to occur later as many JATD patients do reach normal adult height with the exception of patients with end-stage renal disease (Baujat et al. 2013;

Schmidts et al. 2013). Growth hormone treatment is currently not recommended for patients with ciliary chondrodysplasias such as EVC (Baujat & Merrer, 2007) or non-ciliary chondrodysplasias as only temporarily beneficial effects in patients with achondroplasia, hypochondroplasia and nail–cartilage hypoplasia have been noted (Bocca et al. 2004; Harada et al. 2005; Key & Gross, 1996).

Retina

Similarily to renal problems, patients with MSS as well as JATD patients with *IFT140* mutations have a significant risk of developing retinal disease at some point of the disease course. As for renal disease, the age of onset appears variable but can be the first years of life (Perrault et al. 2012; Schmidts et al. 2013). JATD patients with *DYNC2H1* mutations seem to be at a lower risk than IFT140 patients (Schmidts et al. 2013); nevertheless, they require monitoring as ERG abnormalities were noted in up to 50% of cases by Baujat et al. (2013). The risk for JATD patients with mutations in other genes is unclear at the moment. Monitoring should ideally consist of regular ERG examinations as ERG changes usually precede pathological findings in fundoscopy and clinical manifestation of retinal dystrophy (Camuglia et al. 2011).

Kidney

Regular monitoring of renal blood markers and renal ultrasound is recommended, especially for patients with MSS and CED, and JATD cases where IFT140 mutations have been identified while JATD patients with DYNC2H1 mutations are at a lower risk (Baujat et al. 2013; Schmidts et al. 2013). As mutations in WDR19, TTC21B and IFT80 have only been identified in single cases or very few patients to date, the risk for renal disease cannot be predicted for those patients. In case of renal disease supportive therapy is indicated, followed by dialysis and/or renal transplantation if required.

Liver and pancreas

Liver enzymes should be monitored alongside renal retention parameters and accomplished by ultrasound examination. Pancreatic enzymes should be investigated in case of abnormal ultrasound findings or other clinical indications. If ultrasound and/or blood analyses reveal significant abnormalities, MRI examination may be helpful and liver biopsy is recommended. Although published data on the natural course of the disease is very limited, it seems that most JATD and CED patients exhibit a mild course of liver disease (Baujat et al. 2013; de Vries et al. 2010; Schmidts et al. 2013), with only one case of liver transplant and one fatal case in JATD reported in the literature (Hennekam et al. 1983; Yerian et al. 2003). MSS and EVC patients do not seem to develop liver disease frequently (Baujat & Merrer, 2007; Perrault et al. 2012). The clinical course in OFD IV is currently difficult to predict. Ursodeoxycholic acid treatment has been reported to improve liver parameters in JATD patients by Labrune et al. (1999), de Vries et al. (2010) and Baujat et al. (2013).

Heart

Heart defects mainly occur in EVC and should be treated according to cardiological guidelines, conservatively or surgically, including endocarditis prophylaxis for dental surgery. Heart defects can shorten life expectancy in EVC (Baujat & Merrer, 2007; Digilio et al. 1995; Katsouras et al. 2003).

Gastroenterology and nutrition

Severe gastro-oesophageal reflux and 'swallowing problems' are reported by many JATD families during infancy, although it is not currently recognised as an associated feature in the literature. The pathophysiological mechanism is currently unclear; possibly the small thorax provokes increased intra-abdominal pressure. Conservative reflux therapy might be successful, however, on-going failure-to-thrive problems might make placement of a naso-gastric feeding tube or even percutaneous endoscopic gastrostomy necessary.

No data is currently available about risk of 'lifestyle diseases' such as obesity, type2 diabetes and hypertension in ciliary chondrodysplasia patients later in life but patients should be encouraged to lead a healthy lifestyle given the fact that some patients may be less able to exercise due to their thorax phenotype and possible other skeletal problems leading to increased rates of secondary complications, as is observed for other congenital skeletal dysplasias such as achondroplasia (Wright & Irving, 2012).

Dysplastic teeth

Neonatal teeth in EVC might cause feeding problems in which case they should be removed. Older patients with EVC and CED with craniofacial involvement may also require dental treatment or dental surgery (Baujat & Merrer, 2007).

Neurodevelopment

Neurodevelopment seems to usually be normal in Jeune syndrome and MSS patients while CED patients more frequently exhibit problems; however, a subset of Jeune syndrome patients with clinical and radiological features of Joubert syndrome (Lehman et al. 2010) and MSS patients with ataxia have also been reported (Giedion, 1979; Mainzer et al. 1970). Therefore, in case of neurological symptoms and/or developmental delay, evaluation by paediatric neurology specialist and MRI examination of the head is recommended.

Genetic counselling

Given the genetic nature of ciliary chondrodysplasias, genetic counselling is indicated in all cases. Recurrence risk is usually 25% for subsequent pregnancies for all ciliary chondrodysplasias with the exception of Weyers acrofacial dysostosis where it is 50% if one of the parents is carrier and very low in cases of *de novo* mutations. Prenatal genetic testing can be offered to families where the causative mutation has already been identified and families can be informed about the possibility of pre-implantation diagnostics as well as sperm or egg donation. As significant phenotypical variability has been observed between patients from different families with mutations in the same gene as well as between siblings carrying the same mutations (Schmidts et al. 2013), it may be difficult to predict the exact phenotype for subsequent pregnancies based on genetic findings.

Psychological impact

Ciliary chondrodysplasias have significant psychological impact on affected families. Late interruptions of pregnancies and loss of children in the neonatal period and in infancy leaves some families traumatised and in fearful anticipation of 'it happening again' in subsequent pregnancies. The genetic basis of these disorders can provoke feelings of guilt in some parents. Regular, over several years, on-going in-patient treatment in tertiary care centres, located a significant distance from the family home, might make it difficult to care for siblings of the affected child and may impose significant organisational and financial strains on families. As these diseases are rare and no larger studies about long-term outcome of patients have been performed to date, a clear prognosis can often not be given to the family and, currently, no patient support group except for EVC exists in the UK. Some associated disease features such as retinal and renal disease might only manifest later in life when respiratory problems are already overcome. Not knowing 'what else will come in the future' imposes additional emotional stress on affected families. For affected patients, external appearance can also impose psychological distress as, although the thorax volume significantly increases with increasing age, unusual external appearance might persist. The thorax volume might not be sufficient in all female patients to ensure gas exchange until term in a pregnancy situation, which might turn pregnancy into a potentially life threatening event respectively might make pregnancy impossible in severe cases (own unpublished data). No data has been published on fertility rates of ciliary chondrodysplasia patients so it remains unclear if fertility is affected, especially in males as the sperm flagellum is quite similarly composed to the cilium (Fliegauf et al. 2007). Psychological support should be offered to all families affected with ciliary chondrodysplasias.

In summary, ciliary chondrodysplasias require multidisciplinary long-term clinical management which ideally is co-ordinated by the discipline most closely involved with the patient. Best care can be achieved in a centre where experience with ciliary chondrodysplasias and with all disciplines located within in the centre. Patients might need (paediatric) respiratory physicians, cardiologists, nephrologists, ophthalmologists, orthopaedics and surgeons, gastroenterologists, dieticians/nutritionists, geneticists and psychologists. This could be undertaken in a similar way as has been established for other rare ciliary diseases such as Bardet–Biedl syndrome and primary ciliary dyskinesia (PCD) in the UK by establishing national and/or regional multidisciplinary specialist clinics (see Table 3.3)

Table 3.3 Clinical follow-up recommendations for ciliary chondrodysplasias

	SRPS	JATD	MSS	EVC	CED	OFD IV
Physical examination	1), 3)	1), 2), 3)	1), 2), 3)	1), 2), 3)	1), 2), 3)	1), 2), 3)
Palliative care	Always	Unlikely				Depending on severity
Pulmonary evaluation	To establish the clinical phenotype	1), 2) or 6), 3)	1), 6), 3)	1), 6), 3)	1), 6), 3)	1), 6) 3)
Renal evaluation		1), 2), 3)	1), 2), 3)	1), 3)	1), 2), 3	1), 2), 3)
Liver tests		1), 2), 3)	1), 2), 3)	1), 3)	1), 2), 3)	1), 2), 3)
Pancreatic evaluation		1), 3)	1), 3)	1), 3)	1), 3)	1), 3)
Ophthalmology		1) A, B, C 5) A 3) B, C	1) A, B, C 5) A 3) B, C	1) A, B, C 3) B, C	1) A, B, C 5) A 3) B, C	1) A, B, C 5) A 3) B, C
Cardiology		1), 3)	1), 3)	1), 3), 4)	1), 3)	1), 3)
Audiology		3)	3)	3)	3)	1), 2) or 5), 3)
Orthopaedic evaluation		1), 3)	1), 3)	1), 3)	1), 3)	1), 3)
Brain MRI		3)	1), 3)	3)	1), 3)	1), 3)

Key: 1) at diagnosis; 2) yearly until age of 15–18 years. 3) on indication 4) according to cardiology guidelines 5) every 2–3 years, 6) every 6 months if pulmonary disease is present; Ophthalmology: A: ERG, B: Fundoscopy, C: vision assessment. Liver and pancreatic tests: abdominal ultrasound and blood tests. Renal evaluation: ultrasound, blood and urine tests. Cardiological evaluation: auscultation of the heart, echocardiography in case of abnormal auscultation. Orthopaedic evaluation: spine MRI at diagnosis or 6 months of age, X-rays of spine and hips at 6–10 years of age and on indication. Exclusion of hip dysplasia by ultrasound is recommended in all patients in infancy. Physical examination should consist of pulmonary auscultation, abdominal examination, neuromuscular assessment, anthropometric measurements (standing and sitting height, limbs segments, weight and thorax perimeter), arterial blood pressure, respiratory parameters and back static examination.

Source: Data from Baujat G and Merrer LM, Ellis van Creveld syndrome, *Orphanet Journal of Rare Diseases*, Volume 2, Issue 27, pp. 1750–1172, Copyright © 2007 Baujat and Le Merrer; licensee BioMed Central Ltd and Baujat G et al., Asphyxiating thoracic dysplasia: clinical and molecular review of 39 families, *Journal of Medical Genetics*, Volume 50, Issue 2, pp. 91–98, Copyright © 2012 by the BMJ Publishing Group Ltd.

Acknowledgement

I am very grateful to Suzanne Rix (Molecular Medicine Unit, Institute of Child Health, University College London) and Melita Irving (Clinical Genetics Department, Guy's Hospital, London) for helpful discussions of this topic; to Hulya Kayserili (İstanbul University, Medical Faculty, Medical Genetics Department, Istanbul, Turkey) and Bernhard Zabel (University Hospital Freiburg, Paediatric Genetics Division, Freiburg, Germany), for the provision of clinical images; and to Action Medical Research UK for providing me with a Research Training Fellowship (RTF-1411) enabling research on JATD.

References

Ades, L. C., Clapton, W. K., Morphett, A., Morris, L. L. & Haan, E. A. 1994. Polydactyly, campomelia, ambiguous genitalia, cystic dysplastic kidneys, and cerebral malformation in a fetus of consanguineous parents: A new multiple malformation syndrome, or a severe form of oral–facial–digital syndrome type IV? *Am J Med Genet* **49**, 211–217.

Allen, A. W., Moon, J. B., Hovland, K. R. & Minckler, D. S. 1979. Ocular findings in thoracic–pelvic phalangeal dystrophy. *Arch Ophthalmol* **97**, 489–492.

Amar, M. J. A., Sutphen, R. & Kousseff, B.G. 1997. Expanded phenotype of cranioectodermal dysplasia (Sensenbrenner syndrome). *Am J Med Genet* **70**, 349–352.

D'Asdia, M. C., Torrente, I., Consoli, F., Ferese, R., Magliozzi, M., Bernardini, L., et al. 2012. Novel and recurrent EVC and EVC2 mutations in Ellis–van Creveld syndrome and Weyers acrofacial dysostosis. *Eur J Med Genet* **56**, 80–87.

Arts, H. H., Bongers, E. M., Mans, D. A., van Beersum, S. E., Oud, M. M., Bolat, E., et al. 2011. C14ORF179 encoding IFT43 is mutated in Sensenbrenner syndrome. *J Med Genet* **48**, 390–395.

Baker, K. & Beales, P. L. 2009. Making sense of cilia in disease: The human ciliopathies. *Am J Med Genet, C, Semin Med Genet* **151C**, 281–295.

Baraitser, M. 1986. The orofaciodigital (OFD) syndromes. *J Med Genet* **23**, 116–119.

Bard, L. A., Bard, P. A., Owens, G. W. & Hall, B. D. 1978 Retinal involvement in thoracic–pelvic phalangeal dystrophy. *Arch Ophthalmol* **96**, 278–281.

Baujat, G. & Merrer, L. M. 2007. Ellis van Creveld syndrome. *Orphanet J Rare Dis* **2**, 1750–1172.

Baujat, G., Huber, C., El Hokayem, J., Caumes, R., Do Ngoc Thanh, C., David, A., et al. 2013. Asphyxiating thoracic dysplasia: Clinical and molecular review of 39 families. *J Med Genet* **50**, 91–98.

Beales, P. L., Bland, E., Tobin, J. L., Bacchelli, C., Tuysuz, B., Hill, J., et al. 2007. IFT80, which encodes a conserved intraflagellar transport protein, is mutated in Jeune asphyxiating thoracic dystrophy. *Nat Genet* **39**, 727–729.

Beemer, F. A., Langer Jr, L. O., Klep-de Pater, J. M., Hemmes, A. M., Bylsm, J. B., Pauli, R. M., et al. 1983. A new short rib syndrome: Report of two cases. *Am J Med Genet* **14**, 115–123.

Beighton, P., Giedion, A., Gorlin, R., Hall, J., Horton, B., Kozlowski, K., et al. 1992. International Working Group on Constitutional Diseases of Bone. International classification of osteochondrodysplasias. *Am J Med Genet* **44**, 223–229.

Bernstein, J., Brough, A. J. & McAdams, A. J. 1974. The renal lesion in syndromes of multiple congenital malformations. Cerebrohepatorenal syndrome; Jeune asphyxiating thoracic dystrophy; tuberous sclerosis; Meckel syndrome. *Birth Defects Orig Artic Ser* **10**, 35–43.

Bernstein, R., Isdale, J., Pinto, M., Zaaijman, J. T. & Jenkins, T. 1985.Short rib–polydactyly syndrome: A single or heterogeneous entity? A re-evaluation prompted by four new cases. *J Med Genet* **22**, 46–53.

Bocca, G., Weemaes, C. M., van der Burgt, I. & Otten, B. J. 2004. Growth hormone treatment in cartilage–hair hypoplasia: Effects on growth and the immune system. *J Pediatr Endocrinol Metab* **17**, 47–54.

Blackburn, M. G. & Belliveau, R. E. 1971. Ellis–van Creveld syndrome: A report of previously undescribed anomalies in two siblings. *Am J Dis Child* **122**, 267–270.

Bredrup, C., Saunier, S., Oud, M. M., Fiskerstrand, T., Hoischen, A., Brackman, D., et al. 2011. Ciliopathies with skeletal anomalies and renal insufficiency due to mutations in the IFT-A gene WDR19. *Am J Hum Genet* **89**, 634–643.

Burn, J., Dezateux, C., Hall, C. M. & Baraitser, M. 1984. Orofaciodigital syndrome with mesomelic limb shortening. *J Med Genet* **21**, 189–192.

Camuglia, J. E., Greer, R. M., Welch, L. & Gole, G. A. 2011. Use of the electroretinogram in a paediatric hospital. *Clin Experiment Ophthalmol* **39**, 506–512.

Cavalcanti, D. P., Huber, C., Sang, K. H., Baujat, G., Collins, F., Delezoide, A. L., et al. 2011. Cormier-Daire V. Mutation in IFT80 in a fetus with the phenotype of Verma–Naumoff provides molecular evidence forJeune–Verma–Naumoff dysplasia spectrum. *J Med Genet* **48**, 88–92.

Chen, H., Yang, S. S., Gonzalez, E., Fowler, M. & Al Saadi, A. 1980. Short rib–polydactyly syndrome, Majewski type. *Am J Med Genet* **7**, 215–222.

Cole, D. G. & Snell, W. J. 2009. SnapShot: Intraflagellar transport. *Cell* **137**, 784.

Curry, C. J. R. & Hall, B. D. 1979. Polydactyly, conical teeth, nail dysplasia, and short limbs: A new autosomal dominant malformation syndrome. *Birth Defects Orig Art Ser* XV(5B), 253–263.

Dagoneau, N., Goulet, M., Genevieve, D., Sznajer, Y., Martinovic, J., Smithson, S., et al. 2009. DYNC2H1 mutations cause asphyxiating thoracic dystrophyand short rib–polydactyly syndrome, type III. *Am J Hum Genet* **84**, 706–711.

Davenport, J. R., Watts, A. J., Roper, V. C., Croyle, M. J., van Groen, T., Wyss, J. M., et al. 2007. Disruption of intraflagellar transport in adult mice leads to obesity and slow-onset cystic kidney disease. *Curr Biol* **17**, 1586–1594.

Davis, E. E., Zhang, Q., Liu, Q., Diplas, B. H., Davey, L. M., Hartley, J., et al. 2011. TTC21B contributes both causal and modifying alleles across the ciliopathy spectrum. *Nat Genet* **43**, 189–196.

Davis, J. T., Davis, J. T., Ruberg, R. L., Leppink, D. M., McCoy, K. S. & Wright, C. C. 1995. Lateral thoracic expansion for Jeune's asphyxiating dystrophy: A new approach. *Ann Thorac Surg* **60**, 694–696.

Davis, J. T., Heistein, J. B., Castile, R. G., Adler, B., Mutabagani, K. H., Villalobos, R. E., et al. 2001 Lateral thoracic expansion for Jeune's syndrome: Midterm results. *Ann Thorac Surg* **72**, 872–878.

Davis, J. T., Long, F. R., Adler, B. H., Castile R.G. & Weinstein S., 2004. Lateral expansion for Jeune syndrome: Evidence of rib healing and new bone formation. *Ann Thorac Surg* **77**, 445–448.

de Vries, J., Yntema, J. L., van Die, C. E., Crama, N., Cornelissen, E. A. & Hamel, B. C. 2010. Jeune syndrome: Description of 13 cases and a proposal for follow-up protocol. *Eur J Pediatr* **169**, 77–88.

Digilio, M. C., Giannotti, A., Pagnotta, G., Mingarelli, R. & Dallapiccola, B. 1995. Joint dislocation and cerebral anomalies are consistently associated with oral–facial–digital syndrome type IV. *Clin Genet* **48**, 156–159.

Eke, T., Woodruff, G. & Young, I. D. 1996. A new oculorenal syndrome: Retinal dystrophy and tubulointerstitial nephropathy in cranioectodermal dysplasia. (Letter) *Br J Ophthalmol* **80**, 490–491.

Elçioglu, N. H. & Hall, C. M. 2002. Diagnostic dilemmas in the short rib–polydactyly syndrome group. *Am J Med Genet* **111**, 392–400.

Ellis, R. W. & van Creveld, S. 1940. A syndrome characterized by ectodermal dysplasia, polydactyly, chondro-dysplasia and congenital morbus cordis: report of three cases. *Arch Dis Child* **15**, 65–84.

Fischer, E., Legue, E., Doyen, A., Nato, F., Nicolas, J. F., Torres, V., et al. 2006. Defective planar cell polarity in polycystic kidney disease. *Nat Genet* **2038**, 21–23.

Fliegauf, M., Benzing, T. & Omran, H. 2007. When cilia go bad: Cilia defects and ciliopathies. *Nat Rev Mol Cell Biol* **8**, 880–893.

Follit, J. A., Xu, F., Keady, B. T. & Pazour, G. J. 2009. Characterization of mouse IFT complex B. *Cell Motil Cytoskeleton* **66**, 457–468.

Follit, J. A., San Agustin, J. T., Xu, F., Jonassen, J. A., Samtani, R., Lo, C. W., et al. 2008. The Golgin GMAP210/TRIP11 anchors IFT20 to the Golgi complex. *PLoS Genet* 4, e1000315.

Fry, A. E., Klingenberg, C., Matthes, J., Heimdal, K., Hennekam, R. C. & Pilz, D. T. 2009. Connective tissue involvement in two patients with features of cranioectodermal dysplasia. *Am J Med Genet* A. **149A**, 2212–2215.

Giedion, A. 1979. Phalangeal cone shaped epiphysis of the hands (PhCSEH) and chronic renal disease—the conorenal syndromes. *Pediatr Radiol* **8**, 32–38.

Gilissen, C., Arts, H. H., Hoischen, A., Spruijt, L., Mans, D. A., Arts, P., et al. 2010. Exome sequencing identifies WDR35 variants involved in Sensenbrenner syndrome. *Am J Hum Genet* **87**, 418–423.

Goetz, S. C., Ocbina, P. J. & Anderson, K. V. 2009. The primary cilium as a Hedgehog signal transduction machine. *Methods Cell Biol* **94**, 199–222.

Hao, L., Efimenko, E., Swoboda, P. & Scholey, J. M. 2011. The retrograde IFT machinery of *C. elegans* cilia: Two IFT dynein complexes? *PLoS One* 6, e20995.

Harada, D., Yamanaka, Y., Ueda, K., Shimizu, J., Inoue, M., Seino, Y., et al. 2005. An effective case of growth hormone treatment on cartilage–hair hypoplasia. Bone **36**, 317–322.

Hennekam, R. C. M., Beemer, F. M., Gerards, L. J. & Cats, B. 1983. Thoracic pelvic phalangeal dystrophy (Jeune syndroom). *Tijdschr Kindergeneeskd* **51**, 95–100.

Herdman, R. C. & Langer, L. O. 1968. The thoracic asphyxiant dystrophy and renal disease. *Am J Dis Child* **116**, 192–201.

El Hokayem, J., Huber, C., Couve, A., Aziza, J., Baujat, G., Bouvier, R., et al. 2012. NEK1 and DYNC2H1 are both involved in short rib–polydactyly Majewski type but not in Beemer Langer cases. *J Med Genet* **49**, 227–233.

Hopper, M. S. C., Boultbe, J. E. & Watson, A. R. 1979. Polyhydramnion associated with congenital pancreatic cysts and asphyxiating thoracic dysplasia: A case report. *S Afr Med J* **56**, 32–33.

Hossain, Z., Ali, S. M., Ko, H. L., Xu, J., Ng, C. P., Guo, K., et al. 2007. Glomerulocystic kidney disease in mice with a targeted inactivation of Wwtr1. *Proc Natl Acad Sci U.S.A.* **104**, 1631–1636.

Huangfu, D., Liu, A., Rakeman, A. S., Murcia, N. S., Niswander, L. & Anderson, K. V. 2003. Hedgehog signalling in the mouse requires intraflagellar transport proteins. *Nature* **426**, 83–87.

Huber, C. & Cormier-Daire, V. 2012. Ciliary disorder of the skeleton. *Am J Med Genet, C, Semin 555 Med Genet* **160**, 165–174.

Ishikawa, H. & Marshall, W. F. 2011. Ciliogenesis: Building the cell's antenna. *Nat Rev Mol Cell Biol* **12**, 222–234.

Jenkins, D. & Beales, P. L. 2012. Genes and mechanisms in human ciliopathies, in *Emery & Rimoin's Principles and Practice of Medical Genetics*, ed. by **Korf, B., Pyeritz, R. & Rimoin, D.** Elsevier.

Jeune, M., Beraud, C. & Carron, R. 1955. Asphyxiating thoracic dystrophy with familial characteristics. *Arch Fr Pediatr* **12**, 886–891.

Jonassen, J. A., Sanagustin, J., Baker, S. P. & Pazour, G. J. 2012. Disruption of IFT complex A causes cystic kidneys without mitotic spindle misorientation. *J Am Soc Nephrol* **23**, 641–651.

Katsanis, N., Ansley, S. J., Badano, J. L., Eichers, E. R., Lewis, R. A., Hoskins, B. E., et al. 2011. Triallelic inheritance in Bardet–Biedl syndrome, a Mendelian recessive disorder. *Science* **293**, 2256–2259.

Katsouras, C. S., Thomadakis, C. & Michalis, L. K. 2003. Cardiac Ellis–van Creveld syndrome. *Int J Cardiol* **87**, 315–316.

Kannu, P., McFarlane, J. H., Savarirayan, R. & Aftimos, S. 2007. An unclassifiable short rib–polydactyly syndrome with acromesomelic hypomineralization and campomelia in siblings. *Am J Med Genet A* **143A**, 2607–2611.

Keogh S. J., McKee S., Smithson S. F., Grier D. & Steward C. G. 2012. Shwachman–Diamond syndrome: a complex case demonstrating the potential for misdiagnosis as asphyxiating thoracic dystrophy (Jeune syndrome). *BMC Pediatr* **3**, 48.

Key, L. L.& Gross, A. 1996. Response to growth hormone in children with chondrodysplasia. *J Pediatr* **128(5 Pt 2)**, S14–7.

Konstantinidou, A. E. 2009. Cranioectodermal dysplasia: A probable ciliopathy. *Am J Med Genet A* **149A**, 2206–2211.

Kovacs, N., Sarkany, I., Mohay, G., Adamovich, K., Ertl, T., Kosztolanyi, G., et al. 2006. High incidence of short rib–polydactyly syndrome type IV in a Hungarian Roma subpopulation. (Letter) *Am J Med Genet* **140A**, 2816–2818.

Krock, B. L., Mills-Henry, I. & Perkins, B. 2009 Retrograde intraflagellar transport by cytoplasmic dynein-2 is required for outer segment extension in vertebrate photoreceptors but not arrestin translocation. *Invest Opthalmol Vis Sci* **50**, 5463–5471.

Kronenberg, H. 2003. Developmental regulation of the growth plate. *Nature* **423**, 332–336.

Labrune, P., Fabre, M. & Trioche, P. 1999. Jeune syndrome and liver disease: Report of three cases treated with ursodeoxycholic acid. *Am J Med Genet* **87**, 324–328.

Lang, G. D., & Young. I. D. 1991. Cranioectodermal dysplasia in sibs. *J Med Genet* **28**, 424.

Langer Jr, L. O. 1968. Thoracic–pelvic–phalangeal dystrophy: Asphyxiating thoracic dystrophy of the newborn, infantile thoracic dystrophy. *Radiology* **91**, 447–456.

Lehman, A. M., Eydoux, P., Doherty, D., Glass, I. A., Chitayat, D., Chung, B. Y., et al. 2010. Co-occurrence of Joubert syndrome and Jeune asphyxiating thoracic dystrophy. *Am J Med Genet* A **152A**, 1411–1419.

Levin, L. S., Perrin, J. C. S., Ose, L., Dors, J. P., Miller, J. D. & McKusick, V. A. 1977. A heritable syndrome of craniosynostosis, short thin hair, dental abnormalities, and short limbs: Cranioectodermal dysplasia. *J Pediatr* **90**, 55–61.

Lienkamp, S. S., Liu, K., Karner, C. M., Carroll, T. J., Ronneberger, O., Wallingford, J. B., et al. 2012. Vertebrate kidney tubules elongate using a planar cell polarity-dependent, rosette-based mechanism of convergent extension. *Nat Genet* **44**, 1382–1387.

Lin, A. E., Doshi, N., Flom, L., Tenenholz, B. & Filkins, K. L. 1991. Beemer–Langer syndrome with manifestations of an orofaciodigital syndrome. *Am J Med Genet* **39**, 247–251.

Lin, F., Hiesberger, T., Cordes, K., Sinclair, A. M., Goldstein, L. S., Somlo, S., et al. 2003. Kidney-specific inactivation of the KIF3A subunit of kinesin-II inhibits renal ciliogenesis and produces polycystic kidney disease. *Proc Natl Acad Sci U.S.A.* **100**, 5286–5291.

Liu, F., Kohlmeier, S, & Wang, C. Y. 2008. Wnt signalling and skeletal development. *Cell Signal* **20**, 999–1009.

Long, F. & Ornitz, D. M. 2013. Development of the endochondral skeleton. *Cold Spring Harbor Perspect Biol* 5, pii: a008334, doi: 10.1101/cshperspect.a008334

Lurie, I. W. 1994.Further delineation of the Beemer–Langer syndrome using concordance rates in affected sibs. *Am J Med Genet* **50**, 313–317.

Mainzer, F., Saldino, R. M., Ozonoff, M. B. & Minagi, H. 1970. Familial nephropathy associated with retinitis pigmentosa, cerebellar ataxia and skeletal abnormalities. *Am J Med* **49**, 556.

Maroteaux, P. & Savart, P. 1964. Asphyxiating thoracic dystrophy. Radiological study and relation to the Ellis–van-Creveld syndrome. *Ann Radiol (Paris)* **7**, 332–338.

Mill, P., Lockhar, P. J., Fitzpatrick, E., Mountford, H. S., Hall, E. A., Reijns, M. A. M., et al. 2011. Human and mouse mutations in WDR35 cause short-rib polydactyly syndromes due to abnormal ciliogenesis. *Am J Hum Genet* **88**, 508–515.

McKusick, V. A., Egeland, J. A., Eldridge, R. & Krusen, D. E. 1964. Dwarfism in the Amish. I. The Ellis–van Creveld syndrome. *Bull Johns Hopkins Hosp* **115**, 306–336.

Majewski, F., Pfeiffer, R. A., Lenz, W., Muller, R., Feil, G. & Seiler, R. 1971. Polysyndaktylie, verkuerzte Gliedmassen, und Genitalfehlbildungen: Kennzeichen eines selbstaendigen Syndroms? *Z Kinderheilk* **111**, 118–138.

Marec, B. L., Passarge, E., Dellenbach, P., Kerisit, J., Signargout, J., Ferrand, B., et al. 1973. Les formes neonatales lethales de la dysplasie chondro-ectodermique. *Ann Radiol* **16**, 19–26.

Marszalek, J. R., Liu, X., Roberts, E. A., Chui, D., Marth, J. D., Williams, D. S., et al. 2000. Genetic evidence for selective transport of opsin and arrestin by kinesin-II in mammalian photoreceptors. *Cell* **102**, 175–187.

Merrill, A. E., Merriman, B., Farrington-Rock, C., Camacho, N., Sebald, E. T., Funari, V. A., et al. 2009. Ciliary abnormalities due to defects in the retrograde transport protein DYNC2H1 in short-rib polydactyly syndrome. *Am J Hum Genet* **84**, 542–549.

Moyer, J. H., Lee-Tischler, M. J., Kwon, H. Y., Schrick, J. J., Avner, E. D., Sweeney, W. E., et al. 1994. Candidate gene associated with a mutation causing recessive polycystic kidney disease in mice. *Science* **264**, 1329–1333.

Naumoff, P., Young, L. W., Mazer, J. & Amortegui, A. J. 1977. Short-rib–polydactyly syndrome type III. *Radiology* **122**, 443–447.

Naumoff, P. 1980. Personal Communication. Pittsburgh, Pa. 2/18/1980

Nevin, N. C. & Thomas, P. S. 1989. Orofaciodigital syndrome type IV: Report of a patient. *Am J Med Genet* **32**, 151–154.

Oberklaid, F., Danks, D. M., Mayne, V. & Campbell, P. 1977. Asphyxiating thoracic dysplasia. Clinical, radiological, and pathological information on 10 patients. *Arch Dis Child* **52**, 758–765.

Ocbina, P. J., Eggenschwiler, J. T., Moskowitz, I. & Anderson, K. V. 2011. Complex interactions between genes controlling trafficking in primary cilia. *Nat Genet* **43**, 547–553.

Oliveira da Siva, E., Janovitz, D. & Cavalcanti de Albuquerque, S. 1980. Ellis–van Creveld syndrome: Report of 15 cases in an inbred kindred. *J Med Genet* **17**, 349–356.

Passarge, E. 1983. Familial occurrence of a short rib syndrome with hydrops fetalis but without polydactyly. (Letter) *Am J Med Genet* **14**, 403–405.

Patel, V., Li, L., Cobo-Stark, P., Shao, X., Somlo, S., Lin, F., et al. 2008. Acute kidney injury and aberrant planar cell polarity induce cyst formation in mice lacking renal cilia. *Hum Mol Genet* **17**, 1578–1590.

Pazour, G. J., Baker, S. A., Deane, J. A., Cole, D. G., Dickert, B. L., Rosenbaum, J. L., et al. 2002. The intraflagellar transport protein, IFT88, is essential for vertebrate photoreceptor assembly and maintenance. *J Cell Biol* **157**, 103–114.

Perrault, I., Saunier, S., Hanein, S., Filhol, E., Bizet, A. A., Collins, F., et al. 2012. Mainzer–Saldino syndrome is a ciliopathy caused by IFT140 mutations. *Am J Hum Genet* **90**, 864–870.

Popovic-Rolovic, M., Calic-Perisic, N., Bunjevacki, G. & Negovanovic, D. 1976. Juvenile nephronophthisis associated with retinal pigmentary dystrophy, cerebellar ataxia, and skeletal abnormalities. *Arch Dis Child* **51**, 801–803.

Quinlan, R. J., Tobin, J. L. & Beales, P. L. 2008. Modeling ciliopathies: Primary cilia in development and disease. *Curr Top Dev Biol* **84**, 249–310.

Renier, J. C., Larget-Piet, L., Boasson, M., Berthelot, J. & Fouillet, L. 1975. Dysplasie chondroépidermique d'Ellis–van Creveld: Deux cas dans une même fratrie. *Rev Rhum* **42**, 417–422.

Richardson, M. M., Beaudet, A. L., Wagner, M. L., Malini, S., Rosenberg, H. S. & Lucci, J. A. 1977. Prenatal diagnosis of recurrence of Saldino–Noonan dwarfism. *J Pediatr* **91**, 467–471.

Rix, S., Calmont, A., Scambler, P. J. & Beales, P. L. 2011. An Ift80 mouse model of short rib polydactyly syndromes shows defects in hedgehog signalling without loss or malformation of cilia. *Hum Mol Genet* **20**, 1306–1314.

Robins, D. G., French, T. A. & Chakera, T. M. 1976. Juvenile nephronophthisis associated with skeletal abnormalities and hepatic fibrosis. *Arch Dis Child* **51**, 799–801.

Rosenbaum, J. L. & Witman, G. B. 2002. Intraflagellar transport. *Nat Rev Mol Cell Biol*; **3**, 813–825.

Roubicek, M. & Spranger, J. 1984. Weyers acrodental dysostosis in a family. *Clin Genet* **26**, 587–590.

Ruiz-Perez,V. L., Blair, H. J., Rodriguez-Andres, M. E., Blanco, M. J., Wilson, A., Liu, Y. N., et al. 2007. Evc is a positive mediator of Ihh-regulated bone growth that localises at the base of chondrocyte cilia. *Development* **134**, 2903–2912.

Ruiz-Perez, V. L., Tompson, S., Blair, H., Espinoza-Valdez, C., Lapunzina, P., Silva, E., et al. 2003. Mutations in two nonhomologous genes in a head-to-head configuration cause Ellis–van Creveld syndrome. *Am J Hum Genet* **72**, 728–732.

Ruiz-Perez, V. L., Ide, S. E., Strom, T. M., Lorenz, B., Wilson, D., Woods, K., King, L., et al. 2000. Mutations in a new gene in Ellis–van Creveld syndrome and Weyers acrodental dysostosis. *Nat Genet* **24**, 283–286.

Saldino, R. M.& Noonan, C. D. 1972. Severe thoracic dystrophy with striking micromelia, abnormal osseous development, including the spine, and multiple visceral abnormalities. *Am J Roentgenol* **114**, 257–263.

Schaballie, H., Renard, M., Vermylen, C., Scheers, I., Revencu, N., Regal, L., et al. 2013. Misdiagnosis as asphyxiating thoracic dystrophy and CMV-associated haemophagocytic lymphohistiocytosis in Shwachman–Diamond syndrome. *Eur J Pediatr* **172**, 613–622.

Schmidts, M., Frank, V., Eisenberger, T., Al Turki, S., Antony, D., Rix, S., et al. 2013. Combined NGS approaches identify mutations in the intraflagellar transport gene IFT140 in skeletal ciliopathies with early progressive kidney disease. *Hum Mut* **34**, 714–724.

Schmidts, M., Arts, H. H., Bongers, E. M. H. F., Yap, Z., Oud, M. M., Anthony, D. , et al., and the UK10K Consortium. 2013. Exome sequencing identifies DYNC2H1 mutations as a common cause of asphyxiating thoracic dystrophy (Jeune syndrome) without major polydactyly, renal or retinal involvement. *J Med Gene* **50**, 309–23

Sharoni, E., Erez, E., Chorev, G., Dagan O. & Vidne B.A.,1998 A chest reconstruction in asphyxiating thoracic dystrophy. *J Pediatr Surg* **33**, 1578–1581.

Smits, P., Bolton, A. D., Funari, V., Hong, M., Boyden, E. D., Lu, L., et al. 2010. Lethal skeletal dysplasia in mice and humans lacking the golgin GMAP-210. *New Engl J Med* **362**, 206–216.

Spranger, J. W., Brill, P. W. & Poznanski, A. K. 2002. *Bone Dysplasias: An atlas of genetic disorders of the skeleton*, 2nd edition. ISBN-0-19-521474-9. Oxford University Press, New York.

Spranger, J. W., Grimm, B., Weller, M., Weibenbacher, G., Herrmann, J., Gilbert, E. F. & Krepler, R. 1974. Short rib–polydactyly (SRP) syndromes, types Majewski and Saldino–Noonan. *Z Kinderheilk* **116**, 73–94.

Spranger, J. W., Langer, L. O., Weller, M. H. & Herrmann, J . 1974. Short rib–polydactyly syndromes and related conditions. *Birth Defects Orig Art Ser* X(9), 117–123.

Thiel, C., Kessler, K., Giessl, A., Dimmler, A., Shalev, S. A., von der, H. S., et al. 2011. NEK1 mutations cause short-rib polydactyly syndrome type majewski. *Am J Hum Genet* **88**, 106–114.

Thomas, S., Legendre, M., Saunier, S., Bessieres, B., Alby, C., Bonniere, M., et al. 2012. TCTN3 mutations cause Mohr–Majewski syndrome. *Am J Hum Genet* **91**, 372–378.

Todd, D. W., Tinguely, S. J. & Norberg, W. J. 1986. A thoracic expansion technique for Jeune's asphyxiating thoracic dystrophy. *J Pediatr Surg* **21**, 161–163.

Walczak-Sztulpa, J., Eggenschwiler, J., Osborn, D., Brown, D. A., Emma, F., Klingenberg, C., et al. 2010. Cranioectodermal dysplasia, Sensenbrenner syndrome, is a ciliopathy caused by mutations in the IFT122 gene. *Am J Hum Genet* **86**, 949–956.

Verma, I. C., Bhargava, S. & Agarwal, S. 1975. An autosomal recessive form of lethal chondrodystrophy with severe thoracic narrowing, rhizoacromelic type of micromelia, polydactyly and genital anomalies. *Birth Defects Orig Art Ser* XI(6), 167–174.

Weyers, H. 1952. Ueber eine korrelierte Missbildung der Kiefer und Extremitatenakren (Dysostosis acro-facialis). *Fortschr Roentgenstr* **77**, 562–567.

Wright, M. J. & Irving, M. J. 2012. Clinical management of achondroplasia. *Arch Dis Child* **97**, 129–134.

Yang, S. S., Roth, J. A. & Langer, L. O. 1991. Short rib syndrome Beemer–Langer type with polydactyly: A multiple congenital anomalies syndrome. *Am J Med Genet* **39**, 243–246.

Yerian, L. M., Brady, L. & Hart, J. 2003. Hepatic manifestations of Jeune syndrome (asphyxiating thoracic dystrophy). *Semin Liver Dis* **23**, 195–200.

Zaffanello, M., Diomedi-Camassei. F., Melzi, M. L., Torre, G., Callea, F. & Emma, F. 2006. Sensenbrenner syndrome: A new member of the hepatorenal fibrocystic family. *Am J Med Genet* **140A**, 2336–2340.

Zangwill, K. M., Boal, D. K. B. & Ladda, R. L. 1988. Dandy–Walker malformation in Ellis–van Creveld syndrome. *Am J Med Genet* **31**, 123–129.

Chapter 4

Joubert syndrome and Joubert syndrome-related disorders

Victoria Harrison and Andrea H. Németh

History of the eponym

In 1969, Marie Joubert and colleagues (Joubert et al. 1969) described four siblings with episodic hyperpnoea, abnormal eye movements, ataxia and intellectual disability in association with partial or complete agenesis of the cerebellar vermis. A fifth, unrelated case was also described and in addition to episodic hyperpnoea and agenesis of the cerebellar vermis, this patient had a range of other clinical features including polydactyly.

In 1977, Boltshauser and Isler (Boltshauser & Isler, 1977) reported a further three cases from two families with what they described as *Joubert syndrome*. As a result some authors refer to Joubert–Bolthauser syndrome, although more recently the abbreviation JBTS has become increasingly used. The features common to all eight patients included severe intellectual disability, an abnormal breathing pattern occurring neonatally, hypotonia and a variable degree of cerebellar vermis hypoplasia. Tongue protrusion and abnormal eye movements were each present in four out of eight patients. Microcephaly, an occipital meningoencephalocele, ataxia and polydactyly were also described. Parental consanguinity and/or recurrence in siblings were highly suggestive of an autosomal recessive condition with variable expressivity.

Following these original descriptions, many more cases of Joubert syndrome were reported, with numerous additional clinical features expanding the original phenotype, and in 1992 diagnostic criteria were proposed (Saraiva & Baraitser, 1992). These suggested that cerebellar vermis hypoplasia, hypotonia and developmental delay must be present in addition to abnormal breathing and/or abnormal eye movements. Patients were classified into Joubert syndrome types A or B based on the absence or presence of retinal dystrophy.

As radiological techniques advanced, more details of the underlying brain abnormalities emerged. In 1997, the term 'molar tooth sign' (MTS) was first used to describe the appearance of the mid and hindbrain structures on axial brain imaging (Maria et al. 1997) as seen in Figure 4.1 (Brancati et al. 2010). The primary criteria for classical Joubert syndrome are now considered to be the presence of the MTS, hypotonia evolving into ataxia, and intellectual disability. The presence of retinopathy is variable.

However, it is important to note that although the MTS is diagnostic of classical Joubert syndrome, it is also present in other overlapping disorders now collectively referred to

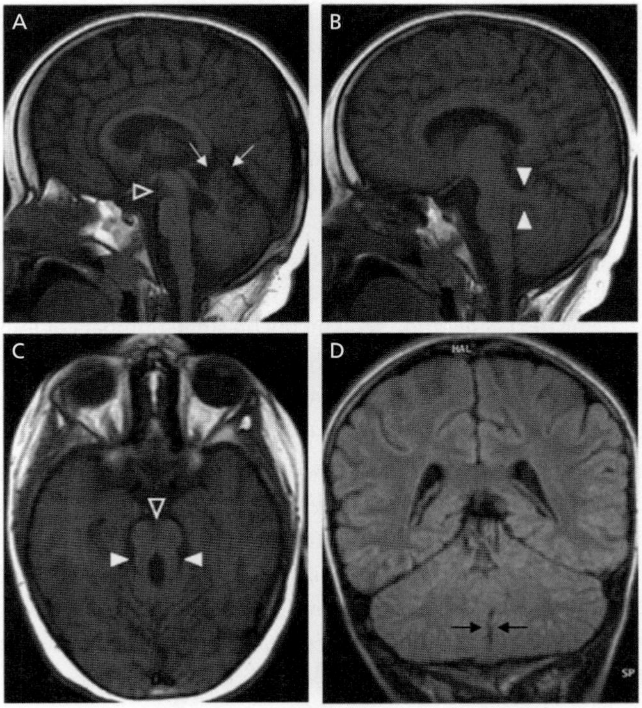

Figure 4.1 Brain MRI sections in patients with JSRD. (A) Mid-sagittal T1-weighted image shows a thin midbrain with corresponding enlargement of the interpeduncular fossa (open arrowhead). There is concurrent superior vermian dysplasia (thin arrows). (B) Parasagittal T1-weighted image shows thickened and maloriented superior cerebellar peduncle (thick arrowheads). (C) Axial T1-weighted image confirms the deepened interpeduncular fossa (open arrowhead) and abnormal superior cerebellar peduncles (thick arrowheads), comprising the 'molar tooth sign'. (D) Coronal FLAIR image shows midline cerebellar cleft (black arrows) indicating agenesis of the inferior vermis.

as Joubert syndrome-related disorders (JSRDs). These conditions have a variety of additional non-neurological features including renal, hepatic and oral–facial abnormalities. They have a variety of eponymous and other names but are unlikely to represent distinctive conditions; rather, they represent the phenotypic variation associated with different underlying cilial defects.

Epidemiology

Joubert syndrome is rare and therefore accurate prevalence data is not available. Figures vary but 1 in 100,000 live births appears to be most commonly used and considered appropriate for the purposes of genetic counselling (Kroes et al. 2008; Parisi, 2009). It is likely that more mildly affected children and adults go unrecognised and the actual prevalence may be higher.

Cases have been reported from many different ethnic backgrounds. Because JBTS/JSRD is almost always recessive, it is frequently found in consanguineous families and this should be specifically asked for when the clinical history is taken. The first family to be described by Joubert and colleagues (Joubert et al. 1969) was French Canadian and the parents of the affected siblings were noted to have a common ancestor nine generations earlier. A number of mutations have subsequently been reported for French Canadian populations and there exists a high prevalence of JBTS within the Saint Lawrence region of Quebec. Studies using next generation sequencing on small numbers of Joubert syndrome families from this and other regions of Quebec have identified compound heterozygous mutations within the genes *C5orf42*, *CC2D2A* and *TMEM231*. This included a number of recurrent mutations in unrelated individuals (Srour et al. 2012a,b).

A single founder mutation has been identified in the Ashkenazi Jewish population. Thirteen patients from eight families were found to have a common region of homozygosity on chromosome 11. Sequencing of the 14 genes within this region identified a single point mutation in *TMEM216* (c.35G>T, R12L). All patients were homozygous for this mutation and all parents were heterozygous. Thirty out of 2766 anonymous, ethnically matched controls were also found to be heterozygous, indicating a possible carrier frequency of 1 in 92 for this population (Edvardson et al. 2010).

Clinical features

Neurology

Hypotonia is present in all children with JBTS and is evident from the neonatal period. There are reduced spontaneous movements and positional plagiocephaly may result. In itself, hypotonia is a non-specific finding common to many neurodevelopmental disorders and it is the presence of additional features that should suggest the diagnosis of JBTS.

There is delay in reaching motor milestones, with one study reporting that 91% of JBTS patients were able to roll over at an average age of 10 months, 73% were able to sit up by approximately 19 months and 50% were able to walk by approximately 47 months (Maria et al. 1999b). Twelve out of 18 patients reviewed by Hodgkins et al. (2004) walked independently between 22 months and 10 years. More severely affected patients may remain wheelchair dependent.

If ambulation is achieved, gait tends to be wide based and unsteady. Fourteen of the 15 JBTS patients clinically assessed by Maria et al. were ataxic and all displayed problems with balance using a variety of tests (unilateral weight bearing, tandem ambulation, Balance Master trials) (Maria et al. 1997).

Intellectual disability and behaviour

Intellectual function is usually in the moderate disability range although it is highly variable and ranges from normal to profound (Hodgkins et al. 2004; Maria et al. 1999b; Parisi,

2009; Valente et al. 2008). Significant expressive speech and motor impairment can make it difficult to accurately assess cognitive function, and receptive abilities are often less severely delayed (Gitten et al. 1998; Hodgkins et al. 2004; Maria et al. 1999a). *Developmental regression in not a feature of JBTS* and if present must prompt urgent re-evaluation of the patient (Maria et al. 1999a).

Gitten and colleagues used a parent report measure (Child Development Inventory) to assess the social, language, adaptive, motor, cognitive and academic development of 32 children with JBTS ranging from 14 to 204 months of age (mean 68.7) (Gitten et al. 1998). The average developmental age was 19 months (63% below chronological age) and was spread across all domains tested. Older children were observed to be further behind their chronological age compared to younger children. There was no significant correlation between the level of disability and severity of cerebellar involvement on magnetic resonance imaging (MRI) scans.

Maria and colleagues reported that 61% of their JBTS cohort said their first words at approximately 26 months (±4.9 months), 75% of which later combined words at approximately 44 months (±17.4 months) (Maria et al. 1999a). In another cohort, 73% of patients aged more than 5 years had intelligible speech and six attended a mainstream school with varying levels of support (Hodgkins et al. 2004).

There is limited data confirming that autism is part of the behavioural phenotype of Joubert syndrome. Three out of 11 children (27%) reported by Ozonoff and colleagues (Ozonoff et al. 1999) were assessed as meeting the DSM-IV criteria for a diagnosis of autism, and a further patient was diagnosed with a pervasive developmental disorder not otherwise specified. However, Takahashi and colleagues (Takahashi et al. 2005) used semi-structured family history interviews and the Autism Behavioural Checklist to assess 31 children with JBTS. None of these children met the clinical criteria for a diagnosis of autism. Lack of eye contact due to oculomotor abnormalities and speech delay due to oromotor dysfunction may mimic the social and communication difficulties seen in children with autism (Doherty, 2009). In the authors' opinion the diagnosis of autism should be carefully reviewed, especially if made prior to the diagnosis of Joubert syndrome.

Other behavioural features reported by some parents of children with JBTS include temper tantrums, hyperactivity, aggression and dependency (Fennell et al. 1999; Hodgkins et al. 2004).

Breathing

Episodes of hyperpnoea with or without periods of apnoea are common in the neonatal period but are not present in all infants with JBTS. Early death attributed to apnoea has been reported (Maria et al. 1999a; Steinlin et al. 1997). If present, breathing abnormalities become less obvious or disappear with increasing age. In older children, they may only be apparent at time of illness or emotional stress (Joubert et al. 1969).

Hodgkins and colleagues reported apnoeic episodes in 13 of their 18 JBTS patients (Hodgkins et al. 2004). In 10, this was only transient lasting up to 3 months. One patient required admission to a paediatric intensive care unit for several weeks because of severe tachypnoeic episodes. One required oxygen at home and another had apnoeic episodes lasting until the age of 18 months.

Eyes

Joubert syndrome is associated with a wide range of ocular and oculomotor findings. Common eye movement abnormalities include delay or failure of saccadic initiation, also known as oculomotor apraxia (often associated with head thrusts or turns), primary position nystagmus (typically see-saw, but can also be pendular) and abnormalities in smooth pursuit (Lambert et al. 1989; Sturm et al. 2010; Tusa & Hove, 1999; Weiss et al. 2009). Nystagmus and oculomotor apraxia (OMA) are often present at birth and may improve with age (Lambert et al. 1989; Parisi, 2009).

Saccades and quick phases of nystagmus are related to an underlying brainstem defect, whereas abnormalities in smooth pursuits and vestibulo-ocular reflex cancellation can be partially explained by cerebellar vermis hypoplasia (Sturm et al. 2010; Tusa & Hove, 1999). Oculomotor neural pathways cross the midline within the brainstem and cerebellum and clinical findings suggest decussation abnormalities that may not be evident on the MRI (Weiss et al. 2009).

Pigmentary abnormalities and retinal dystrophy develop in a subgroup of patients with Joubert syndrome. When this presents early it may be difficult to distinguish from Leber congenital amaurosis (LCA) but patients with retinal dystrophies, especially those with early onset, should be assessed for the presence of other clinical features and investigated accordingly (Doherty, 2009). A later onset pigmentary retinopathy can present from childhood onwards and often has a more variable course (Saraiva & Baraitser, 1992). In JSRDs the phenotype is generally retinitis pigmentosa (predominant loss of rods with bone spicule pigmentation and nyctalopia) in contrast to Bardet–Biedl syndrome (BBS) where the retinal phenotype is more commonly a cone–rod dystrophy (macular degeneration with loss of central colour vision).

Choroidoretinal colobomas have been reported in 19% of patients in one large cohort of Joubert syndrome patients but found in 71% of those with the specific JSRD known as COACH syndrome (see the section 'Range of the phenotype'). If the macula or optic nerves are involved there can be significant visual impairment (Doherty, 2009; Parisi, 2009). Other ophthalmological findings seen in Joubert syndrome include poor visual acuity, ptosis, optic nerve hypoplasia, astigmatism, strabismus, partial third nerve palsy, cataract, ocular fibrosis and optic disc drusen (Hodgkins et al. 2004; Sturm et al. 2010; Tusa & Hove, 1999).

Kidneys

Renal disease is reported to be present in approximately 16% to 30% of patients with JBTS (Doherty, 2009; Maria et al. 1999a; Saraiva & Baraitser, 1992); figures vary widely because of the age-related penetrance. In the past, two different types of kidney involvement have been distinguished:

- *Juvenile nephronophthisis* (NPHP), which often presents in childhood (or later) with defects in urine concentration (salt-losing renal insufficiency). Signs and symptoms include polydipsia, polyuria, anaemia, poor growth, an increase in serum creatinine and echogenic kidneys on renal ultrasound. As the condition progresses to end-stage renal failure the kidneys appear small and scarred.
- *Cystic kidney disease*, which can be identified pre- or post-natally. The cysts are described as multiple, small and cortical, and affected kidneys also have interstitial chronic inflammation and fibrosis (Hodgkins et al. 2004).

Nephronophthisis and cystic dysplasia is now considered to be part of a continuum with the specific renal manifestation varying by stage of renal disease (Parisi & Glass, 1993). A renal disorder resembling autosomal recessive polycystic kidney disease (ARPKD) has also been reported. The clinical features are more typical of ARPKD with enlarged, diffusely microcystic kidneys and early-onset severe hypertension as well as congenital hepatic fibrosis (Gunay-Aygun et al. 2009).

Liver

Liver disease can be asymptomatic or present with raised transaminases, clinical signs and symptoms of portal hypertension (hepatosplenomegaly, oesophageal varices, ascites, upper gastrointestinal bleeding) and/or abnormal imaging (increased echogenicity, cysts, dilated intrahepatic bile ducts). It is not usually apparent at birth, typically occurring from the second decade of life onwards. In rare cases hepatic fibrosis has progressed to end-stage liver failure requiring transplantation (Doherty, 2009; Parisi, 2009).

Dysmorphology

Facial dysmorphism has been described although it is rather non-specific (Maria et al. 1997, 1999a). Findings have included epicanthic folds, upturned nose with prominent nasal bridge and tip, a long face with bitemporal narrowing, a 'trapezoid'-shaped mouth with lower lip eversion, thickened ear lobes, ptosis and elevated eyebrows that may be arched and occasionally low-set and tilted ears. It has been observed that the face tends to become longer and narrower with a more prominent chin as the child ages (Braddock et al. 2007).

The presence of cleft lip and/or palate, a central notch in the upper lip, soft tissue tumours of the tongue and oral frenuli are indicative of a particular subtype of JSRD known as oral–facial–digital type VI syndrome (OFD VI), or Varadi–Papp syndrome, and can lead to the impression of facial dysmorphism (Poretti et al. 2012).

Other

The frequency of polydactyly in JBTS is approximately 10% to 20% (Doherty, 2009; Valente et al. 2008). Postaxial polydactyly is much more common, but pre-axial has also been seen. Mesoaxial polydactyly has been reported in OFD VI (Poretti et al. 2012). Other skeletal features that can be present in JBTS include cone shaped epiphyses, a small thorax, and scoliosis (Parisi, 2009).

Structural heart defects have been previously reported in patients with a range of ciliopathies but are not common in JSRD patients. Abnormalities described in JSRDs include a patent ductus arteriosus (Peker et al. 2009), an atrial septal defect with persistent left superior vena cava (Elmali et al. 2007), and severe congenital aortic stenosis with a bicuspid aortic valve and atrial septal defect (Karp et al. 2012).

Various endocrine abnormalities have been described in JBTS including isolated growth hormone or thyroid hormone deficiency, or even panhypopituitary dysfunction (Parisi, 2009).

Case reports of JBTS patients with atypical additional features include congenital hypothyroidism due to thyroid dysgenesis (Graber et al. 2009), craniocervical spine involvement with progressive scoliosis (Vogel et al. 2012), duodenal atresia (Hodgkins et al. 2004), Hirschsprung's disease and vocal cord paralysis (Maria et al. 1999a).

Range of the phenotype

Joubert syndrome and Joubert syndrome-related disorders are defined by the presence of the MTS and are therefore, by definition, within the spectrum of ciliopathies associated with developmental abnormalities of the cerebellum.

In addition to JBTS, the following conditions are within the JSRD spectrum when the MTS is present:

- CORS: this refers to the **c**erebellar–**o**cular–**r**enal syndrome
- Senior–Løken syndrome (SLS): ocular abnormalities associated with renal dysplasia
- Dekaban–Arima syndrome (DKA): ocular abnormalities associated with hepatic fibrosis
- COACH syndrome: **c**erebellar vermis hypo-/aplasia, **o**ligophrenia, congenital **a**taxia, ocular **c**oloboma, and **h**epatic fibrosis
- Oral–facial–digital (OFD) syndromes: These are a complex group of (at least) nine disorders whose features include midline oral clefts, tongue nodules, hamartomas and digital abnormalities, including brachydactyly, syndactyly and polydactyly. Polycystic kidney disease and central nervous system anomalies including hydrocephalus and cerebellar anomalies are reported in some cases.

One type of OFD syndrome is Varadi–Papp syndrome, named after the Hungarian clinicians who described polydactyly, cleft lip/palate, lingual hamartomas and psychomotor retardation in a gypsy population. Varadi–Papp syndrome is now known as OFD VI. Two cases with the clinical features of OFD VI have been found to have mutations in TMEM216 but many cases remain without a molecular diagnosis (Valente et al. 2010).

OFD type I is X-linked dominant and caused by mutations in *OFD1*. It is usually associated with male lethality in the presence of polycystic kidneys and affected individuals are nearly all females. *OFD1* encodes a cilial protein involved in centrosomal organisation (Ferrante et al. 2001; Romio et al. 2004). Recessive mutations in *OFD1* are associated with X-linked Joubert syndrome with the MTS and have only recently been recognised. The features are variable and include: severe cognitive impairment, variable retinal degeneration, post-axial polydactyly, apnoea, poor growth, dysmorphic features such as mild hirsutism, low-set ears, a broad nasal bridge, prominent phlitrum and maxillary arch and full lips. Other reported features include obesity, optic atrophy, seizures, macrocephaly, frontal bossing, epicanthic folds, renal cystic disease requiring transplantation and polymicrogyria. Frameshifts and an in-frame deletion have been found (Coene et al. 2009; Field et al. 2012; Tsurusaki et al. 2012).

Although distinct conditions, the following are also included within the JSRD spectrum since they can include the molar tooth sign:

- Meckel–Gruber syndrome (MKS) also known as Meckel syndrome: an autosomal recessive disorder, usually lethal and characterised by the presence of occipital encephalocele, polycystic kidney disease, and polydactyly
- Bardet–Biedl syndrome (BBS): an autosomal recessive disorder characterised by post-axial polydactyly, retinal dystrophy, truncal obesity, male hypogonadotropic hypogonadism, female genitourinary malformation, intellectual impairment, and renal abnormalities
- Cogan-type congenital oculomotor apraxia: an autosomal recessive form of congenital oculomotor apraxia characterised by defective horizontal voluntary eye movements with jerkiness. Some individuals have cerebellar vermis hypoplasia, evidence of the MTS and occasionally develop nephronophthisis. The latter patients, since they have the MTS, should be classified as having JSRD
- Leber congenital amaurosis (LCA) is an early onset retinal degeneration often noted in infancy or very early childhood. Visual impairment is found when a poor visual response and lack of tracking becomes evident. Photophobia, nystagmus, sluggish or near-absent pupillary responses, keratoconus, and high hyperopia. Affected children exhibit the oculodigital sign characterised by poking, rubbing and pressing of the eyes. Developmental delay is not uncommon in children with visual impairment but its presence or the occurrence of any neurological signs in a patient with a diagnosis of LCA should prompt the clinician to consider an MRI scan and other investigations pertinent to ciliopathies since BBS and JRSDs may otherwise be overlooked. However, LCA can be present in isolation and caused by mutations in non-cilial genes and, although some authors include LCA within the JSRD spectrum, those patients who do have additional features should probably not have a diagnostic label of LCA, but have this revised to reflect the other features and/or genetic defect.

The key features of each disorder are shown in Table 4.1, but there are numerous exceptions, again reflecting the complex phenotypic variation of cilial disorders.

Since many of the non-neurological features of JSRD develop with age, a precise clinical diagnosis may not be possible until the child is older. Genetic testing, discussed in the

Table 4.1 Key clinical features found in Joubert syndrome and related disorders

	JBTS	CORS	DKA	SLS	COACH	BBS	MKS	OFD6	XLJS	Cogan
MTS	+	+	+/-	+/-	+/-	+/-	+/-	+/-	+	+/-
Hypotonia	+	+	+/-	+/-	+/-	+/-	NA	+/-	+	+
Cognitive impairment	+	+	+/-	+/-	+/-	+/-	NA	+	+/-	+/-
Hyperpnoea/ apnoea	+	+/-	+/-	+/-	+/-	+/-	NA	+/-	+/-	-
Oculomotor apraxia	+	+	-	-	-	-	NA	-	-	+
Retinal degeneration	+/-	+	+	+	-	+	-	-	+/-	-
Polymicrogyria	+/-	-	-	-	-	-	-	-	+/-	-
Cystic renal dystrophy	-	+	+	-	-	+	+	-	+/-	-
Nephronoph-thisis	-	+	+	+	-	+/-	-	-	-	+/-
Coloboma	-	-	-	-	+	-	+/-	-	-	-
Hepatic fibrosis	-	-	-	-	+	+/-	-	-	-	-
Encephalo-coele	-	-	-	-	-	-	+	-	-	-
Polydactyly	-	-	-	-	-	+	+	+	+/-	-
Obesity	-	-	-	-	-	+	NA	-	+/-	-
Tongue hamartoma	-	-	-	-	-	-	-	+	-	-
Oral frenula	-	-	-	-	-	-	-	+	+/-	-
Orofacial clefting	-	-	-	-	-	-	+/-	+	-	-

Adapted with permission from Sattar S and Gleeson, The ciliopathies in neuronal development: a clinical approach to investigation of Joubert syndrome and Joubert syndrome-related disorders, *Developmental Medicine and Child Neurology*, Volume 53, pp. 793–798, Copyright © 2011 The Authors, Developmental Medicine and Child Neurology, Copyright © 2011 Mac Keith Press.

section 'Genetics', is of considerable use in facilitating a specific diagnosis of JBTS or JSRD and the advent of Next Generation Sequencing should assist clinicians in this respect.

In addition to the evident phenotypic heterogeneity of JBTS/JSRDs, the phenotypic spectrum of cilial proteins causing these disorders can include conditions that *do not* have the typical MTS; for example, mutations in the same genes causing JSRDs can cause other ciliopathies, but in the absence of the MTS are not considered within this spectrum:

- Nephronophthisis
- Leber congenital amaurosis

non-progressive intellectual impairment and cerebellar hypoplasia, indicating a developmental defect, combined with progressive later onset disorders of other systems, such as retina, liver and kidney. This clearly indicates differential requirements for cilial processes of different cell types and at different developmental stages, and unravelling these mechanisms is likely to be very important in our understanding of these complex processes that go beyond ciliopathies.

Analysis of JSRDs has also provided very important insights into the evolution of gene regulation, which has wide relevance for genomic studies. Linkage analysis of some JBTS families revealed a locus on chromosome 11 and in about half the cases mutations were found in *TMEM216*, a known JBTS gene. However, the remaining patients, all of whom had similar clinical features to those with *TMEM216* mutations including optic coloboma, retinal dysplasia, nephronophthisis and occasional occipital encephalocoele, did not have identifiable mutations in *TMEM216*. Subsequent analysis revealed mutations in *TMEM138*, a neighbouring gene, of previously unknown significance. This gene has no homology with *TMEM216*, but is arranged in a head-to-tail conformation on chromosome 11. Further extensive analysis showed that the alignment of the two genes adjacent to each other occurred during the transition from amphibia to reptiles ~340 million years ago. This arrangement allows for co-ordinated expression of the two genes, mediated by regulatory intergenic sequences. Both genes are required for ciliogenesis possibly by marking vesicles en route from the Golgi to the base of cilium. *TMEM216* was shown to be localised to post-Golgi vesicles along microtubules, as well as the Golgi apparatus surrounding the base of cilium, whereas *TMEM138* localised to adjacent but non-overlapping distinct vesicles. Morpholino experiments in zebrafish went on to demonstrate that prior to the evolutionary linking of the two genes they had separate functions including involvement in distinct brain and cardiac developmental pathways (Lee et al. 2012b). In addition to providing insights into the regulation of two cilial genes this work provides a model for understanding the evolution of gene regulation by non-coding genomic regions.

Genetics

Classical Joubert syndrome is an autosomal recessive condition, and, as such, parents should be counselled for a 25% offspring recurrence risk. Affected individuals rarely reproduce, although if they are going to there is a high risk of recurrence in consanguineous couples and presence of consanguinity should be noted when assessing a family. The exceptions to autosomal recessive inheritance are the rare cases of X-linked JSRD (*OFD1* gene) and care must be taken to consider this, as the genetic counselling issues are very different. There are suggestions that some cases could be caused by *de novo* dominant mutations but this is so far unproven.

As with many of the ciliopathies, JBTS/JRSDs are clinically and genetically heterogeneous. JSRDs are associated with an ever-increasing number of genes that each account for only a small proportion of cases (ranging from <1% to approximately 10%).

At the time of writing there are 20 genes that are known to cause Joubert syndrome-related disorders (see Table 4.2) (Arts et al. 2007; Baala et al. 2007; Bielas et al. 2009; Cantagrel et al. 2008; Chaki et al. 2012; Coene et al. 2009; Dafinger et al. 2011; Davis et al. 2011; Delous et al. 2007; Edvardson et al. 2010; Ferland et al. 2004; Garcia-Gonzalo et al. 2011; Gorden et al. 2008; Huang et al. 2011; Lee et al. 2012a,b; Noor et al. 2008a,b; Parisi et al. 2004; Srour et al. 2012a,b; Thomas et al. 2012; Valente et al. 2006). A 21st gene producing a ciliary protein, *TTC21B*, has been described as JBTS11 but mutations have only been found in the heterozygous state. Two mutations in this gene were found in Jeune asphyxiating thoracodystrophy and NPHP but the significance of the heterozygous changes in the JBTS patients is still unclear (Davis et al. 2011). The genes with the highest individual contribution to JBTS are *AHI1* (~10%), *CC2D2A* (~10%), *TMEM67* (~10%), *RPGRIP1L* (~2–4%) and *TMEM216* (~3%) (Doherty, 2009; Kroes et al. 2008; Parisi et al. 2006; Valente et al. 2008) while the other genetic causes are rarer (Brancati et al. 2009; Castori et al. 2005; Field et al. 2012; Parisi et al. 2006; Putoux et al. 2011; Sang et al. 2011) (see Table 4.2). It is estimated that approximately 50% of cases have mutations in known genes, suggesting that there may be other JSRD genes to be identified (Parisi & Glass, 1993). Recently a non-cilial gene, *ZNF423*, which encodes a zinc finger protein, has been reported to cause JBTS in three families, one of whom had clear JBTS and the other two in whom there was cerebellar vermis hypoplasia, renal abnormalities, and in one case tongue tumours (Chaki et al. 2012). Of the three cases, two had heterozygous mutations and the authors propose that these are dominant negative mutations. However, parental DNA was not available to confirm that these are *de novo* dominant (as opposed to rare benign heterozygous polymorphisms, or single recessive mutations with the second as yet unidentified). ZNF423 interacts with PARP1, which in turn recruits MRE11 and ATM to site of DNA damage. ZNF423 also interacts directly with CEP290, and if confirmed this data would provide the first evidence of functional links between DNA repair and cilial pathways (Chaki et al. 2012).

Many different mutation types have been described in genes causing JBTS/JSRD. Almost all mutations identified in *AHI1* and *CEP290* have been null/truncating mutations, whereas only mis-sense mutations have been described in *INPP5E* and *ARL13B* suggesting that null alleles for some of the genes may be lethal (Lee & Gleeson, 2011). The proportion of each genetic subtype and the types of variants reported are shown in Table 4.2 (Parisi & Glass, 1993).

There are some genotype–phenotype correlations although they are of limited value for genetic testing strategies. For example, Joubert with retinal dystrophy is associated with mutations in *AHI1*, *CEP290*, *TMEM216*, *TMEM138*, *INPP5E* and *CEP41*, whereas Joubert syndrome with liver disease is associated with mutations in *TMEM67*, *CC2D2A*, *RPGRIP1L*, *CEP290* and *INPP5E*. In practice, as can be seen from Table 4.3, there is considerable overlap.

Prior to the advent of Next Generation sequencing (NGS), a method for performing high-throughput sequencing of numerous genes, serial testing strategies using Sanger sequencing were reasonable and are still likely to be employed in many centres for some time. If serial sequencing of genes is performed then genotype phenotype correlations may help guide the choice of genes: for example in a patient without a retinal dystrophy then *AHI1* would not be the first choice but instead *CC2D2A* (since most patients with *AHI1* mutations do have a retinal dystrophy), or if renal or liver disease were present then *TMEM67* would be a reasonable first choice. However, the difficulties with this strategy are that not all features are present from an early age, serial testing can be very expensive and time consuming and, when parents are making reproductive decisions, a molecular diagnosis may be requested before all clinical features are apparent and the choice of genes to be tested is unclear. For these reasons JBTS/JSRDs are ideal candidates for developing parallel testing of genes using NGS. However, there is little current data on the use of NGS for JBTS/JSRD. We have unpublished data suggesting that it is highly effective, and there is one report of exome sequencing as a method for diagnosis (Tsurusaki et al. 2012) but this is not in widespread use yet. In that paper mutations were found in *INPP5E*, probably a rare cause of JBTS, illustrating the advantages of massively parallel sequencing. A more likely interim strategy will be targeted capture and sequencing of known Joubert or cilial genes and this is being developed in some centres http://www.ouh.nhs.uk/services/referrals/genetics/genetics-laboratories/default.aspx

In addition to the primary pathogenic mutations in various genes there is also increasing evidence for the effect of genetic modifiers (also known as epistatic effects) in JSRDs. In 2007, Tory and colleagues (Tory et al. 2007) reported a cohort of 13 unrelated patients with nephronophthisis and at least one JBTS-related neurological feature associated with homozygous or compound heterozygous mutations in *NPHP1*. Additional mutations that were predicted to be damaging were found in *CEP290 (NPHP6)* and *AHI1* and the frequency of the latter variant (R830W) was statistically significant when compared to normal controls and *NPHP1*-associated nephronophthisis without neurological involvement, suggesting that the *NPHP1* phenotype (and perhaps others) can be modified by other genes. The effect of *AHI1* (R830W) was confirmed in a second study (Louie et al. 2010) and was also demonstrated in mouse models, where the presence of a heterozygous deletion of *AHI1* increased the severity of retinal degeneration caused by homozygous deletions of *NPHP1* (Louie et al. 2010). A similar effect has been shown for the A229T allele of RPGRIP1L, which is highly over-represented in patients with JSRDs who have a retinal phenotype (Khanna et al. 2009; Louie et al. 2010).

In addition to over-representation of certain SNPs in Joubert associated genes, there have been some reports of addition pathogenic mutations in patients with JBTS. A heterozygous *KIF7* truncating mutation was reported in a patient with two *TMEM67* mutations, and a heterozygous *CEP290* mutation was found in a patient with two *CEP41* mutations, possibly suggesting 'mutational load' effects (Dafinger et al. 2011; Lee et al. 2012a). Such effects have been reported in BBS as well, although the significance of these additional mutations is debated (Abu-Safieh et al. 2012; Katsanis et al. 2001).

From a practical perspective, much further work is required before such data on genetic modifiers or additional mutations could be used in clinical practice, for example to identify those patients at highest risk of developing retinal or renal disease. Such work is not without clinical interest: it is possible that some types of JRSD's could be suitable for gene therapy; therefore identifying those patients at risk could be of benefit.

Underlying ciliary defect

Cilia are antenna-like structures, which protrude from the cell surface and are composed of a microtubule cytoskeleton known as the axoneme and are anchored to the cell by the centrosome-derived basal body. Cilia are classified into two types: primary (non-motile) and motile. The axoneme of primary cilia contains nine doublet microtubules (9 + 0 axoneme) but lacks the two central microtubules found in motile cilia (9 + 2 axoneme). Motile cilia are present in a variety of tissues including the respiratory tract, the oviduct, some cells of the choroid plexus and ependymal lining of the ventricle, whereas primary (non-motile) cilia are present in virtually all brain cells as well as many peripheral tissues including the epithelium of renal tubules, bile ducts and retinal photoreceptors (Valente et al. 2008). Cilia have a specialised mechanism, known as intraflagellar transport (IFT), which moves proteins in an anterograde and retrograde manner up and down the cilia. To date, all but one (ZNF423) of the proteins involved in JSRDs have been shown to function directly within this primary cilium complex and appear to affect cilial biogenesis or stability. For example, *ARL13B* mutations result in ciliary shortening and deformity (Li et al. 2010), *AHI1* knock-down impairs ciliogenesis but also affects trafficking from the Golgi to the base of the primary cilium (Hsiao et al. 2009), mutations in *RPGRIP1L* appear to affect cilium assembly and stability (Coene et al. 2011), *KIF7* mutations affect microtubular dynamics and Golgi morphology (Dafinger et al. 2011) and CEP290 is required for cilary localisation of Rab8 which is in turn required for cilary biogenesis (Kim et al. 2008).

Primary cilia, sometimes known as sensory cilia, have a broad range of functions including sensing a wide range of extracellular signals which are transduced to affect cell proliferation, polarity, nerve growth, differentiation and tissue maintenance. Although much is still to be understood about the role of cilia in signalling it is clear that they are involved in key developmental signalling cascades involving Wnt/planar polarity pathways, sonic hedgehog and phosphatidylinositol.

In addition to the specific roles of individual proteins, the concept of the ciliary proteome ('ciliome') or ciliary networks has arisen (Inglis et al. 2006; Sang et al. 2011). This has allowed the identification of novel JBTS loci through analysis of novel proteins known to interact with known to cause JBTS. An interesting new finding has been the identification of recessive mutations in *ATXN10*, which are reported to be associated with nephronophthisis, seizures and cerebral atrophy (Sang et al. 2011). Dominant mutations in *ATXN10* are associated with a rare form of spinocerebellar ataxia type 10, a neurodegenerative disorder, and if confirmed this association could link late onset neurodegeneration with cilial function.

Physiological effect of the ciliary defect

An enormous amount of progress has been made in recent years in understanding the basic cellular mechanisms involved in the function of primary cilia. However, the link between these basic cellular processes and clinical phenotypes has been much harder to unravel.

The simplest mechanism to understand is that of photoreceptor degeneration. Photoreceptors are highly specialised cells, which convert light signals into an electrical output. This process of phototransduction occurs in the highly specialised primary cilium of the photoreceptor, known as the outer segment. The outer segment is composed of stacks of disc membranes, which are linked to the inner segment by the connecting cilium. The disc membranes contain the crucial proteins required for phototransduction such as rhodopsin. All these proteins are translated in the inner segment and must be transported to the outer segment via the connecting cilium. The connecting cilium has the typical structure of a primary cilium with a 9 + 0 microtubule arrangement, a basal body and ciliary rootlet. Disc membranes form by evagination of the plasma membrane at the apical end of the connecting cilium and pinch off to form the outer segment. The discs at the most apical end are shed and engulfed by the retinal pigment epithelium and new discs are added at the base of the outer segment. This process of disc renewal continues throughout life, as does protein transport via the IFT mechanism through the connecting cilium, including 2000 molecules of opsin every minute. Impaired transport of opsin is known to cause photoreceptor degeneration, which is therefore an understandable consequence of defects in cilial biogenesis and stability.

Another physiological effect that illustrates cilial disruption involves non-canonical Wnt/planar cell polarity (PCP) signalling as a possible mechanism for the development of polycystic kidney disease. This suggests that during renal morphogenesis malorientation of the mitotic spindle leads to incorrect orientation of tubule growth, for example in a lateral rather than longitudinal direction, leading to dilatation of the tubule or cystic formation (Hildebrandt et al. 2011).

Less straightforward to understand are the roles of ciliary proteins in brain development. Early development of the cerebellum relies on proliferation of granule cell progenitors within the outer external granule layer. This process is driven by Sonic hedgehog (Shh) signals derived from the underlying Purkinje cells. The final structure of the mature cerebellum requires cell interactions and inward radial migration of external granule layer cells to form the inner granular layer. Shh-dependent granular cell progenitor proliferation is impaired in foetuses with JBTS; however, this appears to affect global cerebellar development rather than vermis hypoplasia specifically (Aguilar et al. 2012). Defective Wnt signalling may explain vermian hypoplasia due to incomplete cerebellar midline fusion (Lancaster et al. 2011).

Clinical management

A diagnosis of Joubert syndrome should prompt a multidisciplinary review of the patient (or parents if the diagnosis is made after termination of pregnancy). From the discussion above it will be evident that this diagnosis needs to be evaluated very carefully, and the wide range of clinical features associated with JSRDs should be sought to ensure that the clinical diagnosis is correct. Ideally, the family should be seen in a specialist clinic where paediatricians/neurologists and geneticists are present with additional easy access to renal, hepatic and ophthalmic specialists. Other specialists that a patient may need to see include respiratory physicians, rheumatologists, cardiologists, endocrinologists and specialist surgical teams.

Often the patient will be cared for in a general hospital setting and it is important that one specialist takes the lead in arranging follow-up care. This is particularly important since the renal, hepatic and ophthalmic complications of the condition can present many years after the initial diagnosis and early identification is essential. Careful hand-over from paediatrics to adult teams is required so patients are not lost in the long term. The Joubert syndrome and related disorders website has an information sheet for clinicians caring for patients (http://www.jsrdf.org/). A clinical checklist is also provided in Table 4.4.

Table 4.4 Clinical checklist for investigation and management of JBTS/JSRD

MRI to identify MTS and other brain malformations	◆ Thin (3-mm thickness) axial cuts through posterior fossa from midbrain to the pons plus standard axial, coronal, and sagittal cuts
Clinical history	◆ Three-generation family tree including any consanguinity and miscarriages ◆ Developmental profile: age at smiling, crawling, walking, social development ◆ Symptoms of renal impairment: excessive thirst, polyuria, fatigue, nausea, anorexia, dizziness, weakness, weight loss, fever ◆ Symptoms of visual impairment: Lack of visual attentiveness, lack of fixing and following, clumsiness, falls, poor colour discrimination, difficulty in dark or low illumination, complaints of tunnel vision or missing patches, difficulty at school with reading, or blackboards ◆ Symptoms of CNS abnormalities: headache, seizures, nausea, vomiting ◆ Other: behavioural disorder, problems of social interaction
Clinical examination	◆ Height, weight, head circumference, blood pressure ◆ Cranium: bony abnormalities suggesting encephalocoeles etc ◆ Face: tongue nodules/hamartomas/frenulae, facial dysmorphology including clefts ◆ Limbs: polydactyly, syndactyly, brachydactyly, bony dysplasia ◆ Chest: respiratory rate and regularity ◆ Abdomen: hepatosplenomegaly ◆ Eyes: presence of squint, ptosis, eye movement abnormalities, colobomas of iris, fundoscopy for retinitis pigmentosa, macular abnormalities ◆ Neurology: ataxic features (e.g. dysmetria, dysdiadochokinesis, nystagmus), presence of reduced or increased tone, increasing head circumference or symptoms of hydrocephalus ◆ External genitalia and endocrine: penile size in a male, any growth deficiency, menstrual irregularity, obesity

Table 4.4 (continued) Clinical checklist for investigation and management of JBTS/JSRD

Routine investigations	◆ Blood urea, electrolytes, creatinine ◆ Blood bilirubin and liver enzymes ◆ Full blood count including platelets and clotting ◆ Urine for blood and protein, early morning urine for concentrating ability ◆ Genetic testing* ◆ Annual abdominal ultrasound (liver and kidneys) ◆ Speech and swallowing by speech and language therapist ◆ Developmental profile by Community Paediatric Team or other
Specialist investigations where appropriate	◆ Polysomnography to be considered early, especially if history or apnoea or tachypnoea ◆ ERG, VEPs, OCT ◆ Renal biopsy when appropriate ◆ Abdominal MRI and biopsy where appropriate
Specialist management	◆ Surgical correction of squint or ptosis when appropriate ◆ Correction of refractive errors ◆ Referral to low vision specialist teams if present ◆ Monitoring and treatment for apnoea, especially in newborn period, occasionally mechanical ventilation required ◆ Nasogastic tubes or gastrostomy for severe feeding difficulties ◆ Early physiotherapy for motor delay ◆ Neurosurgical review for hydrocephalus, encephalocoele ◆ Nephronophthisis or other renal impairment likely to require dialysis or transplantation, specialist management of complications of end-stage renal disease (e.g. anemia, hypertension etc.) ◆ Treatment of portal hypertension by gastroenterologist/hepatologist, avoid of hepatotoxic drugs as indicated ◆ Surgical correction of polydactyly, clefting, tongue hamartomas, cardiac disease
Mandatory specialist review:	◆ Paediatrics: until hand-over to adult teams ◆ Developmental assessments: at least yearly until school age ◆ Ophthalmology (neuro and medical retina): yearly ◆ Nephrology: yearly ◆ Clinical genetics: until molecular diagnosis and for all reproductive issues
Other specialist review for any relevant features	◆ Endocrinology ◆ Neurosurgery ◆ Plastic surgery

*This should be routine, but if not available obtain urgent advice from clinical genetics.

In addition to a clinical diagnosis, a genetic diagnosis should be sought whenever possible. This is particularly important in families who may wish to understand reproductive risks or consider prenatal diagnosis in future pregnancies, but may also be important in alerting the clinicians to likely long-term sequelae. Until recently genetic testing has been very difficult, since most tests have been available only in research laboratories. However, Next Generation sequencing technologies offer the possibility of parallel sequencing of multiple genes. In the past two years the number of known JRSD genes has doubled, but it remains unknown what proportion of cases this accounts for. The current estimates are that ~50% of patients have mutations in known genes, but this will not be clear without further research (Parisi & Glass, 1993; Sattar & Gleeson, 2011).

Once a diagnosis of Joubert syndrome has been made, a series of baseline investigations should be arranged to identify any additional clinical features that may be present (see Table 4.4).

From the neonatal period

The use of apnoea monitoring equipment should be considered, especially in the neonatal period. Supplementary oxygen may be used and in severe cases of respiratory dysfunction tracheostomy and/or mechanical ventilation have been required. Patients with symptoms suggestive of obstructive sleep apnoea should be considered for formal sleep studies and possible nocturnal non-invasive ventilation.

An ophthalmology assessment may form part of the initial investigations but if not, should be performed at the time of diagnosis and at least annually thereafter. Retinal abnormalities can be progressive and repeat electro-retinography should be requested if there is clinical concern. Surgery may be required to treat ptosis and strabismus, and glasses may be required to correct refractive errors. Retinal degeneration can be associated with cystic macular oedema, which aggravates the visual deterioration and can be treated with acetazolamide. Regular assessments by a medical retina specialist ophthalmologist are therefore recommended.

Baseline renal investigations, including urea and electrolytes should be performed at diagnosis. A baseline renal ultrasound scan should also be performed looking for structural renal abnormalities such as cystic dysplastic kidneys. Even if no structurally abnormalities are present at the time of diagnosis, renal function should continue to be monitored regularly. The most common renal abnormality in JSRDs is NPHP. In juvenile NPHP, polydipsia and polyuria caused by a reduced urinary concentrating ability and sodium wasting occur. The decreased urinary concentrating defect is demonstrated by a low urinary osmolality (<400 mosm/kg in the first urine sample in the morning). Urinary sodium wasting may result in hyponatraemia and hypovolaemia, with subsequent growth reduction. In the clinic patients should have growth and blood pressure regularly monitored, although blood pressure is usually normal before the onset of renal failure. Similarly, urine should be tested for the presence of haematuria or proteinuria although they are generally absent or minimal. Once renal impairment develops, more typical features include anaemia, a metabolic acidosis, nausea, anorexia and weakness. Renal ultrasound may reveal normal-sized kidneys, but renal parenchymal hyperechogenicity and loss of corticomedullary differentiation are often observed. At later stages, small cysts are present in the medulla. Renal biopsy shows severe tubular damage on light microscopy. Nephronophthisis can progress to end-stage renal failure (ESRF) and require dialysis and/or transplantation. ESRF may result in additional complications such as anaemia, hypertension and renal bone disease.

Infantile NPHP differs from that presenting in early childhood, not only by its early onset but also by the histopathological features. Whereas cystic dilatations of the collecting ducts are seen in these patients, the typical changes in the tubular basement membranes seen in juvenile NPHP are usually absent. Ultrasonography usually shows moderately enlarged kidneys and severe hypertension is common.

Liver

Although hepatic fibrosis is less common than retinal or renal disorders it can be life threatening and follow up by a hepatologist is warranted in any patient with clinical evidence of liver disease (e.g. hepato(spleno)megaly, raised serum liver enzymes). Patients require yearly monitoring by their hepatologist for any complications associated with portal hypertension including variceal bleeding, hypersplenism, ascending cholangitis, and, to a lesser extent, biliary stones, cholangiocarcinoma, and hepatocellular carcinoma, even if they remain asymptomatic. Fever with upper quadrant pain and/or elevation of transaminases or biliary markers [e.g. gamma glutamyl transpeptidase (GGT)] may be signs of cholangitis and should lead to prompt evaluation and initiation of antimicrobial therapy; a small subset of patients may manifest the disease as unexplained recurrent sepsis with gram negative organisms. The platelet and white cell count should be monitored; they decline as portal hypertension worsens.

Liver function tests and an ultrasound scan should be performed following diagnosis and annually thereafter. If abnormal, referral to a specialist should be considered and hepatotoxic medication avoided. Liver fibrosis can be progressive and result in end-stage liver failure. Porto-systemic shunting and transplantation have been required in rare but serious cases.

Other systems

Hydrocephalus in JBTS is rare but a rapidly increasing head size, bulging fontanelle or symptoms suggestive of raised intracranial pressure should prompt investigation and referral to a neurosurgeon. Congenital cardiac defects are also rare but potentially serious and a one-off electrocardiogram and echocardiogram should be performed if there is any clinical concern, with review by a cardiologist if any abnormalities are identified.

Older children

Regular assessments of a child's development and growth should take place. Scoliosis is common in children with hypotonia and should be monitored, especially during periods of more rapid growth (e.g. puberty). Children should be assessed by a speech and language therapist. nasogastric (NG) feeding and/or gastrostomy may be required if swallowing is considered unsafe. Seizures have been observed in patients with JBTS and if suspected should be investigated and treated by a paediatric neurologist. Treatment of ongoing seizures is with standard antiepileptic medication. Polydactyly does not usually have functional implications but can be corrected surgically. Investigation of pituitary function and possible referral to an endocrinologist should be considered if there is any clinical concern. Physiotherapy, occupational therapy and orthotics may be beneficial. The use of assistive devices and/or sign language can aid communication given the particular difficulties with expressive language.

Prenatal

Couples at a 25% risk of having a child with JBTS syndrome may request prenatal diagnosis. Care also should be taken to search for X-linked cases where the recurrence risks are different. If the mutations in the family are known, then chorionic villus sampling or amniocentesis can be offered at the appropriate time. When a molecular diagnosis has not been made prior to a pregnancy, ultrasound or MRI may identify abnormalities suggestive of recurrence. Ultrasound findings of a posterior fossa abnormality or another JBTS related feature (e.g. polydactyly, cystic kidneys) in an at-risk couple is highly suspicious. However, due to the reduced sensitivity of ultrasound as well as clinical variability, a normal or equivocal scan should prompt consideration of foetal MRI. The MTS sign has been identified as early as 17–18 weeks gestation but the features often appear more prominent at later gestations. The ratio of measurements at the pontomesencephalic junction may enhance the accuracy of the foetal MRI scan when the appearance of the MTS is less certain (Saleem & Zaki, 2010; Saleem et al. 2011). However, it is likely that foetal MRI may not detect all cases and genetic testing of the foetus is preferable when possible.

Where the mutations are known, the option of pre-implantation genetic diagnosis should be discussed with at-risk couples and a referral can be made to a specialist centre if this is to be pursued.

In the absence of a family history, identification of a posterior fossa abnormality on routine ultrasound scanning may occur. Joubert syndrome should be considered in the differential diagnosis and extra-cranial features should be specifically looked for. A foetal MRI scan may help to further clarify the nature of the abnormality and if the pregnancy is continued, clinical and radiological assessments should follow in the post-natal period. If termination of pregnancy is performed then a post-mortem with storage of DNA should be done whenever possible. This may allow a precise molecular diagnosis to assist genetic counselling in future pregnancies.

Conclusions

JBTS and JSRDS are complex ciliopathies, which require specialist investigation and management. The genetic basis of these conditions are being identified and aid with diagnosis and genetic counselling. The underlying cilial defects are being unravelled although much more research is required to understand the link between the ciliary defects and the physiological consequences.

References

Abu-Safieh, L., Al-Anazi, S., Al-Abdi, L., Hashem, M., Alkuraya, H., Alamr, M., et al. 2012. In search of triallelism in Bardet–Biedl syndrome. *Eur J Hum Genet* **20**, 420–427.

Ali, B. R., Silhavy, J. L., Akawi, N. A., Gleeson, J. G. & Al-Gazali, L. 2012. A mutation in KIF7 is responsible for the autosomal recessive syndrome of macrocephaly, multiple epiphyseal dysplasia and distinctive facial appearance. *Orphanet J Rare Dis* **7**, 27.

Aguilar, A., Meunier, A., Strehl, L., Martinovic, J., Bonniere, M., Attie-Bitach, T., et al. 2012. Analysis of human samples reveals impaired SHH-dependent cerebellar development in Joubert syndrome/ Meckel syndrome. *Proc Natl Acad Sci U.S.A.* **109**, 16951–16956.

Arts, H. H., Doherty, D., Van Beersum, S. E., Parisi, M. A., Letteboer, S. J., Gorden, N. T., et al. 2007. Mutations in the gene encoding the basal body protein RPGRIP1L, a nephrocystin-4 interactor, cause Joubert syndrome. *Nat Genet* **39**, 882–888.

Baala, L., Romano, S., Khaddour, R., Saunier, S., Smith, U. M., Audollent, S., et al. 2007. The Meckel–Gruber syndrome gene, MKS3, is mutated in Joubert syndrome. *Am J Hum Genet* **80**, 186–194.

Bachmann-Gagescu, R., Ishak, G. E., Dempsey, J. C., Adkins, J., O'Day, D., Phelps, I. G., et al. 2012. Genotype–phenotype correlation in CC2D2A-related Joubert syndrome reveals an association with ventriculomegaly and seizures. *J Med Genet* **49**, 126–137.

Bielas, S. L., Silhavy, J. L., Brancati, F., Kisseleva, M. V., Al-Gazali, L., Sztriha, L., et al. 2009. Mutations in INPP5E, encoding inositol polyphosphate-5-phosphatase E, link phosphatidyl inositol signaling to the ciliopathies. *Nat Genet* **41**, 1032–1036.

Boltshauser, E. & Isler, W. 1977. Joubert syndrome: Episodic hyperpnea, abnormal eye movements, retardation and ataxia, associated with dysplasia of the cerebellar vermis. *Neuropadiatrie*, **8**, 57–66.

Braddock, S. R., Henley, K. M. & Maria, B. L. 2007. The face of Joubert syndrome: A study of dysmorphology and anthropometry. *Am J Med Genet Part A*, **143A**, 3235–3242.

Brancati, F., Dallapiccola, B. & Valente, E. M. 2010. Joubert syndrome and related disorders. *Orphanet J Rare Dis* **5**, 20.

Brancati, F., Iannicelli, M., Travaglini, L., Mazzotta, A., Bertini, E., Boltshauser, E., et al. 2009. MKS3/TMEM67 mutations are a major cause of COACH syndrome, a Joubert syndrome related disorder with liver involvement. *Hum Mut*, **30**, E432–42.

Cantagrel, V., Silhavy, J. L., Bielas, S. L., Swistun, D., Marsh, S. E., Bertrand, J. Y., et al. 2008. Mutations in the cilia gene ARL13B lead to the classical form of Joubert syndrome. *Am J Hum Genet* **83**, 170–179.

Castori, M., Valente, E. M., Donati, M. A., Salvi, S., Fazzi, E., Procopio, E., et al. 2005. NPHP1 gene deletion is a rare cause of Joubert syndrome related disorders. *J Med Genet* **42**, e9.

Chaki, M., Airik, R., Ghosh, A. K., Giles, R. H., Chen, R., Slaats, G. G., et al. 2012. Exome capture reveals ZNF423 and CEP164 mutations, linking renal ciliopathies to DNA damage response signaling. *Cell*, **150**, 533–548.

Coene, K. L., Mans, D. A., Boldt, K., Gloeckner, C. J., Van Reeuwijk, J., Bolat, E., et al. 2011. The ciliopathy-associated protein homologs RPGRIP1 and RPGRIP1L are linked to cilium integrity through interaction with Nek4 serine/threonine kinase. *Hum Mol Genet* **20**, 3592–3605.

Coene, K. L., Roepman, R., Doherty, D., Afroze, B., Kroes, H. Y., Letteboer, S. J., et al. 2009. OFD1 is mutated in X-linked Joubert syndrome and interacts with LCA5-encoded lebercilin. *Am J Hum Genet* **85**, 465–481.

Dafinger, C., Liebau, M. C., Elsayed, S. M., Hellenbroich, Y., Boltshauser, E., Korenke, G. C., et al. 2011. Mutations in KIF7 link Joubert syndrome with Sonic Hedgehog signaling and microtubule dynamics. *J Clin Invest* **121**, 2662–2667.

Davis, E. E., Zhang, Q., Liu, Q., Diplas, B. H., Davey, L. M., Hartley, J., et al. 2011. TTC21B contributes both causal and modifying alleles across the ciliopathy spectrum. *Nat Genet* **43**, 189–196.

Delous, M., Baala, L., Salomon, R., Laclef, C., Vierkotten, J., Tory, K., et al. 2007. The ciliary gene RPGRIP1L is mutated in cerebello-oculo-renal syndrome (Joubert syndrome type B) and Meckel syndrome. *Nat Genet* **39**, 875–881.

Doherty, D. 2009. Joubert syndrome: Insights into brain development, cilium biology, and complex disease. *Sem Pediatr Neurol* **16**, 143–154.

Edvardson, S., Shaag, A., Zenvirt, S., Erlich, Y., Hannon, G. J., Shanske, A. L., et al. 2010. Joubert syndrome 2 (JBTS2) in Ashkenazi Jews is associated with a TMEM216 mutation. *Am J Hum Genet* **86**, 93–97.

Elmali, M., Ozmen, Z., Ceyhun, M., Tokatlioglu, O., Incesu, L. & Diren, B. 2007. Joubert syndrome with atrial septal defect and persistent left superior vena cava. *Diagn Intervent Radiol* **13**, 94–96.

Fennell, E. B., Gitten, J. C., Dede, D. E. & Maria, B. L. 1999. Cognition, behavior, and development in Joubert syndrome. *J Child Neurol* **14**, 592–596.

Ferland, R. J., Eyaid, W., Collura, R. V., Tully, L. D., Hill, R. S., Al-Nouri, D., et al. 2004. Abnormal cerebellar development and axonal decussation due to mutations in AHI1 in Joubert syndrome. *Nat Genet* **36**, 1008–1013.

Ferrante, M. I., Giorgio, G., Feather, S. A., Bulfone, A., Wright, V., Ghiani, M., et al. 2001. Identification of the gene for oral–facial–digital type I syndrome. *Am J Hum Genet* **68**, 569–576.

Field, M., Scheffer, I. E., Gill, D., Wilson, M., Christie, L., Shaw, M., et al. 2012. Expanding the molecular basis and phenotypic spectrum of X-linked Joubert syndrome associated with OFD1 mutations. *Eur J Hum Genet* **20**, 806–809.

Garcia-Gonzalo, F. R., Corbit, K. C., Sirerol-Piquer, M. S., Ramaswami, G., Otto, E. A., Noriega, T. R., et al. 2011. A transition zone complex regulates mammalian ciliogenesis and ciliary membrane composition. *Nat Genet* **43**, 776–784.

Giordano, L., Vignoli, A., Pinelli, L., Brancati, F., Accorsi, P., Faravelli, F., et al. 2009. Joubert syndrome with bilateral polymicrogyria: Clinical and neuropathological findings in two brothers. *Am J Med Genet Part A* **149A**, 1511–1515.

Gitten, J., Dede, D., Fennell, E., Quisling, R. & Maria, B. L. 1998. Neurobehavioral development in Joubert syndrome. *J Child Neurol* **13**, 391–397.

Gleeson, J. G., Keeler, L. C., Parisi, M. A., Marsh, S. E., Chance, P. F., Glass, I. A., et al. 2004. Molar tooth sign of the midbrain–hindbrain junction: Occurrence in multiple distinct syndromes. *Am J Med Genet, Part A* **125A**, 125–134; discussion 117.

Gorden, N. T., Arts, H. H., Parisi, M. A., Coene, K. L., Letteboer, S. J., Van Beersum, S. E., et al. 2008. CC2D2A is mutated in Joubert syndrome and interacts with the ciliopathy-associated basal body protein CEP290. *Am J Hum Genet* **83**, 559–571.

Graber, J. J., Lau, H. & Sathe, S. 2009. Teaching neuroImages: Molar tooth sign with hypotonia, ataxia, and nystagmus (Joubert syndrome) and hypothyroidism. *Neurology*, **73**, e106.

Gunay-Aygun, M., Parisi, M. A., Doherty, D., Tuchman, M., Tsilou, E., Kleiner, D. E., et al. 2009. MKS3-related ciliopathy with features of autosomal recessive polycystic kidney disease, nephronophthisis, and Joubert syndrome. *J Pediatr* **155**, 386–92 e1.

Hildebrandt, F., Benzing, T. & Katsanis, N. 2011. Ciliopathies. *New Engl J Med* **364**, 1533–1543.

Hodgkins, P. R., Harris, C. M., Shawkat, F. S., Thompson, D. A., Chong, K., Timms, C., et al. 2004. Joubert syndrome: Long-term follow-up. *Dev Med Child Neurol* **46**, 694–699.

Hsiao, Y. C., Tong, Z. J., Westfall, J. E., Ault, J. G., Page-Mccaw, P. S. & Ferland, R. J. 2009. Ahi1, whose human ortholog is mutated in Joubert syndrome, is required for Rab8a localization, ciliogenesis and vesicle trafficking. *Hum Mol Genet* **18**, 3926–3941.

Huang, L., Szymanska, K., Jensen, V. L., Janecke, A. R., Innes, A. M., Davis, E. E., et al. 2011. TMEM237 is mutated in individuals with a Joubert syndrome related disorder and expands the role of the TMEM family at the ciliary transition zone. *Am J Hum Genet* **89**, 713–730.

Inglis, P. N., Boroevich, K. A. & Leroux, M. R. 2006. Piecing together a ciliome. *Trends Genet* **22**, 491–500.

Joubert, M., Eisenring, J. J., Robb, J. P. & Andermann, F. 1969. Familial agenesis of the cerebellar vermis. A syndrome of episodic hyperpnea, abnormal eye movements, ataxia, and retardation. *Neurology*, **19**, 813–825.

Karp, N., Grosse-Wortmann, L. & Bowdin, S. 2012. Severe aortic stenosis, bicuspid aortic valve and atrial septal defect in a child with Joubert syndrome and Related Disorders (JSRD)—a case report and review of congenital heart defects reported in the human ciliopathies. *Eur J Med Genet* **55**, 605–610.

Katsanis, N., Ansley, S. J., Badano, J. L., Eichers, E. R., Lewis, R. A., Hoskins, B. E., et al. 2001. Triallelic inheritance in Bardet–Biedl syndrome, a Mendelian recessive disorder. *Science* **293**, 2256–2259.

Khanna, H., Davis, E. E., Murga-Zamalloa, C. A., Estrada-Cuzcano, A., Lopez, I., Den Hollander, A. I., et al. 2009. A common allele in RPGRIP1L is a modifier of retinal degeneration in ciliopathies. *Nat Genet* **41**, 739–745.

Kim, J., Krishnaswami, S. R. & Gleeson, J. G. 2008. CEP290 interacts with the centriolar satellite component PCM-1 and is required for Rab8 localization to the primary cilium. *Hum Mol Genet* **17**, 3796–3805.

Kroes, H. Y., Van Zon, P. H., Fransen Van De Putte, D., Nelen, M. R., Nievelstein, R. J., Wittebol-Post, D., et al. 2008. DNA analysis of AHI1, NPHP1 and CYCLIN D1 in Joubert syndrome patients from the Netherlands. *Eur J Med Genet* **51**, 24–34.

Lambert, S. R., Kriss, A., Gresty, M., Benton, S. & Taylor, D. 1989. Joubert syndrome. *Arch Ophthalmol* **107**, 709–713.

Lancaster, M. A., Schroth, J. & Gleeson, J. G. 2011. Subcellular spatial regulation of canonical Wnt signalling at the primary cilium. *Nat Cell Biol* **13**, 700–707.

Lee, J. E. & Gleeson, J. G. 2011. Cilia in the nervous system: Linking cilia function and neurodevelopmental disorders. *Curr Opin Neurol* **24**, 98–105.

Lee, J. E., Silhavy, J. L., Zaki, M. S., Schroth, J., Bielas, S. L., Marsh, S. E., et al. 2012a. CEP41 is mutated in Joubert syndrome and is required for tubulin glutamylation at the cilium. *Nat Genet* **44**, 193–199.

Lee, J. H., Silhavy, J. L., Lee, J. E., Al-Gazali, L., Thomas, S., Davis, E. E., et al. 2012b. Evolutionarily assembled cis-regulatory module at a human ciliopathy locus. *Science*, **335**, 966–969.

Li, Y., Wei, Q., Zhang, Y., Ling, K. & Hu, J. 2010. The small GTPases ARL-13 and ARL-3 coordinate intraflagellar transport and ciliogenesis. *J Cell Biol* **189**, 1039–1051.

Louie, C. M., Caridi, G., Lopes, V. S., Brancati, F., Kispert, A., Lancaster, M. A., et al. 2010. AHI1 is required for photoreceptor outer segment development and is a modifier for retinal degeneration in nephronophthisis. *Nat Genet* **42**, 175–180.

Maria, B. L., Boltshauser, E., Palmer, S. C. & Tran, T. X. 1999a. Clinical features and revised diagnostic criteria in Joubert syndrome. *J Child Neurol* **14**, 583–590; discussion 590–1.

Maria, B. L., Hoang, K. B., Tusa, R. J., Mancuso, A. A., Hamed, L. M., Quisling, R. G., et al. 1997. "Joubert syndrome" revisited: Key ocular motor signs with magnetic resonance imaging correlation. *J Child Neurol* **12**, 423–430.

Maria, B. L., Quisling, R. G., Rosainz, L. C., Yachnis, A. T., Gitten, J., Dede, D., et al. 1999b. Molar tooth sign in Joubert syndrome: Clinical, radiologic, and pathologic significance. *J Child Neurol* **14**, 368–376.

Noor, A., Windpassinger, C., Patel, M., Stachowiak, B., Mikhailov, A., Azam, M., et al. 2008a. Addendum. CC2D2A, encoding a coiled-coil and C2 domain protein, causes autosomal-recessive mental retardation with retinitis pigmentosa. *Am J Hum Genet* **83**, 656.

Noor, A., Windpassinger, C., Patel, M., Stachowiak, B., Mikhailov, A., Azam, M., et al. 2008b. CC2D2A, encoding a coiled-coil and C2 domain protein, causes autosomal–recessive mental retardation with retinitis pigmentosa. *Am J Hum Genet* **82**, 1011–1018.

Ozonoff, S., Williams, B. J., Gale, S. & Miller, J. N. 1999. Autism and autistic behavior in Joubert syndrome. *J Child Neurol* **14**, 636–641.

Parisi, M. & Glass, I. 1993. Joubert Syndrome and Related Disorders. 2003 Jul 9 [Updated 2013 Apr 11]. In: **Pagon RA, Bird TD, Dolan CR, et al.**, editors. GeneReviews™ [Internet]. Seattle (WA): University of Washington, Seattle; 1993. Available from: http://www.ncbi.nlm.nih.gov/books/NBK1325 /

Parisi, M. A. 2009. Clinical and molecular features of Joubert syndrome and related disorders. *Am J Med Genet, Part C, Sem Med Genet* **151C**, 326–340.

Parisi, M. A., Bennett, C. L., Eckert, M. L., Dobyns, W. B., Gleeson, J. G., Shaw, D. W., et al. 2004. The NPHP1 gene deletion associated with juvenile nephronophthisis is present in a subset of individuals with Joubert syndrome. *Am J Hum Genet* **75**, 82–91.

Parisi, M. A., Doherty, D., Eckert, M. L., Shaw, D. W., Ozyurek, H., Aysun, S., et al. 2006. AHI1 mutations cause both retinal dystrophy and renal cystic disease in Joubert syndrome. *J Med Genet* **43**, 334–339.

Peker, E., Kirimi, E., Sal, E., Ceylan, A., Ustyol, L. & Caksen, H. 2009. Joubert syndrome associated with patent ductus arteriosus in a newborn infant. *Genet Couns* **20**, 289–292.

Poretti, A., Boltshauser, E., Loenneker, T., Valente, E. M., Brancati, F., Il'yasov, K., et al. 2007. Diffusion tensor imaging in Joubert syndrome. *AJNR, Am J Neuroradiol* **28**, 1929–1933.

Poretti, A., Vitiello, G., Hennekam, R. C., Arrigoni, F., Bertini, E., Borgatti, R., et al. 2012. Delineation and diagnostic criteria of oral–facial–digital syndrome type VI. *Orphanet J Rare Dis* **7**, 4.

Putoux, A., Thomas, S., Coene, K. L., Davis, E. E., Alanay, Y., Ogur, G., et al. 2011. KIF7 mutations cause fetal hydrolethalus and acrocallosal syndromes. *Nat Genet* **43**, 601–606.

Romio, L., Fry, A. M., Winyard, P. J., Malcolm, S., Woolf, A. S. & Feather, S. A. 2004. OFD1 is a centrosomal/basal body protein expressed during mesenchymal–epithelial transition in human nephrogenesis. *JASN, J Am Soc Nephrol* **15**, 2556–2568.

Saleem, S. N. & Zaki, M. S. 2010. Role of MR imaging in prenatal diagnosis of pregnancies at risk for Joubert syndrome and related cerebellar disorders. *AJNR, Am J Neuroradiol* **31**, 424–429.

Saleem, S. N., Zaki, M. S., Soliman, N. A. & Momtaz, M. 2011. Prenatal magnetic resonance imaging diagnosis of molar tooth sign at 17 to 18 weeks of gestation in two fetuses at risk for Joubert syndrome and related cerebellar disorders. *Neuropediatrics*, **42**, 35–38.

Sang, L., Miller, J. J., Corbit, K. C., Giles, R. H., Brauer, M. J., Otto, E. A., et al. 2011. Mapping the NPHP-JBTS-MKS protein network reveals ciliopathy disease genes and pathways. *Cell*, **145**, 513–528.

Saraiva, J. M. & Baraitser, M. 1992. Joubert syndrome: A review. *Am J Med Genet* **43**, 726–731.

Sattar, S. & Gleeson, J. G. 2011. The ciliopathies in neuronal development: A clinical approach to investigation of Joubert syndrome and Joubert syndrome-related disorders. *Dev Med Child Neurol* **53**, 793–798.

Senocak, E. U., Oguz, K. K., Haliloglu, G., Topcu, M. & Cila, A. 2010. Structural abnormalities of the brain other than molar tooth sign in Joubert syndrome-related disorders. *Diagn Intervent Radiol* **16**, 3–6.

Srour, M., Hamdan, F. F., Schwartzentruber, J. A., Patry, L., Ospina, L. H., Shevell, M. I., et al. 2012a. Mutations in TMEM231 cause Joubert syndrome in French Canadians. *J Med Genet* **49**, 636–641.

Srour, M., Schwartzentruber, J., Hamdan, F. F., Ospina, L. H., Patry, L., Labuda, D., et al. 2012b. Mutations in C5ORF42 cause Joubert syndrome in the French Canadian population. *Am J Hum Genet* **90**, 693–700.

Steinlin, M., Schmid, M., Landau, K. & Boltshauser, E. 1997. Follow-up in children with Joubert syndrome. *Neuropediatrics* **28**, 204–211.

Sturm, V., Leiba, H., Menke, M. N., Valente, E. M., Poretti, A., Landau, K., et al. 2010. Ophthalmological findings in Joubert syndrome. *Eye* **24**, 222–225.

Takahashi, T. N., Farmer, J. E., Deidrick, K. K., Hsu, B. S., Miles, J. H. & Maria, B. L. 2005. Joubert syndrome is not a cause of classical autism. *Am J Med Genet, Part A* **132**, 347–351.

Thomas, S., Legendre, M., Saunier, S., Bessieres, B., Alby, C., Bonniere, M., et al. 2012. TCTN3 mutations cause Mohr–Majewski syndrome. *Am J Hum Genet* **91**, 372–378.

Tory, K., Lacoste, T., Burglen, L., Moriniere, V., Boddaert, N., Macher, M. A., et al. 2007. High NPHP1 and NPHP6 mutation rate in patients with Joubert syndrome and nephronophthisis: Potential epistatic effect of NPHP6 and AHI1 mutations in patients with NPHP1 mutations. *JASN, J Am Soc Nephrol* **18**, 1566–1575.

Tsurusaki, Y., Kosho, T., Hatasaki, K., Narumi, Y., Wakui, K., Fukushima, Y., et al. 2012. Exome sequencing in a family with an X-linked lethal malformation syndrome: Clinical consequences of hemizygous truncating OFD1 mutations in male patients. *Clin Genet* **83**, 135–144.

Tusa, R. J. & Hove, M. T. 1999. Ocular and oculomotor signs in Joubert syndrome. *J Child Neurol* **14**, 621–627.

Valente, E. M., Brancati, F. & Dallapiccola, B. 2008. Genotypes and phenotypes of Joubert syndrome and related disorders. *Eur J Med Genet* **51**, 1–23.

Valente, E. M., Logan, C. V., Mougou-Zerelli, S., Lee, J. H., Silhavy, J. L., Brancati, F., et al. 2010. Mutations in TMEM216 perturb ciliogenesis and cause Joubert, Meckel and related syndromes. *Nat Genet* **42**, 619–625.

Valente, E. M., Silhavy, J. L., Brancati, F., Barrano, G., Krishnaswami, S. R., Castori, M., et al. 2006. Mutations in CEP290, which encodes a centrosomal protein, cause pleiotropic forms of Joubert syndrome. *Nat Genet* **38**, 623–625.

Vogel, T. W., Dlouhy, B. J. & Menezes, A. H. 2012. Craniovertebral junction abnormality in a case of Joubert syndrome. *Child's nervous system: ChNS: Off J Int Soc Pediatr Neurosurg* **28**, 1109–1112.

Weiss, A. H., Doherty, D., Parisi, M., Shaw, D., Glass, I. & Phillips, J. O. 2009. Eye movement abnormalities in Joubert syndrome. *Invest Ophthalmol Vis Sci* **50**, 4669–4677.

Yachnis, A. T. & Rorke, L. B. 1999. Neuropathology of Joubert syndrome. *J Child Neurol* **14**, 655–659; discussion 669–672.

Chapter 5

Bardet–Biedl syndrome

Philip L. Beales and Elizabeth Forsythe

History of the eponym

In 1866 John Laurence and Robert Moon described a family of four siblings with retinal dystrophy, obesity, spastic paraparesis and cognitive deficit (Laurence & Moon, 1995). George Bardet and Artur Biedl later (1920s) reported separately on further similarly affected individuals with the triad of retinal degeneration, post-axial polydactyly and obesity (Bardet, 1995; Biedl, 1995). Shortly after the condition was coined the Laurence–Moon–Bardet–Biedl syndrome. Later, the syndrome was divided into two entities: the Laurence–Moon syndrome and the Bardet–Biedl syndrome owing to the presence of spastic paraparesis in the Laurence and Moon family; but there is considerable phenotypic overlap suggesting that they may be allelic (Beales et al. 1999). Bardet–Biedl syndrome (BBS) is now the standard term in common usage.

Prevalence and inheritance

BBS is a pleiotropic genetic disorder with significant interfamilial and intrafamilial variation (Riise et al. 1997; Waters & Beales, 1993). Inheritance is traditionally considered autosomal recessive although notable exceptions exist whereby BBS may behave as an oligogenic disorder (Katsanis et al. 2001). The prevalence of BBS varies markedly between populations; ranging from 1 in 160,000 in northern European populations (Waters & Beales, 1993) to 1 in 13,500 and 1 in 17,500, respectively, in isolated communities in Kuwait and Newfoundland where a higher level of consanguinity prevails (Farag & Teebi, 1989; Green et al. 1989). Studies on the families in Newfoundland affected with BBS revealed the presence of at least eight different BBS mutations at six BBS loci and therefore the high prevalence on this island cannot be attributed to a single founder (Webb et al. 2009).

Clinical features

The BBS phenotype evolves slowly throughout the first decade of life, although there is considerable variability. As a result, most patients are diagnosed in late childhood or early adulthood (Beales et al. 1999).

Post-axial polydactyly

Post-axial polydactyly is usual (~70%), and may be the only obvious dysmorphic feature at birth (Beales et al. 1999). This may affect all four limbs or only upper or lower limbs and may occur together with brachydactyly and/or syndactlyly.

Rod–cone dystrophy

The most common diagnostic handle prompting investigation for BBS is the development of rod–cone dystrophy. Primary loss of rod photoreceptors is followed by later demise of cone photoreceptors (Hamel, 2007). This presents as an atypical retinal dystrophy often with early macular involvement (Baker & Beales, 2009). Most affected children have normal vision during their early years but typically between 6 and 8 years there is gradual onset of night blindness, followed by photophobia and loss of central and colour vision (Hamel, 2007). Many variants of the ophthalmological phenotype in BBS have been described and some patients develop the converse sequence of pathological events with early loss of cone photoreceptors followed by rod photoreceptors (Beales et al. 1999). Electroretinography is the investigation of choice and may show early changes within the first two years of life although significant changes are rarely visible before the age of five (Baker & Beales, 2009). Symptoms usually develop in the first decade of life and most patients are legally blind by the second or third decade (Adams et al. 2007) although moderate forms of the disease do exist. Other eye abnormalities such as cataracts and refractive errors are also prevalent in BBS. One should not forget that diabetic retinopathy can also contribute to eye disease in those patients with early-onset and uncontrolled type 2 diabetes.

Obesity

Obesity is another major clinical finding and the incidence is reported to be 72–86% in the BBS population (Beales et al. 1999; Hjortshoj et al. 2010; Moore et al. 2005; Riise et al. 1997; Rooryck & Lacombe, 2008; Tobin & Beales, 2007). Birth weight is usually within the normal range although there is evidence of a skewed distribution towards the upper centiles (Putoux et al. 2011). One third of those with a normal birth weight develop obesity by the age of 1 year (Putoux et al. 2011). Although adult obesity tends to be truncal, it appears to be widespread and diffuse in childhood.

The development of type 2 diabetes mellitus is prevalent amongst patients (Beales et al. 1999; Hjortshoj et al. 2010; Moore et al. 2005; Riise et al. 1997; Rooryck & Lacombe, 2008; Tobin & Beales, 2007). It may be related to the level of obesity and it is often found in association with other signs of metabolic syndrome such as hypertension, hyperlipidaemia and hypertriglyceridaemia.

Hypogonadism

Hypogonadism may manifest as delayed puberty or hypogenitalism in males and genital abnormalities in females (Beales et al. 1999; Deveault et al. 2011; Moore et al. 2005). This may occur independently or in conjunction with biochemical hypogonadism. Female babies may be born with abdominal distension, a consequence of hydrometrocolpos whereby fluid builds up in the uterus secondary to an imperforate hymen. A wide variety of genital malformations have been observed in females with BBS. Whilst males are almost invariably infertile, it is important not to assume that most females with BBS will be infertile and they should not therefore forego contraceptive advice.

Developmental delay and cognitive deficit

Developmental delay and cognitive deficit are common in BBS. Delay is often global but may be specific to certain areas of development (Baker & Beales, 2009). In a cohort of 109 patients, cognitive deficit was reported in 62% and half of these patients required special schooling (Beales et al. 1999). Children with BBS are often reported to have labile behaviour with outbursts of frustration (Beales et al. 1999; Deveault et al. 2011; Moore et al. 2005). Many prefer a fixed routine and may display obsessive compulsive behaviour and lack of social dominance (Barnett et al. 2002). Others have a more severe behavioural phenotype, develop autistic spectrum disorder or psychosis (Barnett et al. 2002).

Renal abnormalities

Renal abnormalities are a major cause of morbidity and mortality in BBS (O'Dea et al., 1996). The renal phenotype is variable but classically manifests with cystic tubular disease and anatomical malformations that include duplex urinary systems, horseshoe kidneys and hydronephrosis (Putoux et al. 2011). The latter can often manifest during the second trimester, evident on routine anomaly ultrasound scans. Urinary concentration defects are prevalent even in patients with near normal renal function and apparent normal renal structure, so it is important to inquire of the amount of liquid intake and how often the patient passes urine during the day and night (Marion et al. 2011).

Speech deficit

Speech deficit has been reported in 60% of patients (Barnett et al. 2002). This mainly consists of high-pitched nasal speech and children often do not develop intelligible speech before the age of 4 years. It has been suggested that substitution of the first consonant of a word may be characteristic (Beales et al. 1999; Barnett et al. 2002). Speech difficulties may be complicated by hearing loss, which is reported in 17–21% of patients (Beales et al. 1999; Deveault et al. 2011). Most patients suffer conductive hearing loss secondary to chronic otitis media (Beales et al. 1999). Speech delay in children with BBS is generally responsive to speech therapy (Barnett et al. 2002). This may be related to structural palatal changes (high-arched palate is common) or velopalatal insufficiency and so formal speech and language assessment is recommended early.

Other organ systems and abnormalities

Involvement of other organ systems such as the heart and gastrointestinal system are also observed. The type of cardiac abnormalities observed in BBS is highly variable. An echocardiography study of 22 individuals from three highly consanguineous Bedouin families revealed a frequency of heart defects of 50% (Elbedour et al. 1994). Beales and colleagues (Beales et al. 1999) found a frequency of only 7% in a study based on 109 patients. Cardiac abnormalities include valvular stenoses, patent ductus arteriosis and cardiomyopathies (Beales et al. 1999). Hepatic involvement ranges from fibrosis to cystic dilatation of the bile duct, intrahepatic and extrahepatic tracts (Baker & Beales, 2009).

Hirschsprung's disease has been documented in BBS but the incidence of this association is unclear (de Pontual et al., 2009). Constipation is more commonly encountered in children and young adults (personal observations).

Dental crowding and a high-arched palate are common. Other abnormalities include hypodontia, malocclusion and enamel hypoplasia (Waters & Beales, 1993). Anosmia has been described following observations in a mouse model. In a study of 19 BBS patients nine had anosmia/hyposmia (Kulaga et al. 2004).

Many affected individuals suffer from a degree of clumsiness and 40% of one cohort described signs of ataxia and poor coordination (Beales et al. 1999). Dysdiadochokinesia and past pointing are common (79%) as are difficulties with tandem walking and the Fogg test (Beales et al. 1999).

Overlap with other ciliopathies

There is a significant overlap in clinical features within the ciliopathy disease spectrum. Although the classical features associated with BBS are well documented, there is significant phenotypic overlap with other ciliopathies such as Alström, McKusick–Kauffman, Joubert and Meckel syndromes. Alström syndrome is usually differentiated from BBS by the presence of hearing loss and absence of polydactyly and significant learning difficulties. Patients with McKusick–Kauffman syndrome have a high prevalence of urogenital anomalies but normally lack the obesity, rod–cone dystrophy and learning difficulties characteristic of BBS. Diagnostic difficulties arise as these features are age-dependent in BBS (Slavotinek & Biesecker, 2000).

Carriers of BBS do not appear to be at increased risk of hypertension or diabetes (Beales et al. 1999; Webb et al. 2009) although conflicting evidence exists regarding the possible increased risk of obesity in carriers (Beales et al. 1999; Croft et al. 1995; Guo & Rahmouni, 2011; Webb et al. 2009). An increase in renal cancers and malformations (Beales et al. 2000) as well as retinal dysfunction (Kim et al. 2007) has been reported in obligate carriers.

Of note, genome-wide association studies in non-BBS lean and obese Caucasian patients revealed a significant association (albeit weak) between single-nucleotide polymorphisms in *BBS2*, *BBS4* and *BBS6* and common obesity (Benzinou et al. 2006).

Screening and diagnosis

Unless the diagnosis is suspected based on antenatal imaging revealing polydactyly and structural renal abnormalities, BBS is usually not diagnosed before the patient starts to develop the visual problems characteristic of rod–cone dystrophy. Although there are some

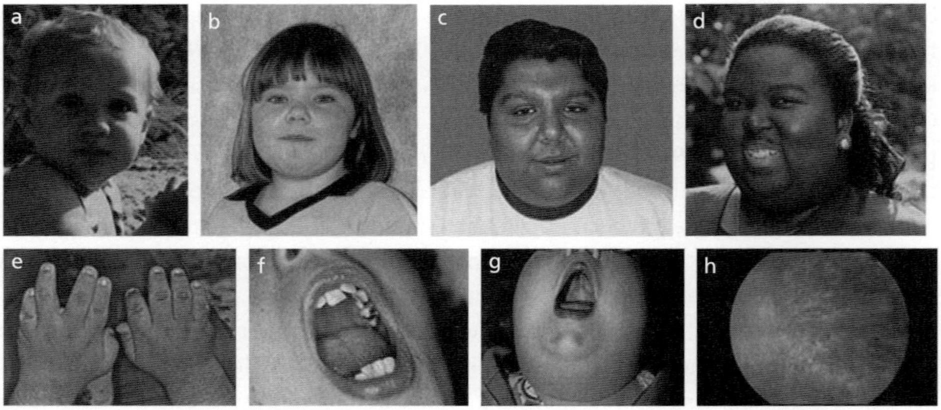

Figure 5.1 Images of patients demonstrating the dysmorphic features associated with BBS. (a,b,c,d) Demonstrate typical facial features. These are often subtle and are not always present. Features include deep-set eyes, hypertelorism, downward slanting palpebral fissures, a flat nasal bridge, small mouth, malar hypoplasia and retrognathia. (e) Brachydactyly and scars from excision of accessory digits. (f) Dental crowding. (g) High arched palate. (h) Fundoscopy demonstrating rod–cone dystrophy.

distinctive dysmorphic features such as hypertelorism, midface hypoplasia and retrognathia these are inconsistently present and can be subtle (Lorda-Sanchez et al. 2001; Tobin et al. 2008) Figure 5.1 demonstrates the variation of facial features seen in BBS. Modified diagnostic criteria by Beales and colleagues (Beales et al. 1999) suggest that either four primary features or three primary and two secondary features are required to make a clinical diagnosis. Table 5.1 summarises the modified major and minor clinical features and their incidence in BBS.

Molecular confirmation of the diagnosis can be obtained in nearly 80% via direct sequencing of 16 BBS genes. Billingsley and colleagues (Billingsley et al. 2011) have suggested a practical approach to mutational screening where sequencing is prioritised in accordance with the frequency of pathogenic mutations between and within BBS genes. BBS with other ciliopathies lend themselves to panel testing using Next Generation sequencing approaches currently under development that will provide an efficient and cost-effective service.

Clinical management

A multidisciplinary approach is required to effectively manage this pleiotropic condition. Although research is in progress, there is still no targeted treatment for BBS. Complications associated with BBS should be treated symptomatically as in the general population. Figure 5.2a outlines the management of patients with BBS. Figure 5.2b outlines

Table 5.1 Diagnostic features and prevalence in BBS

Feature	Frequency
Primary features	
Rod–cone dystrophy	93%
Polydactyly	63–81% All four limbs: 21% Upper limbs only: 9% Lower limbs only: 21%
Obesity	72–92%
Genital anomalies	59–98%
Renal anomalies	53%
Learning difficulties	61%
Secondary features	
Speech delay	54–81%
Developmental delay	50–91%
Diabetes mellitus	6–48%
Dental anomalies	51%
Congenital heart disease	7%
Brachydactyly/syndactyly	46–100%/8–95%
Ataxia/poor coordination	40–86%
Anosmia/hyposmia	60%

Four primary features or three primary features and two secondary features are required for a clinical diagnosis of Bardet– Biedl syndrome.

Reprinted from Elizabeth Forsythe and Philip L Beales, Bardet–Biedl syndrome, *European Journal of Human Genetics*, Volume 21, pp. 8–13, Copyright © 2013 Macmillan Publishers Limited. Source: Data from Beales et al., New criteria for improved diagnosis of Bardet–Biedl syndrome: results of a population survey, *Journal of Medical Genetics*, Volume 36, Issue 6, pp. 437–446, Copyright © 1999 by the BMJ Publishing Group Ltd; Rooryck and Lacombe, Bardet–Biedl syndrome, *Ann Endocrinol (Paris)*,Volume 69, Number 6, pp. 463–71, Copyright © 2008 Elsevier Masson SAS and Putoux et al, Phenotypic variability of Bardet–Biedl syndrome: focusing on the kidney, *Pediatric Nephrology*, Volume 27, Issue 1, pp. 7–15, Copyright © 2011 IPNA.

the appropriate assessment pathway of patients with suspected BBS who do not fulfil the modified diagnostic criteria for BBS.

Blood pressure should be measured every 6 months or more often if there is evidence of hypertension. Antihypertensives and lipid-lowering medication should be prescribed as appropriate. It is recommended that all patients have at least one baseline renal ultrasound to rule out any obvious malformations. Focussed questioning regarding any symptoms of diabetes insipidus is helpful as this often-overlooked entity is frequently seen in patients with BBS (Marion et al. 2011). Patients with proven diabetes insipidus often do not respond to vasopressin therapy owing to end-organ unresponsiveness (nephrogenic diabetes insipidus). Patients with renal impairment should be referred to a nephrologist.

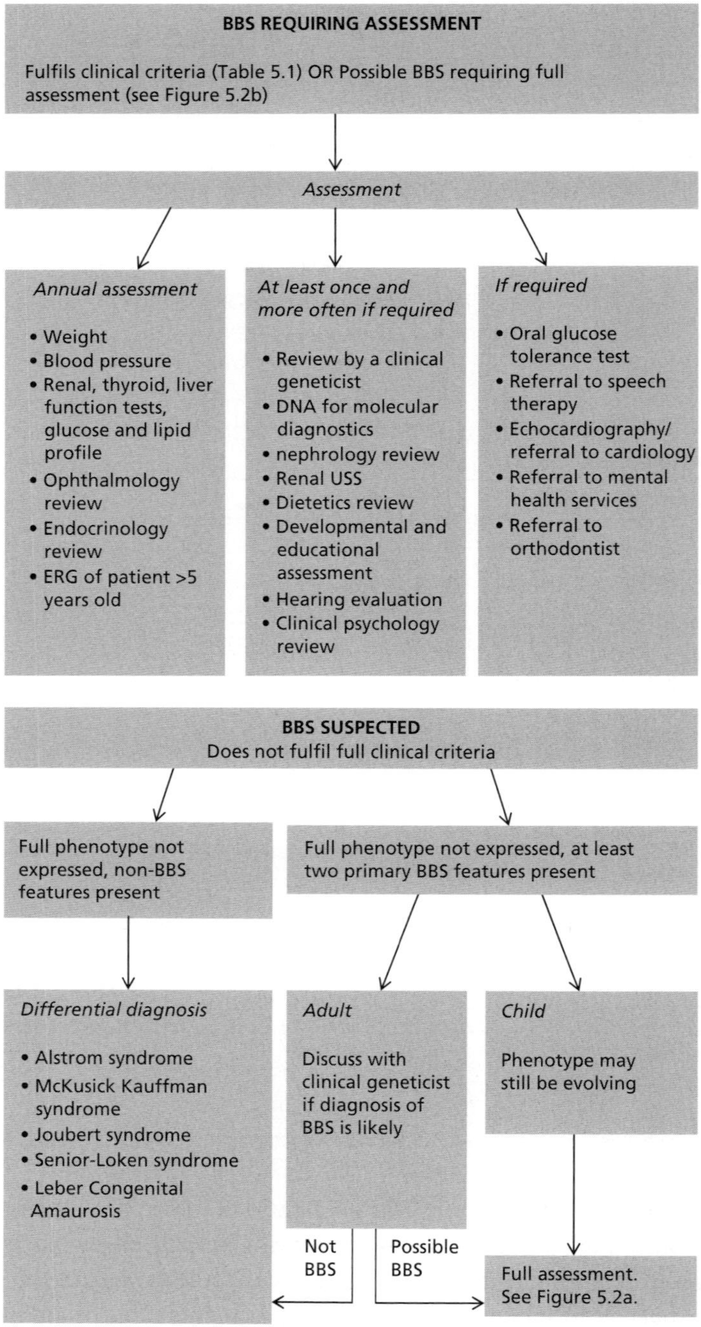

Figure 5.2 Management of patients with suspected and confirmed BBS. (a) An outline of the management of patients with clinically diagnosed BBS. (b) An outline of the appropriate assessment pathway of patients with suspected BBS who do not fulfil the modified clinical criteria for BBS.

Adapted from Elizabeth Forsythe and Philip L Beales, Bardet–Biedl syndrome, *European Journal of Human Genetics*, Volume 21, pp. 8–13, Copyright © 2013 Macmillan Publishers Limited.

A detailed ophthalmological assessment including electrophysiology (e.g. electroretinogram and visually evoked responses) and imaging (e.g. fundal photography and optical coherence tomography) is required to determine the onset, progression and level of rod–cone dystrophy and to screen for other visual defects such as refractive error, diabetic retinopathy or cataracts. Visual aids and mobility training can significantly improve the quality of life for those who are visually impaired.

Effective weight management is imperative to avoid associated morbidity such as metabolic syndrome to which patients with BBS may be particularly susceptible. Exercise and dietetics review can assist in providing a weight loss strategy.

Regular developmental and educational assessment is required to ensure that patients gain optimally from their learning environment. It is noteworthy that although early assessment is helpful in placing a child in the best educational environment, it is not always indicative of full adult potential (Beales et al. 1999). Many patients benefit from a review by a clinical psychologist to help with anxiety, depression or behavioural issues that are more common in BBS.

Endocrinological assessment must include investigation for any signs and symptoms of diabetes mellitus with subsequent oral glucose tolerance testing if appropriate. Assessment of thyroid function, lipid profile and development of secondary sexual characteristics is important. If appropriate, further pituitary function testing can be done, and hormone replacement therapy instigated.

Depending on individual need, referral to an orthodontist for assessment of dental crowding/hypodontia or a cardiologist for investigation of any structural cardiac abnormalities may be appropriate.

Genetics

The last decade has witnessed an unprecedented level of research interest into this once obscure syndrome resulting in the discovery of at least 16 BBS genes accounting for approximately 80% of clinically diagnosed BBS patients (Table 5.2) and the elucidation of the underlying patho-aetiology, primary cilia dysfunction. The majority of pathogenic mutations are found in *BBS1* and *BBS10* accounting for 23.2% and 20%, respectively (Waters & Beales, 1993), although some regional variation in prevalence exists (Hjortshoj et al. 2010). Some genes appear to have greater ethnic specific frequency than others although no mutated genes are found exclusively in a single ethnic population. In northern European individuals two main mutations; *BBS1* M390R and *BBS10* C91LfsX5, are the most commonly encountered alleles. Mutations in *BBS4*, *BBS5* and *TTC8* are mainly seen in patients of Middle Eastern and North African descent (Billingsley et al. 2011). Although traditionally considered an autosomal recessive condition, there are several reported cases of a 'triallelic' mode of inheritance where three mutations in BBS genes are required for the phenotype to manifest itself, or alternatively where a third disease locus acts as a disease modifier (Badano et al. 2003b; Beales et al. 2003; Eichers et al. 2004; Katsanis et al. 2001, 2002).

Table 5.2 Bardet–Biedl syndrome genes

Gene	Frequency	Locus	Function
BBS1	23%	11q13	BBSome protein
BBS2	8%	16q21	BBSome protein
BBS3/ARL6	0.4%	3p12–p13	GTPase
BBS4	2%	15q22.3–q23	BBSome protein
BBS5	0.4%	2q31	BBSome protein
BBS6/MKKS	6%	20p12	Part of chaperonin complex
BBS7	2%	4q27	BBSome protein
BBS8/TTC8	1%	14q32.1	BBSome protein
BBS9/B1	6%	7p14	BBSome protein
BBS10	20%	12q21.2	Part of chaperonin complex
BBS11/TRIM32	0.1%	9q31–q34.1	E3 ubiquitin ligase
BBS12	5%	4q27	Part of chaperonin complex
BBS13/MKS1	4.5%	17q23	Centriole migration
BBS14/CEP290/NPHP6	1%	12q21.3	Basal body: RPGR interaction
BBS15/WDPCP	1%	2p15	Basal body: localisation of septins and ciliogenesis
BBS16/SDCCAG8	1%	1q43	Basal body: interacts with OFD I

Reprinted from Elizabeth Forsythe and Philip L Beales, Bardet–Biedl syndrome, *European Journal of Human Genetics*, Volume 21, pp. 8–13, Copyright © 2013 Macmillan Publishers Limited.

Source: Data from Waters and Beales, Bardet–Biedl syndrome, Copyright © 1993; Moor et al, Clinical and genetic epidemiology of Bardet–Biedl syndrome in Newfoundland: a 22-year prospective, population-based, cohort study, *American Journal of Medical Genetics Part A*, Volume 132, pp. 352–60, Copyirght © 2005 Wiley–Liss, Inc.; Kim et al., Retinal dysfunction in carriers of Bardet–Biedl syndrome, *Ophthalmic Genetics*, Volume 28, pp. 163–8, Copyright © 2007 Informa Plc.; Heon et al., Ocular phenotypes of three genetic variants of Bardet–Biedl syndrome, *American Journal of Medical Genetics*, Volume 132A, pp. 283–287, Copyright © 2004 Wiley–Liss, Inc.; Nachury et al., A core complex of BBS proteins cooperates with the GTPase Rab8 to promote ciliary membrane biogenesis, *Cell*, Volume 129, pp. 1201–13, Copyright © 2007 Elsevier Inc.; and Jin and Nachury,The BBSome, *Current Biology*, Volume 19, pp. 472–473, Copyright © 2007 Elsevier Inc.

As well as the phenotypic overlaps exhibited between ciliopathies there is emerging evidence suggesting that genes mutated in BBS and other ciliopathies exhibit some genetic overlap (Zaghloul & Katsanis, 2009). For example, mutations in *BBS2*, *BBS4* and *BBS6* have been identified in patients with Meckel syndrome (Karmous-Benailly et al. 2005). Similarly, mutations in *MKS1* which normally lead to Meckel syndrome may be associated with a BBS phenotype and mutations in *MKS3* have been identified in patients with BBS and Joubert syndrome (Gerdes et al. 2009). It appears that the ciliopathy phenotype may reflect both the specific mutated locus and the total mutational load (Gerdes et al. 2009; Leitch et al. 2008).

Genotype–phenotype correlations

Genotype–phenotype correlations are poor although one study suggests a milder phenotype associated with the common M390R mutation found in *BBS1* (Hjortshoj et al. 2010).

Other studies have suggested that specific ocular phenotypes (Heon et al. 2005; Riise et al. 2002) and more severe digital abnormalities (Heon et al. 2005) may be linked to *BBS2*, *BBS3* and *BBS4* mutations. Large-scale studies have not supported these findings, and predictions about correlation are further compounded by the significant interfamilial and intrafamilial phenotypic variability. This supports the hypothesis that the BBS proteins interact in a common cellular process thus making the genotypes clinically indistinguishable (Zaghloul & Katsanis, 2009).

The biology of Bardet–Biedl syndrome

Cilia are highly conserved cellular structures projecting from the apical surface of most vertebral cells. They fall into two classes: motile and immotile (primary) cilia (Baker & Beales, 2009; Tobin & Beales, 2007) (See Chapter 1 for further details about cilia). Figure 5.3 outlines the structure of the cilium and some of the protein complexes defined therein. Immotile (primary) cilia are thought to function mainly as a sensory organelle regulating signal transduction pathways (Baker & Beales, 2009; Tobin & Beales, 2007; Waters & Beales, 2011a). Defects in immotile cilia are characterised clinically by retinitis pigmentosa, polydactyly, situs inversus, learning difficulties, and cystic kidneys, liver and pancreas (Gerdes et al. 2009). BBS is a disease of primary cilia dysfunction.

The cilium is anchored in a basal body, a specialised centriole that acts as a microtubule organising centre. Ciliogenesis and maintenance is orchestrated via the basal body acting in concert with the BBSome which, in turn, is modulated by a chaperonin complex and members of the *Rab* family of proteins (Waters & Beales, 2011b). Together these proteins facilitate intraflagellar transport. This cellular process of bidirectional movement of particles facilitates the formation and maintenance of the cilium (see Figure 5.3).

BBS1, *BBS2*, *BBS4*, *BBS5*, *BBS7*, *BBS8* and *BBS9* form the BBSome complex (Jin & Nachury, 2009; Loktev et al. 2008; Nachury et al. 2007) and *BBS6*, *BBS10* and *BBS12* form the chaperonin complex (Seo et al. 2010) (Figure 5.3). The remaining known disease-causing genes have variable predicted functions, as outlined in Table 5.2. No phenotypic difference has been ascertained between patients harbouring disease-causing mutations in genes associated with the BBSome compared to individuals with mutations in genes associated with the chaperonin complex.

The molecular pathogenesis of the pleiotropic effects of BBS has been elucidated only in part. The obesity observed in BBS is multifactorial in origin and there is evidence for both a central neurogenic and a peripheral adipogenic cause. Energy metabolism appears to be similar in BBS patients and matched obese patients in the general population (Grace et al. 2003). Mouse models demonstrate an association between BBS deficiency, increased food intake (neurogenic model) and decreased physical activity; and there is evidence of leptin resistance (Sheffield, 2010). Leptin is a satiety hormone that acts by binding to leptin receptors in the hypothalamus. BBS mutant mice have increased leptin levels. Co-immunoprecipitation experiments show that the *BBS1* protein interacts directly

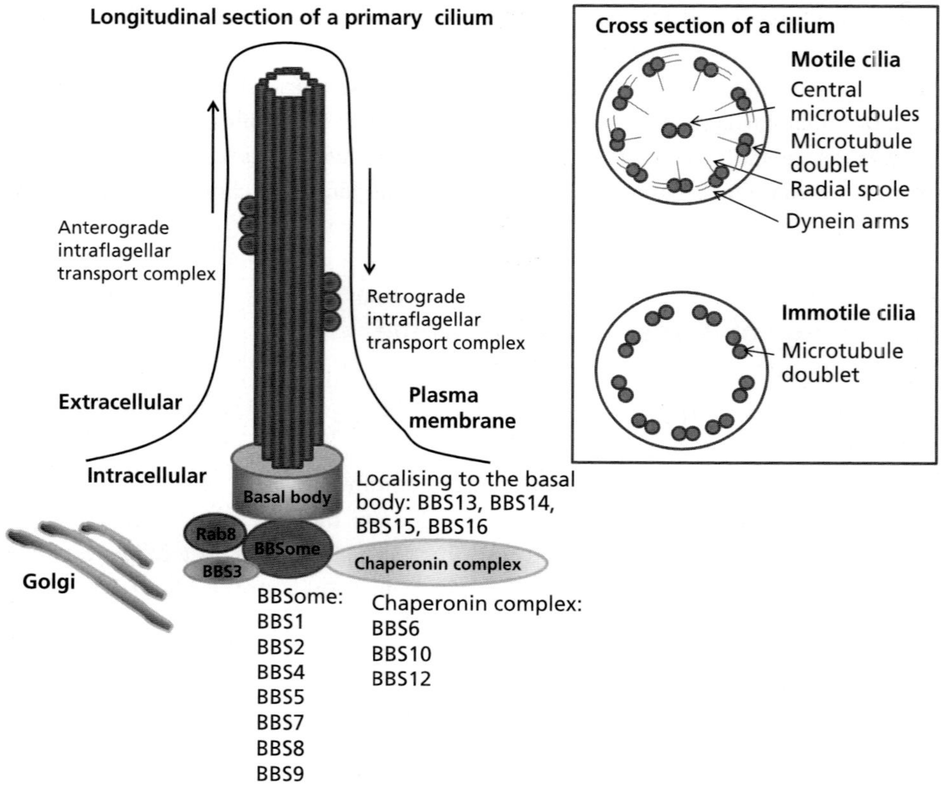

Figure 5.3 Structure of a cilium. Schematic diagram illustrating a longitudinal section of a primary cilium including the localisation of the proteins encoded by the disease-causing genes in Bardet–Biedl syndrome. The chaperonin complex mediates assembly of the BBSome which interacts with *BBS3* (a GTPase) and members of the *Rab* family to organise ciliogenesis and maintenance of the cilium. The cross section view demonstrates the configuration of motile cilia (9 + 2) and immotile cilia (9 + 0).

with leptin receptors and that the common M390R mutation disrupts this interaction (Rahmouni et al. 2008; Seo et al. 2009).

Marion and colleagues found that primary cilia are transiently present in differentiating pre-adipocytes and contain receptors for Wnt and hedgehog pathways (Marion et al. 2009). These highly conserved signalling pathways are integral to normal development and act as modulators of adipogenesis. Inhibition of *BBS10* and *BBS12* induces adipogenesis (adipocentric model) and suggested that this is secondary to disruption of Wnt signalling, which could not be detected after BBS knockdown (Marion et al. 2009).

Rod–cone dystrophy is thought to be a consequence of abnormal trafficking across the defective modified cilia connecting the inner and outer segments of photoreceptors leading to apoptosis (Mockel et al. 2011; Nishimura et al. 2004; Sheffield, 2010).

Receptors for sonic hedgehog signalling are found on cilia in the developing limb buds (Mockel et al. 2011). Intraflagellar transport proteins are thought to modulate this pathway (Gerdes et al. 2009) and dysregulation has been associated with the limb malformations observed in ciliopathies (Bimonte et al. 2011).

The mTor signalling inhibitor rapamycin rescues the renal cysts found in BBS zebrafish morphants implicating signalling pathways upstream of mTor in BBS renal pathology (Tobin & Beales, 2008).

Hyposmia/anosmia may be attributed to defective ciliated olfactory epithelium and subfertility due in part to defective cilia/flagella in sperm cells/oviducts.

BBS was first linked to ciliary dysfunction following the identification of *BBS8* which localises to ciliated structures and to basal bodies and centrosomes in cells (Ansley et al. 2003). Since then several animal models have confirmed these findings and have furthered current understanding of the function of BBS proteins. Mouse and zebrafish models have identified the role of BBS proteins in Wnt signalling (Ross et al. 2005; Gerdes et al. 2009). BBS proteins have been demonstrated to regulate intraflagellar transport and lipid homeostasis in worms and modulate intracellular trafficking and centrosomal function in zebrafish (Blacque & Leroux, 2006). Mouse models share many of the phenotypic features observed in humans affected with BBS and therefore serve as an excellent mammalian model of the disease. Simons and colleagues (Simons et al. 2011) demonstrated this with successful gene therapy preventing photoreceptor death in the *Bbs4*-null murine model. Furthermore, progress in understanding the biological basis for obesity in mouse models may lead to targeted drug therapy for the management of obesity in BBS patients (Guo & Rahmouni, 2011; Seo et al. 2009).

Genetic counselling

BBS is generally inherited in an autosomal recessive manner. Although there are multiple cases of triallelic inheritance these cases are difficult to identify and thought to account for less than 10% of all cases (Waters & Beales, 1993). It is therefore appropriate to counsel patients and families according to traditional autosomal recessive recurrence risk. Hence any sibling of an affected child has a 25% risk of being affected, 50% risk of being an asymptomatic carrier and 25% of being unaffected and not a carrier.

In families where the disease-causing mutations are known, pre-implantation genetic diagnosis or prenatal testing may be possible. Alternatively, in at-risk families where the mutation is unknown targeted second trimester sonography may be applied for visualisation of post-axial polydactyly and renal malformations indicative of a diagnosis of BBS (Cassart et al. 2004; Dar et al. 2001).

Patients and parents should be advised of the heterogeneous nature of the condition. The considerable interfamilial and intrafamilial variation hampers predictions about individual educational attainments, visual deterioration or other difficulties associated with BBS. Although infertility is the norm, this cannot be assumed as there are several reports of both men and women with BBS who have had children (Beales et al. 1999).

Conclusions

Since the first gene for BBS was identified over a decade ago there have been extensive developments within the field. Sixteen disease-causing genes have now been discovered and more have yet to be discovered. Our understanding of their functional properties has facilitated insight into the molecular mechanisms underlying ciliary phenotypes in general and BBS in particular. In the coming years it is likely that other disease-causing genes will be identified, and that there will be further improvement of the clinical diagnostic services allowing for faster diagnosis and prenatal testing.

The ability to more accurately predict the level of disability an affected individual is likely to experience may be improved by furthering our understanding of the molecular processes leading to phenotypic variation. Understanding the epigenetic factors that may account for intrafamilial variation and other modifiers of the condition will be imperative in this process.

Elucidation of the molecular pathogenesis of the clinical features of BBS and research into therapeutics may yield novel treatment options that target organ-specific aspects of the condition, such as renal cysts or rod–cone degeneration or have a more general modulating effect on several aspects of the condition.

Acknowledgement

This chapter is based, with permission, on Elizabeth Forsythe and Philip L. Beales, Bardet–Biedl syndrome, *European Journal of Human Genetics*, Volume 21, pp. 8–13, Copyright © 2013 Macmillan Publishers Limited, doi:10.1038/ejhg.2012.115, first published online 20 June 2012.

References

Adams, N. A., Awadein, A. & Toma, H. S. 2007. The retinal ciliopathies. *Ophthalmic Genet* **28**, 113–125.

Ansley, S. J., Badano, J. L., Blacque, O. E., Hill, J., Hoskins, B. E., Leitch, C. C., et al. 2003. Basal body dysfunction is a likely cause of pleiotropic Bardet–Biedl syndrome. *Nature* **425**(6958), 628–633.

Badano, J. L., Kim, J. C., Hoskins, B. E., Lewis, R. A., Ansley, S. J., Cutler, D. J., et al. 2003a. Heterozygous mutations in BBS1, BBS2 and BBS6 have a potential epistatic effect on Bardet–Biedl patients with two mutations at a second BBS locus. *Hum Mol Genet* **12**, 1651–1659.

Badano, J. L., Kim, J. C., Hoskins, B. E., Lewis, R. A., Ansley, S. J., Cutler, D. J., et al. 2003b. Heterozygous mutations in BBS1, BBS2 and BBS6 have a potential epistatic effect on Bardet–Biedl patients with two mutations at a second BBS locus. *Hum Mol Genet* **12**, 1651–1659.

Baker, K. & Beales, P. L. 2009. Making sense of cilia in disease: The human ciliopathies. *Am J Med Genet, C, Semin Med Genet* **151C**, 281–295.

Bardet, G. 1995. On congenital obesity syndrome with polydactyly and retinitis pigmentosa (a contribution to the study of clinical forms of hypophyseal obesity). 1920. *Obes Res*, **3**, 387–399.

Barnett, S., Reilly, S., Carr, L., Ojo, I., Beales, P. L. & Charman, T. 2002. Behavioural phenotype of Bardet–Biedl syndrome. *J Med Genet* **39**(12), e76.

Beales, P. L., Badano, J. L., Ross, A. J., Ansley, S. J., Hoskins, B. E., Kirsten, B., et al. 2003. Genetic interaction of BBS1 mutations with alleles at other BBS loci can result in non-Mendelian Bardet–Biedl syndrome. *Am J Hum Genet* **72**, 1187–1199.

Beales, P. L., Elcioglu, N., Woolf, A. S., Parker, D. & Flinter, F. A. 1999. New criteria for improved diagnosis of Bardet–Biedl syndrome: Results of a population survey. *J Med Genet* **36**, 437–446.

Beales, P. L., Reid, H. A., Griffiths, M. H., Maher, E. R., Flinter, F. A. & Woolf, A. S. 2000. Renal cancer and malformations in relatives of patients with Bardet–Biedl syndrome. *Nephrol Dial Transplant*, **15**, 1977–1985.

Beales, P. L., Warner, A. M., Hitman, G. A., Thakker, R. & Flinter, F. A. 1997. Bardet–Biedl syndrome: A molecular and phenotypic study of 18 families. *J Med Genet* **34**, 92–98.

Benzinou, M., Walley, A., Lobbens, S., Charles, M. A., Jouret, B., Fumeron, F., et al. 2006. Bardet–Biedl syndrome gene variants are associated with both childhood and adult common obesity in French Caucasians. *Diabetes*, **55**, 2876–2882.

Biedl, A. 1995. A pair of siblings with adiposo-genital dystrophy. 1922. *Obes Res*, **3**, 404.

Billingsley, G., Deveault, C. & Heon, E. 2011. BBS mutational analysis: A strategic approach. *Ophthalmic Genet* **32**, 181–187.

Bimonte, S., De Angelis, A., Quagliata, L., Giusti, F., Tammaro, R., Dallai, R., et al. 2011. Ofd1 is required in limb bud patterning and endochondral bone development. *Dev Biol* **349**, 179–191.

Blacque, O. E. & Leroux, M. R. 2006. Bardet–Biedl syndrome: An emerging pathomechanism of intracellular transport. *Cell Mol Life Sci*, **63**, 2145–2161.

Cassart, M., Eurin, D., Didier, F., Guibaud, L. & Avni, E. F. 2004. Antenatal renal sonographic anomalies and postnatal follow-up of renal involvement in Bardet–Biedl syndrome. *Ultrasound Obstet Gynecol* **24**, 51–54.

Croft, J. B., Morrell, D., Chase, C. L. & Swift, M. 1995. Obesity in heterozygous carriers of the gene for the Bardet–Biedl syndrome. *Am J Med Genet* **55**, 12–15.

Dar, P., Sachs, G. S., Carter, S. M., Ferreira, J. C., Nitowsky, H. M. & Gross, S. J. 2001. Prenatal diagnosis of Bardet–Biedl syndrome by targeted second-trimester sonography. *Ultrasound Obstet Gynecol* **17**, 354–356.

De Pontual, L., Zaghloul, N. A., Thomas, S., Davis, E. E., Mcgaughey, D. M., Dollfus, H., et al. 2009. Epistasis between RET and BBS mutations modulates enteric innervation and causes syndromic Hirschsprung disease. *Proc Natl Acad Sci U.S.A.* **106**, 13921–13926.

Deveault, C., Billingsley, G., Duncan, J. L., Bin, J., Theal, R., Vincent, A., et al. 2011. BBS genotype–phenotype assessment of a multiethnic patient cohort calls for a revision of the disease definition. *Hum Mutat.* **32**(6):610–619.

Eichers, E. R., Lewis, R. A., Katsanis, N. & Lupski, J. R. 2004. Triallelic inheritance: A bridge between Mendelian and multifactorial traits. *Ann Med* **36**, 262–272.

Elbedour, K., Zucker, N., Zalzstein, E., Barki, Y. & Carmi, R. 1994. Cardiac abnormalities in the Bardet–Biedl syndrome: Echocardiographic studies of 22 patients. *Am J Med Genet* **52**, 164–169.

Farag, T. I. & Teebi, A. S. 1989. High incidence of Bardet Biedl syndrome among the Bedouin. *Clin Genet* **36**, 463–464.

Gerdes, J. M., Davis, E. E. & Katsanis, N. 2009. The vertebrate primary cilium in development, homeostasis, and disease. *Cell* **137**, 32–45.

Grace, C., Beales, P., Summerbell, C., Jebb, S. A., Wright, A., Parker, D. et al. 2003. Energy metabolism in Bardet–Biedl syndrome. *Int J Obes Relat Metab Disord* **27**, 1319–1324.

Green, J. S., Parfrey, P. S., Harnett, J. D., Farid, N. R., Cramer, B. C., Johnson, G., et al. 1989. The cardinal manifestations of Bardet–Biedl syndrome, a form of Laurence–Moon–Biedl syndrome. *N Engl J Med* **321**, 1002–1009.

Guo, D. F. & Rahmouni, K. 2011. Molecular basis of the obesity associated with Bardet–Biedl syndrome. *Trends Endocrinol Metab*, **22**, 286–293.

Hamel, C. P. 2007. Cone rod dystrophies. *Orphanet J Rare Dis* **2**, 7.

Heon, E., Westall, C., Carmi, R., Elbedour, K., Panton, C., Mackeen, L., et al. 2005. Ocular phenotypes of three genetic variants of Bardet–Biedl syndrome. *Am J Med Genet A* **132A**, 283–287.

Hjortshoj, T. D., Gronskov, K., Philp, A. R., Nishimura, D. Y., Riise, R., Sheffield, V. C., et al. 2010. Bardet–Biedl syndrome in Denmark—report of 13 novel sequence variations in six genes. *Hum Mutat* **31**, 429–436.

Jin, H. & Nachury, M. V. 2009. The BBSome. *Curr Biol* **19**, R472–473.

Karmous-Benailly, H., Martinovic, J., Gubler, M. C., Sirot, Y., Clech, L., Ozilou, C., et al. 2005. Antenatal presentation of Bardet–Biedl syndrome may mimic Meckel syndrome. *Am J Hum Genet* **76**, 493–504.

Katsanis, N., Ansley, S. J., Badano, J. L., Eichers, E. R., Lewis, R. A., Hoskins, B. E., et al. 2001. Triallelic inheritance in Bardet–Biedl syndrome, a Mendelian recessive disorder. *Science*, **293**, 2256–2259.

Katsanis, N., Eichers, E. R., Ansley, S. J., Lewis, R. A., Kayserili, H., Hoskins, B. E., et al. 2002. BBS4 is a minor contributor to Bardet–Biedl syndrome and may also participate in triallelic inheritance. *Am J Hum Genet* **71**, 22–29.

Kim, L. S., Fishman, G. A., Seiple, W. H., Szlyk, J. P. & Stone, E. M. 2007. Retinal dysfunction in carriers of Bardet–Biedl syndrome. *Ophthalmic Genet* **28**, 163–168.

Kulaga, H. M., Leitch, C. C., Eichers, E. R., Badano, J. L., Lesemann, A., Hoskins, B. E., et al. 2004. Loss of BBS proteins causes anosmia in humans and defects in olfactory cilia structure and function in the mouse. *Nat Genet* **36**, 994–998.

Laurence, J. Z. & Moon, R. C. 1995. Four cases of "retinitis pigmentosa" occurring in the same family, and accompanied by general imperfections of development. 1866. *Obes Res*, **3**, 400–403.

Leitch, C. C., Zaghloul, N. A., Davis, E. E., Stoetzel, C., Diaz-Font, A., Rix, S., et al. 2008. Hypomorphic mutations in syndromic encephalocele genes are associated with Bardet–Biedl syndrome. *Nat Genet* **40**, 443–448.

Loktev, A. V., Zhang, Q., Beck, J. S., Searby, C. C., Scheetz, T. E., Bazan, J. F., et al. 2008. A BBSome subunit links ciliogenesis, microtubule stability, and acetylation. *Dev Cell* **15**, 854–865.

Lorda-Sanchez, I., Ayuso, C., Sanz, R. & Ibañez, A. 2001 Does Bardet–Biedl syndrome have a characteristic face? *J Med Genet* **38**, E14.

Marion, V., Schlicht, D., Mockel, A., Caillard, S., Imhoff, O., Stoetzel, C., et al. 2011. Bardet–Biedl syndrome highlights the major role of the primary cilium in efficient water reabsorption. *Kidney Int* **79**, 1013–1025.

Marion, V., Stoetzel, C., Schlicht, D., Messaddeq, N., Koch, M., Flori, E., et al. 2009. Transient ciliogenesis involving Bardet–Biedl syndrome proteins is a fundamental characteristic of adipogenic differentiation. *Proc Natl Acad Sci U.S.A.* **106**, 1820–1825.

Mockel, A., Perdomo, Y., Stutzmann, F., Letsch, J., Marion, V. & Dollfus, H. 2011. Retinal dystrophy in Bardet–Biedl syndrome and related syndromic ciliopathies. *Prog Retin Eye Res* **30**, 258–274.

Moore, S. J., Green, J. S., Fan, Y., Bhogal, A. K., Dicks, E., Fernandez, B. A., Stefanelli, M., et al. 2005. Clinical and genetic epidemiology of Bardet–Biedl syndrome in Newfoundland: A 22-year prospective, population-based, cohort study. *Am J Med Genet A* **132**, 352–360.

Nachury, M. V., Loktev, A. V., Zhang, Q., Westlake, C. J., Peranen, J., Merdes, A., et al. 2007. A core complex of BBS proteins cooperates with the GTPase Rab8 to promote ciliary membrane biogenesis. *Cell* **129**, 1201–1213.

Nishimura, D. Y., Fath, M., Mullins, R. F., Searby, C., Andrews, M., Davis, R., et al. 2004. Bbs2-null mice have neurosensory deficits, a defect in social dominance, and retinopathy associated with mislocalization of rhodopsin. *Proc Natl Acad Sci U.S.A.* **101**, 16588–16593.

O'Dea, D., Parfrey, P. S., Harnett, J. D., Hefferton, D., Cramer, B. C. & Green, J. 1996. The importance of renal impairment in the natural history of Bardet–Biedl syndrome. *Am J Kidney Dis* **27**, 776–783.

Putoux, A., Attie-Bitach, T., Martinovic, J. & Gubler, M. C. 2011. Phenotypic variability of Bardet–Biedl syndrome: Focusing on the kidney. *Pediatr Nephrol* **27**, 7–15.

Rahmouni, K., Fath, M. A., Seo, S., Thedens, D. R., Berry, C. J., Weiss, R., et al. 2008. Leptin resistance contributes to obesity and hypertension in mouse models of Bardet–Biedl syndrome. *J Clin Invest* **118**, 1458–1467.

Riise, R., Andreasson, S., Borgastrom, M. K., Wright, A. F., Tommerup, N., Rosenberg, T., et al. 1997. Intrafamilial variation of the phenotype in Bardet–Biedl syndrome. *Br J Ophthalmol* **81**, 378–385.

Riise, R., Tornqvist, K., Wright, A. F., Mykytyn, K. & Sheffield, V. C. 2002. The phenotype in Norwegian patients with Bardet–Biedl syndrome with mutations in the BBS4 gene. *Arch Ophthalmol* **120**, 1364–1367.

Rooryck, C. & Lacombe, D. 2008. Bardet–Biedl syndrome. *Ann Endocrinol (Paris)*, **69**, 463–471.

Ross, A. J., May-Simera, H., Eichers, E. R., Kai, M., Hill, J., Jagger, D. J., et al. 2005. Disruption of Bardet–Biedl syndrome ciliary proteins perturbs planar cell polarity in vertebrates. *Nat Genet* **37**, 1135–1140.

Seo, S., Baye, L. M., Schulz, N. P., Beck, J. S., Zhang, Q., Slusarski, D. C., et al. 2010. BBS6, BBS10, and BBS12 form a complex with CCT/TRiC family chaperonins and mediate BBSome assembly. *Proc Natl Acad Sci U.S.A.* **107**, 1488–1493.

Seo, S., Guo, D. F., Bugge, K., Morgan, D. A., Rahmouni, K. & Sheffield, V. C. 2009. Requirement of Bardet–Biedl syndrome proteins for leptin receptor signaling. *Hum Mol Genet* **18**, 1323–1331.

Sheffield, V. C. 2010. The blind leading the obese: The molecular pathophysiology of a human obesity syndrome. *Trans Am Clin Climatol Assoc* **121**, 172–181; discussion 181–182.

Simons, D. L., Boye, S. L., Hauswirth, W. W. & Wu, S. M. 2011. Gene therapy prevents photoreceptor death and preserves retinal function in a Bardet–Biedl syndrome mouse model. *Proc Natl Acad Sci U.S.A.* **108**, 6276–6281.

Slavotinek, A. M. & Biesecker, L. G. 2000. Phenotypic overlap of McKusick–Kaufman syndrome with Bardet–Biedl syndrome: A literature review. *Am J Med Genet* **95**, 208–215.

Tobin, J. L. & Beales, P. L. 2007. Bardet–Biedl syndrome: Beyond the cilium. *Pediatr Nephrol* **22**, 926–936.

Tobin, J. L. & Beales, P. L. 2008. Restoration of renal function in zebrafish models of ciliopathies. *Pediatr Nephrol* **23**, 2095–2099.

Tobin, J. L., Di Franco, M., Eichers, E., May-Simera, H., Garcia, M., Yan, J., et al. 2008. Inhibition of neural crest migration underlies craniofacial dysmorphology and Hirschsprung's disease in Bardet–Biedl syndrome. *Proc Natl Acad Sci U.S.A.* **105**, 6714–6719.

Waters, A. M. & Beales, P. L. 2003. Bardet–Biedl syndrome, in *GeneReviews*™, ed. by Pagon, R. A., Adam, M. P., Bird, T. D., et al. University of Washington: Seattle. Available at: http://www.ncbi.nlm.nih.gov/books/NBK1363/

Waters, A. M. & Beales, P. L. 2011. Ciliopathies: An expanding disease spectrum. *Pediatr Nephrol* **26**, 1039–1056.

Webb, M. P., Dicks, E. L., Green, J. S., Moore, S. J., Warden, G. M., Gamberg, J. S., et al. 2009. Autosomal recessive Bardet–Biedl syndrome: First-degree relatives have no predisposition to metabolic and renal disorders. *Kidney Int*, **76**, 215–223.

Zaghloul, N. A. & Katsanis, N. 2009. Mechanistic insights into Bardet–Biedl syndrome, a model ciliopathy. *J Clin Invest* **119**, 428–437.

Chapter 6

Leber congenital amaurosis and other non-syndromic retinal ciliopathies

Thomas D. Kenny, Philip L. Beales
and Ronald Roepman

History of the eponym

Leber congenital amaurosis (LCA) was recognised and described for the first time in 1869 by Theodor Karl Gustav Leber. LCA was initially introduced as the congenital form of retinitis pigmentosa (RP) (Leber, 1869). Historically, the name LCA also became associated with congenital retinal blindness found in syndromes such as Joubert syndrome, peroxisomal diseases, Batten disease (neuronal ceroid lipofuscinosis) and others. Most vision researchers currently consider LCA the most severe retinal dystrophy without major systemic features (den Hollander et al. 2008).

Epidemiology

Leber congenital amaurosis (LCA, MIM 204000) is the most severe form of early-onset retinal blindness and typically becomes evident in the first year of life (den Hollander et al. 2008), usually at around the age of 6 weeks, when parents note the oscillations of the eyes (nystagmus) or the absence of fixation.

The frequency varies between 1:30,000 (Koenekoop, 2004) and 1:81,000 (Stone, 2007). LCA is generally inherited in an autosomal recessive manner. Although rare it accounts for ~5% of all inherited retinopathies and approximately 20% of children attend schools for the blind (Koenekoop, 2004).

Three longitudinal studies of visual function in LCA patients were performed before the genome era, precluding useful genotype–phenotype correlations. Snellen visual acuity, grating acuities, dark-adapted visual thresholds, and flash visual evoked potentials were performed in a total of 90 LCA patients (Brecelj & Stirn-Kranjc, 1999; Fulton et al. 1996; Heher et al. 1992). Visual deterioration was observed in 15%, stability in 75%, and improvement in 10% of the patients.

Clinical features

LCA represents a group of hereditary retinal diseases characterised and unified by the following constellation of four clinical features: severe and early visual loss, sensory nystagmus, amaurotic pupils, and absent electrical signals on electro-retinogram (ERG) (Franceschetti & Dieterle, 1954; Leber, 1869).

Poor visual function is accompanied by nystagmus, photophobia, absent pupillary responses, hyperopia, extinguished or severely reduced rod and cone signals on ERG, and a highly variable retinal appearance.

Manifestation of visual function and visual acuity in LCA patients ranges widely, usually from 20/200 to light perception or even no light perception. Visual acuities of 20/50 have been described in patients with *CRB1*, *LRAT* and *RPE65* mutations, but acuities in that range do not appear to remain stable. Most patients with LCA have stable or relatively stable residual visual function, but in some LCA patients this deteriorates until they lose all visual function, while LCA patients seem to improve only rarely. It was mentioned by Koenekoop and co-workers, who have found improvements in visual acuity, visual field, and cone ERG b-wave amplitudes in an LCA patient with a mutation in the *CRX* gene that was followed for 12 years (Koenekoop et al. 2002). Although these are rare exceptions, it is important to assess as this could have significant advantages for any future therapy.

LCA patients with *CRB1*, *LCA5* and *RPE65* mutations show mild improvements in their visual function, but then decline after a period of stability (Lorenz et al. 2000; Yzer et al. 2003). LCA patients with *CEP290* and *GUCY2D* mutations appear to have a very significant loss of vision, but then remain stable, while LCA patients who harbour *AIPL1* and *RPGRIP1* mutations have progressive loss of vision (Dharmaraj et al. 2000; Koenekoop et al. 2007). Other studies suggested that visual outcome is distinct and clinically recognisable in patients with mutations in *RPE65* versus those with *GUCY2D* mutations (Lorenz et al. 2000; Perrault et al. 1999). The first group is characterised by measurable acuities and nyctalopia, while the latter group is defined by poor vision, photoaversion and lack of nyctalopia.

Range of the phenotype

The phenotypic variability found in LCA patients is striking; heterogeneity is found in retinal appearance, refractive errors, photoaversion, nyctalopia and the oculodigital sign. Associated features such as keratoconus and cataracts, manifest visual function and longitudinal changes in visual function are also variable. The retinal appearances vary considerably (Figure 6.1). The full phenotypic range of retinal aspects associated with LCA still needs to be determined and correlated with the various genotypes, but currently ranges from essentially normal retinal appearance, to mild retinal vessel attenuation, pseudopapilledema of the optic disc, maculopathy, macular coloboma, bone spicule pigmentation, nummular pigmentation, salt-and-pepper pigmentation, yellow confluent peripheral spots, white retinal spots, marbleised retinal changes, preserved para-arteriolar RPE (PPRPE) and Coats' reaction. It appears that gene-specific phenotypic features exist in LCA (Dharmaraj et al. 2004; Galvin et al. 2005; Koenekoop et al. 2007). These genotype–phenotype correlations are found in the retinal appearances and longitudinal changes in

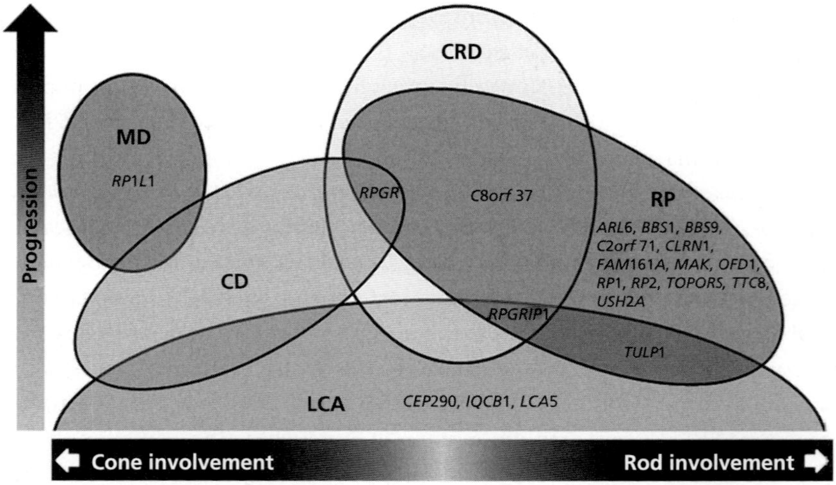

Figure 6.1 A schematic representation of the phenotypic and genetic overlap among non-syndromic retinal ciliopathies. The involvement of cone and rod photoreceptors and progression of diseases are shown in order to match the genetic overlap. In end-stage disease, CD can hardly be distinguished from CRD. Patients with RP initially display tunnel vision owing to rod defects that very often progresses to complete blindness when the cones are also affected. In patients with LCA, the defects can occur in both types of photoreceptors, or in RPE cells, and therefore clinical and molecular genetic overlap with CD, CRD or RP can be expected. At early stage, MD can show an overlap with CD. *RPGR* is also associated with MD. CD, cone dystrophy; CRD, cone–rod dystrophy; LCA, Leber congenital amaurosis; MD, macular dystrophy; RP, retinitis pigmentosa

Reproduced from Estrada-Cuzcano et al., Non-syndromic retinal ciliopathies: translating gene discovery into therapy, *Human Molecular Genetics*, Volume 21, pp. 111–124, Copyright © 2012 with permission of Oxford University Press. Originally adapted from den Hollander, A.I., Black, A., Bennett, J. & Cremers, F.P., Lighting a candle in the dark: advances in genetics and gene therapy of recessive retinal dystrophies, *The Journal of Clinical Investigation*, Issue 120, pp. 3042–3053, Copyright © 2010 American Society for Clinical Investigation.

visual function (Koenekoop et al. 2007). Macular colobomas are a prominent and frequent retinal feature found in LCA, but the term is a misnomer as they are not developmental colobomas. They likely represent complete loss of retinal tissue in the fovea. Colobomas are not found in all genetic types of LCA and may easily be confused with the scars of ocular toxoplasmosis.

LCA patients often have high refractive errors (from hyperopia to myopia) and most patients are high hyperopes (Heher et al. 1992), suggesting that congenital blindness significantly affects the emmetropisation process or that the defective retinal genes also play a role in the determination of the size of the infant eye. It has been suggested that the degree of hyperopia may indicate the presence or absence of associated features in LCA (Wagner et al. 1985) but this association was not born out in subsequent studies (Dagi et al. 1990). Photoaversion (photophobia) can be a prominent feature in LCA (Dagi et al. 1990), as can nyctalopia (night blindness) (Lorenz et al. 2000), and these symptoms may be gene specific (Perrault et al. 1999). The oculodigital sign of Franceschetti can be a very important and sometimes disturbing feature of LCA, although it is not pathognomonic for the disease. The sign consists of a repetitive, deep pushing of the knuckle or finger into the eye and socket. The exact molecular phenomenon is not clear but may be related to the production of phosphenes, which produce sparks of light that may satisfy the patients. The oculodigital sign may also be regarded as a repetitive stereotypic behaviour, also known as blindism. The oculodigital phenomenon can be harmful as the deep-set eyes of LCA patients may be caused by the persistent pushing, resulting in orbital fat atrophy. Deep-set eyes (enophthalmos) may become a prominent facial feature of LCA patients. It has also been suggested that keratoconus, the thinning and ectasia of the cornea, is caused by the oculodigital phenomenon.

Genetics

To date, more than 500 mutations have been identified in LCA and juvenile RP genes (www.retnet.org). The inheritance pattern of LCA in most cases is autosomal recessive, and due to restricted family size results in many isolated cases. However, a small number of dominant cases have been reported. Dominantly inherited LCA has been associated with a 12-bp deletion in the *CRX* gene (Sohocki et al. 1998). In addition, two *de novo* frameshift mutations in *CRX* have been described in two unrelated LCA patients (Freund et al. 1998), and two *de novo* missense mutations in the IMPDH1 gene have been detected in two other unrelated patients (Bowne et al. 2006). All four patients were isolated cases, and therefore a dominant mode of inheritance could not be confirmed. The mutations were not present in the patient's parents, indicating that they arose as new mutations, or were perhaps the result of a germ-line mosaicism. A similar situation was identified for the transcription factor OTX2, where a single case was described where a rare, heterozygous, nonsense mutation causes LCA in combination with pituitary dysfunction (Henderson et al. 2009). The finding of single *de novo* mutations indicates that not all isolated LCA cases are recessive.

A systematic mutation analysis of all known LCA genes in large LCA cohorts has not been performed, but a comprehensive analysis of one or more LCA genes in patient cohorts has been carried out (Booij et al. 2005; Dharmaraj et al. 2000; Hanein et al. 2004; Lotery et al. 2000; Sitorus et al. 2003) and others have analysed patient groups for all known mutations in the known LCA genes using the LCA mutation chip (Henderson et al. 2007; Simonelli et al. 2007; Vallespin et al. 2007; Yzer et al. 2006; Zernant et al. 2005). Together, mutations in these 24 genes account for approximately 80% of all LCA cases. The most frequently mutated genes are *CEP290* (15%), *GUCY2D* (12%) and *CRB1* (10%).

Several founder mutations have been identified that are frequent in certain populations but absent in others. The intronic *CEP290* mutation p.Cys998X was found in 20% of LCA patients of worldwide origin in two reports (den Hollander et al. 2006; Perrault et al. 2007). All patients that carried this mutation were of European ancestry. The mutation was found less frequently in a group of Italian LCA patients (4%; Simonelli et al. 2007) and Spanish LCA patients (8%; Vallespin et al. 2007), suggesting that it is an ancient mutation that arose in the north of Europe several centuries ago. Another, less frequent, founder mutation (p.Lys1575X) in the *CEP290* gene was detected in the north of France (Perrault et al. 2007). Mutations in the *GUCY2D* gene are more frequently found in patients from Mediterranean countries than in patients of worldwide origin (Hanein et al. 2002). However, the *GUCY2D* mutation p.Arg768Trp was found relatively frequent in north-western Europe (Yzer et al. 2006), and the c.2943delG mutation represents an ancient founder mutation in the Finnish population (Hanein et al. 2002). The most frequent *CRB1* allele, p.Cys948Tyr, is found in patients of worldwide origin (den Hollander et al. 1999). The p.Trp278X mutation in the *AIPL1* gene accounts for approximately half of all *AIPL1* alleles, and may represent a founder mutation in the Pakistani population (Sohocki et al. 2000a,b).

Can genotype–phenotype correlations be discerned for LCA? Certain retinal appearances and changes in longitudinal aspects of visual function appear to be gene specific (Koenekoop et al. 2007). For example, a PPRPE appearance of the retina is strongly suggestive of *CRB1* mutations. Transient visual function improvements are suggestive of *RPE65* mutations, and a relatively preserved retinal appearance suggests *GUCY2D* mutations. Genotype–phenotype correlations such as these found in LCA appear to be much scarcer in the much more prevalent RP disease phenotype. It is tempting to speculate that this important difference is due to the fact that LCA is a developmental retinal dystrophy, which may arrest at different time points depending on the gene defect (giving rise to phenotypic differences), while RP is an 'acquired' retinal dystrophy and inexorably leads to slow but progressive cell death. The mechanistic cause for this difference is currently not known and explanations await further investigations.

Underlying molecular defects

Phototransduction (*AIPL1, GUCY2D, CNGA3*)

The aryl hydrocarbon receptor protein-like 1 (AIPL1) is a protein containing three conserved tetratricopeptide repeat (TPR) domains. The TPR motif is a degenerate, 34-residue sequence comprising a pair of anti-parallel α-helices. TPR domains function as molecular scaffolds mediating protein interactions. A primate-specific poly-proline-rich sequence with unknown function is present at the C-terminus in human AIPL1. At early stages AIPL1 is expressed in the central and peripheral retina, which coincides with rod and cone photoreceptor development (van der Spuy et al. 2003). However, in the adult retina, expression is restricted to rod photoreceptors (van der Spuy et al. 2002). AIPL1 acts as a specialised chaperone by enhancing the farnesylation of cGMP-PDE-α, promoting cGMP-PDE folding, and assembling the subunit complex. In humans, AIPL1 mutations lead to early onset and severe LCA.

GUCY2D encodes a membrane guanylatecyclase RetGC-1, the enzyme involved in the resynthesis of cGMP required for the recovery of the dark state after phototransduction. GUCY2D is expressed specifically in the retina, where it localises to the nuclei and inner segments of rod and cone photoreceptors. Many patients carry GUCY2D protein truncating mutations on both alleles, which are expected to result in the total absence of cyclase activity (Perrault et al. 1999). Functional analysis of LCA mutations *in vitro* showed that missense mutations in the catalytic domain result in complete inability to hydrolyse GTP to cGMP (Rozet et al. 2001). Some missense mutations in the extracellular domain do not affect catalytic activity and likely result in misfolding of the mutant protein and subsequent degradation in the endoplasmatic reticulum (Rozet et al. 2001).

The alpha-subunit of the cone photoreceptor cGMP-gated cation channel (CNGA3) is also a key component of the phototransduction pathway. Initially, mutations in the *CNGA3* gene were identified in patients with total colour blindness or complete achromatopsia (Kohl et al. 1998). Whole exome sequencing also identified mutations in this gene in LCA patients (Wang et al. 2011).

Retinoid cycle (*RDH12, LRAT, RPE65*)

Photoactivation of rhodopsin and cone pigments causes isomerisation of 11-*cis*-retinal to all-*trans*-retinal, which is recycled in a pathway termed the visual (retinoid) cycle. Three genes (*LRAT*, *RDH12* and *RPE65*) encoding proteins that play important roles in the visual cycle were found to be mutated in patients with LCA (Janecke et al. 2004; Morimura et al. 1998; Thompson et al. 2000).

RDH12 is expressed in human photoreceptor inner segments and ONL (Haeseleer et al. 2002; Jacobson et al. 2007; Kurth et al. 2007). In the visual cycle, RDH12 can catalyse reduction of all-*trans*-retinal and 11-*cis*-retinal to their corresponding retinols. Studies suggest that decreased 11-*cis*-retinal production (Haeseleer et al. 2002; Jacobson et al. 2007; Kurth et al. 2007; McBee et al. 2001) can be a cause of the degeneration caused by RDH12 mutations.

The retinal pigment epithelium (RPE) is a monolayer of cells apposed to the outer surface of the retinal photoreceptor cells and is involved in many aspects of photoreceptor cell maintenance including the retinoid visual cycle (photoreceptor outer segment disc phagocytosis and recycling). Two key components of this visual cycle are LRAT and RPE65.

LRAT is a protein that catalyses the synthesis of retinyl esters, thereby drawing retinal from the circulation to storage depots such as lipid droplets of hepatic stellate cells and the retinosome structures in the RPE (Imanishi et al. 2004a,b).

RPE65 is a microsomal protein with isomerase activity (Jin et al. 2005; Moiseyev et al. 2006; Redmond et al. 2005). Isomerohydrolase activity assays showed that the enzymatic activity of RPE65 requires LRAT co-expression.

Photoreceptor development and structure (*CRX, OTX2, CRB1*)

The cone–rod homeobox gene *CRX* is a member of a highly conserved gene family, and encodes a homeobox transcription factor (Furukawa et al. 1997). *CRX* is related to the products of other homeobox genes, which all play crucial roles at various stages of eye development (Chow & Lang, 2001). *CRX* is the earliest expressed photoreceptor marker in the retina, and is also expressed in pinealocytes in the pineal gland and regulates photoentrainment (Furukawa et al. 1999). *CRX* is essential for the differentiation and maintenance of photoreceptor cells. It acts synergistically with the eye-specific transcription factors neural leucine-zipper (NRL) and retinal homeobox protein RX in the transactivation of photoreceptor-specific genes, regulating the high level of expression of photoreceptor outer segment proteins (Chen et al. 1997; Freund et al. 1997; Furukawa et al. 1997; Kimura et al. 2000; Mitton et al. 2000). Recently, CRX was found to interact with different histone acetyl-transferases, suggesting that a possible mechanism for CRX-mediated transcriptional activation is to recruit histone acetyl-transferases to photoreceptor gene chromatin for histone acetylation, thereby inducing and maintaining appropriate chromatin configurations for transcription (Peng & Chen, 2007).

OTX2 is also a transcription factor that is of major importance for photoreceptor development; it controls retinal cell fate and transactivates CRX (Nishida et al. 2003). Although rare, a *de novo*, dominant mutation has been described in a patient suffering from LCA and pituitary dysfunction (Henderson et al. 2009).

The intracellular domain of CRB1 has a highly conserved role in organising a macromolecular protein scaffold (Richard et al. 2006) This scaffold is located at the sub-apical region, a region just apical to the adherens junctions at the outer limiting membrane in the retina (Kantardzhieva et al. 2005, 2006; van de Pavert et al. 2004). Adherens junctions are the sites where the cytoplasmic face of the plasma membrane is attached to actin filaments. They are found in photoreceptors and Müller cells, as well as in epithelial cells, where they separate the apical and basolateral membrane.

Ion channels (*KCNJ13, CaBP4*)

A combination of homozygosity mapping and exome sequencing identified a homozygous nonsense mutation in KCNJ13, encoding a potassium channel subunit Kir7.1. All LCA patients showed a distinct and unusual retinal appearance and a similar early onset of visual loss. This suggested both impaired retinal development and progressive retinal degeneration, involving both rod and cone pathways (Sergouniotis et al. 2011).

CaBP4 is a neuronal, Ca^{2+} binding protein with similarity to calmodulin. CaBP4 associates directly with the C-terminal domain of the Ca(v)1.4 alpha(1)-subunit and shifted the activation of the Ca(v)1.4 channel to hyperpolarised voltages in transfected cells (Haeseleer et al. 2004). A null mutation in the CaBP4 gene causes LCA (Aldahmesh, 2010).

Transport across the photoreceptor connecting cilium (*TULP1, RPGRIP1, CEP290, IQCB1, MYO7A, BBS4, ALMS1, lebercilin, SPATA7*)

TULP1 is a member of the Tubby-like protein (TULP) family (Ikeda et al. 1999, 2002a,b). These proteins have an important role in the development and function of the central nervous system (Carroll et al. 2004; Hong et al. 2001; Ikeda et al. 2002a). The TULP1 protein contains a C-terminal 'Tubby domain' that is conserved among the TULP family, and contains a phosphatidylinositol-binding region that may anchor the protein to the cell membrane (Xi et al. 2005). Its N-terminal half contains a nuclear localisation signal and transcriptional activation activity, and the Tubby domain exhibits DNA-binding activity, which indicates a potential transcription factor activity (Boggon et al. 1999).

The gene is predominantly expressed in the photoreceptors of the retina (Ikeda et al. 1999; North et al. 1997). The protein was also detected at lower levels in the brain, in the paraventricular nuclei of the hypothalamus (Ikeda et al. 2000). In the retina, the protein localises predominantly to the photoreceptor cells, both rods and cones (Hagstrom et al. 1999; Ikeda et al. 2000). In agreement with this, the absence of Tulp1 was found to severely affect both light- and dark-adapted ERG responses (Hagstrom et al. 1999). In contrast to its suggested function as a transcription factor, no significant signal could be detected in the photoreceptor nuclei (Hagstrom et al. 2001). Over-expression of the protein in cell lines, however, revealed localisation of a subset of the protein to the nucleoli (Xi et al. 2005). This may indicate that the levels of TULP1 in the nuclei are too low to be detected, or its epitope is masked in the nucleus. Information about gene targets of TULP1 as a transcription factor remains elusive. TULP1 was also identified in retinal neuroblasts as early as 8 fetal weeks, suggesting an important role of TULP1 in retinal differentiation (Milam et al. 2000). This is in line with the early onset of the retinal degeneration in LCA. A low-level staining of the differentiating, as well as adult ganglion cells may also be important in that respect (Milam et al. 2000).

The RP GTPase regulator interacting protein 1 (RPGRIP1) directly binds to the RP GTPase regulator (RPGR) with its C-terminal RPGR interacting domain (RID) (Boylan & Wright, 2000; Roepman et al. 2000). Disease-associated missense mutations in the RCC1-like domain of RPGR disrupted the interaction between RPGRIP1 and RPGR, suggesting that this defect could underly the pathogenesis of RP (Roepman et al. 2000). LCA associated mutations in the RID of RPGRIP1 could lead to a gain- and loss-of-binding to RPGR (Lu et al. 2005). RPGRIP1 contains two coiled-coil domains that are homologous to those found in proteins involved in vesicular trafficking (Boylan & Wright, 2000; Roepman et al. 2000), and a bipartite nuclear localisation signal that could facilitate shuttling to the nucleus of some isoforms (Roepman et al. 2005). Localisation to the connecting cilium (Zhao et al. 2003), and its binding to NPHP4 which is disrupted due to mutations in either the *RPGRIP1* or *NPHP4* gene (Roepman et al. 2005) indicates a central role in transport across the connecting cilium.

CEP290 is a centrosomal and basal body protein. It was first identified in a proteomic analysis of the human centrosome (Andersen et al. 2003). The protein is strongly conserved throughout evolution, and contains several predicted motifs, including 13 coiled-coil domains (Sayer et al. 2006). A bipartite nuclear localisation signal would explain its partial localisation to the nucleus, where it binds to and activates transcription factor ATF4 (Sayer et al. 2006). Besides its localisation to the centrosome of dividing cells and to the nucleus, the protein localises to the basal bodies at the base of the cilia in many different cell types, including the photoreceptor connecting cilium (Chang et al. 2006; Sayer et al. 2006; Valente et al. 2006). Using immunoprecipitation from retinal extracts, it was demonstrated that CEP290 exists in the same complex with several microtubule-based transport proteins, including RPGR (Chang et al. 2006). Many of these proteins were also detected in similar co-immunoprecipitation experiments with RPGR (den Hollander et al. 2007; Khanna et al. 2005). The dynamics of this complex, including the direct interactions of the protein members, remain to be identified.

The IQCB1/NPHP5 protein that was initially found to be localising to cilia and causal for Senior–Løken syndrome (Otto et al. 2005), but that was later also found to be mutated in LCA patients (Estrada-Cuzcano et al. 2011). It may interact with calmodulin and RPGR (Otto et al. 2005), but its function in the photoreceptors, and if this is transport related, has not been well studied.

MYO7A was initially identified as the protein involved in Usher syndrome type 1B (Hasson et al. 1995), and, besides, in hair cells, the protein was also detected at the photoreceptor connecting cilium. In an exome sequencing effort to unravel the mutations in an LCA cohort pre-screened for mutations in the known genes, mutations of MYO7A were also detected in these patients (Wang et al. 2011).

BBS4 is one of seven proteins of the BBSome, a protein complex detected at the basal body that is essential for ciliary membrane biogenesis. BBS4 was previously found to be involved in Bardet–Biedl syndrome (Mykytyn et al. 2001). It was later found that BBS4 targets cargo to the pericentriolar region and is required for microtubule anchoring and cell cycle progression (Kim et al. 2004). Recently, a homozygous missense mutation, identified by exome sequencing in the *BBS4* gene, was suggested to be causative for LCA in a consanguineous family (Wang et al. 2011). This may indicate that other BBS proteins as well may be involved in this disorder, and these may in the future more rapidly be discovered by the rapidly advancing sequencing techniques.

Mutations in *ALMS1* were originally described in Alström syndrome (Collin et al. 2002). The protein encoded by this gene has been implicated in ciliary function, cell cycle control, and intracellular transport, and associates with α-actinin and components of the endosome recycling pathway (Collin et al. 2012a). In the same cohort used to detect IQCB1, CNGA3 and MYO7A mutations, association of ALMS1 with LCA could also be confirmed (Wang et al. 2011).

The *LCA5* gene is almost ubiquitously expressed during early embryonic development, while at later stages its expression shifts towards ciliated tissues. The *LCA5* gene product lebercilin was found to be present in the ciliary proteome database (Gherman et al. 2006) and localisation to cilia of cultured cells and to the photoreceptor connecting cilia was confirmed using anti-lebercilin antibodies in immunohistochemistry and immunoelectron microscopy (EM) analyses (den Hollander et al. 2007). In non-ciliated cells, lebercilin localised to the mother centriole of the centrosome and to microtubules. An increased expression of recombinant lebercilin in some cells induced microtubule bundling, indicating a role of the protein in microtubule dynamics. In ciliated cells, lebercilin localised to the basal bodies and transition zone of the cilia when the protein expression was low (den Hollander et al. 2007). Upon increased expression, the protein also decorated the full ciliary axoneme and microtubule cytoskeleton of the cells. Tandem affinity purification of lebercilin initially revealed interesting details of the link between lebercilin and ciliary and centrosomal function (den Hollander et al. 2007). More detailed quantitative proteomics analyses, and the generation of a knock-out mouse model, enabled the identification that lebercilin associates with the intraflagellar transport machinery, a connection that is lost with the mutant lebercilin proteins (Boldt et al. 2011).

Finally, SPATA7 is a protein associated with LCA that was only very recently found to localise to the photoreceptor cilium (R. Roepman, personal communication). Its function in that region remains still to be unveiled.

With these ciliary proteins, the role of disrupted ciliary processes in the molecular pathogenesis of LCA is emphasised, pinpointing a growing group of LCA subtypes as ciliopathies.

Miscellaneous (IMPDH1, MERTK, RD3, NMNAT1)

The 'miscellaneous' group of LCA-associated genes and proteins is rapidly growing, as next generation sequencing is boosting the speed of gene discovery. Many of these genes encode proteins that not known to belong to the previously mentioned structures or processes.

IMPDH1 functions as a tetramer composed of four identical subunits, which catalyses the rate-limiting step of *de novo* guanine synthesis. It converts inosine monophosphate into xanthosine monophosphate with the reduction of nicotinamide adenine dinucleotide (NAD). The autosomal dominant RP-associated mutations are not located in the catalytic domain and do not reduce the IMPDH1 enzyme activity (Aherne et al. 2004; Mortimer & Hedstrom, 2005). Mutations are located in the second cystathionine beta synthase (CBS) domain of IMPDH1 Although the role of the CBS domains in IMPDH1 is not yet known, it was shown that it can bind single-stranded nucleic acids and therefore might play a role in transcription, translation, post-translational modification, localisation or other aspects of RNA metabolism (McLean et al. 2004). IMPDH1 mutations associated with autosomal dominant RP and *de novo* LCA significantly reduce the nucleic acid binding affinity and specificity (Bowne et al. 2006; Mortimer & Hedstrom, 2005). Although the disease mechanism associated with IMPDH1 mutations is not yet understood, it might involve a perturbation in RNA metabolism in photoreceptor cells.

Nearly all MERTK mutations identified in patients with early onset retinal dystrophy represent loss-of-function alleles that are predicted to result in truncated MERTK protein lacking the intracellular tyrosine kinase domain (Gal et al. 2000; McHenry et al. 2004; Tschernutter et al. 2006). Only one missense mutation has been identified, which presumably causes loss of function due to decreased protein stability (McHenry et al. 2004).

RD3 shows sub-nuclear localisation adjacent to premyelocytic leukaemia gene product (PML) bodies. PML bodies are implicated in diverse biological functions, including DNA repair, antiviral response, apoptosis, proteolysis, gene regulation, and tumour suppression. The precise role of RD3 in the retina and its disease mechanism remain to be elucidated.

NMNAT1 is a protein with a dual function. The activity of the nuclear nicotinamide mononucleotide adenyltransferase NMNAT1 is essential in NAD^+ synthesis. However, its second role is that it acts as a chaperone that protects against neuronal activity-induced degeneration. It is likely that this disrupted neuroprotective role leads to the LCA phenotype that was simultaneously published by four groups in the journal *Nature Genetics* in 2012 (Chiang et al. 2012; Falk et al. 2012; Koenekoop et al. 2012; Perrault et al. 2012).

Other non-syndromic retinal ciliopathies

At least 158 genes have been associated with inherited retinal dystrophies, one-third of which encode proteins that localise to the cilium (Estrada-Cuzcano et al. 2012).

Retinitis pigmentosa

RP is the most common inherited retinal degeneration with a worldwide prevalence of ~1:4000 individuals (Haim, 2002). RP is initially characterised by rod photoreceptor dysfunction, giving rise to night blindness, followed by progressive mid-peripheral vision loss and development of tunnel vision. At an advanced stage when cones are also affected, blindness ensues. The disease is genetically heterogeneous and displays all Mendelian patterns of inheritance as well as examples of digenic inheritance (Daiger et al. 2007).

Cone dystrophy

Cone dystrophy (CD) is a progressive disorder of the cones with a prevalence of ~1:30,000 to 1:40,000 (Michaelides et al. 2006). Initially, patients have normal cone function, but develop visual loss and colour vision disturbance in the first or second decade (Thiadens et al. 2011). Macular abnormalities can be present, and the optic nerve may show a variable degree of temporal pallor. On ERG, cone responses progressively deteriorate and rod responses are initially normal. The visual acuity generally diminishes to legal blindness before the third or fourth decade of life.

Cone–rod dystrophy

Cone–rod dystrophy (CRD) has an estimated prevalence of 1:30,000 to 1:40,000 (Michaelides et al. 2006), and also displays all types of Mendelian inheritance. CRD is characterised by a primary loss of cone photoreceptors subsequently followed by the loss of rod photoreceptors (Thiadens et al. 2011). The disease in most cases becomes apparent during primary school years with symptoms that include photo-aversion, decreased visual acuity with or without nystagmus (pendular or roving eye movements), colour vision defects and decreased sensitivity of the central visual field. Since rods are also involved, night blindness and peripheral vision loss may occur. The diagnosis of CRD is mainly based on ERG recordings, in which cone responses are equal to or more severely reduced than rod responses (Hamel, 2007).

Macular dystrophy

Macular dystrophy (MD) affects the central area of the retina, the macula, leading to loss of colour and sharp vision. Inherited forms of the disease are transmitted by autosomal dominant and autosomal recessive patterns of inheritance. Various forms of MD are recognised based on particular clinical features, such as a dark choroid typical of Stargardt disease and a yellow 'egg-yolk' appearance of the macula seen in Best disease, although the retinal appearance is essentially normal in occult MD.

Genetic heterogeneity

Although CD, CRD, LCA, MD and RP are described as distinct clinical entities, they are not always easily distinguished and can have overlapping genetic causes (den Hollander et al. 2010) (Figure 6.2). To date, 74 genes have been associated with these conditions (including LCA) and more are expected. Of these, 21 genes (28%) encode ciliary proteins. Several gene mutations have so far only been described in one or a few families. It has been estimated that mutations in 16 ciliary genes (*ARL6, BBS1, BBS9, C2orf71, C8orf37, CLRN1, FAM161A, MAK, OFD1, RP1, RP2, RPGR, TOPORS, TTC8, TULP1, USH2A*) represent at least 36% of the genetic causes in RP.

Common mutations

Among the most frequent causes of ciliary RP are mutations in the Usher syndrome gene, *USH2A* (particularly p.Cys759Phe) (Rivolta et al. 2000). In CD and CRD, only 4% of the cases are due to mutations in three ciliary genes (*C8orf37, RPGR, RPGRIP1*), which is somewhat unexpected given the high percentage of mutations in ciliary genes identified in LCA and RP. It is possible that the number of ciliary genes is underrepresented due to the high percentage of unsolved cases in CD and CRD (90 and 60%, respectively) (Estrada-Cuzcano et al. 2012).

Genotype–phenotype correlations

Clear-cut correlations have not been established between ciliary genotypes and retinal phenotypes despite partial overlap of the clinical features (Figure 6.1, Table 6.1). However, among genes that cause both syndromic and non-syndromic retinal ciliopathies (*ARL6, BBS1, BBS9, CLRN1, CEP290, IQCB1, OFD1, TTC8/BBS8, USH2A*), there are some interesting examples of genotype–phenotype correlations. In *TTC8/BBS8*, a splice site mutation, leading to the in-frame skipping of a 30 bp retina-specific exon, was identified in a family with non-syndromic RP, whereas most BBS mutations are predicted to significantly impact mRNA stability and/or protein function (Riazuddin et al. 2010). Interestingly, a deletion in *BBS9* (p.Glu148_Val234del) has been described to cause non-syndromic RP in two affected members of the family, whereas the third affected member showed all the primary features of BBS (Abu-Safieh et al. 2012) suggesting that additional factors may contribute to the disease in the latter individual. A deep intronic mutation in *OFD1* was identified in a large family with RP that inserts a cryptic exon into the mRNA but reduced levels of normally spliced gene product remain (Webb et al. 2012). This disease mechanism of reduced expression of syndromic ciliopathy genes causing isolated retinal dystrophy is reminiscent of an intronic mutation in *CEP290* that causes LCA (den Hollander et al. 2006), suggesting that reduced dosage of correctly spliced ciliopathy genes may be a common disease mechanism in retinal degeneration.

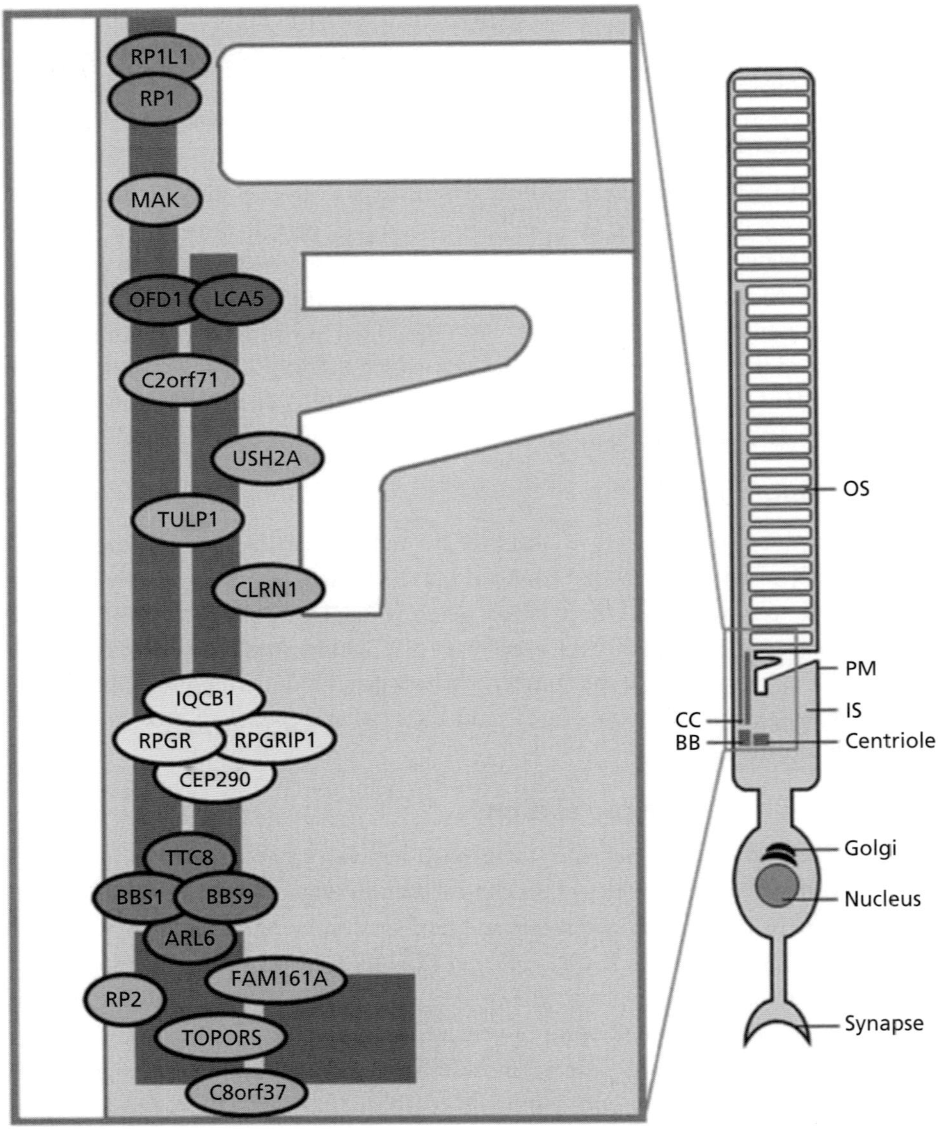

Figure 6.2 A model of the localisation of non-syndromic retinal ciliopathy proteins in the ciliary compartment of the photoreceptor. For each protein, the corresponding gene name has been used. At least four protein complexes have been reported in literature which localise to the connecting cilium. The BBSome consisting of BBS1, BBS9 and TTC8, and recruited by ARL6, forms a coat at the base of the CC. A complex consisting of nephronophthisis-associated proteins CEP290/IQCB1 connected to RPGR by RPGRIP1 localises to the CC, such as OFD1 bound to lebercilin. In the OS axonemes, localisation of RP1L1 and RP1 is observed. BB, basal body; CC, connecting cilium; IS, inner segment; OS, outer segment; PM, plasma membrane.

Reproduced from Estrada-Cuzcano et al., Non-syndromic retinal ciliopathies: translating gene discovery into therapy, *Human Molecular Genetics*, Volume 21, pp. 111–124, Copyright © 2012 with permission of Oxford University Press.

Table 6.1 Non-syndromic retinal ciliopathy genes

Entrez gene symbol	Entrez gene identifier	Localisation of the encoded protein in photoreceptors	Pattern of inheritance	Non-syndromic disease association	Syndromic disease association
ARL6	84,100	Photoreceptor and ganglion cell layer, nerve fiber layer	AR	RP	BBS
BBS1	582	BB, CC, synapse of the plexiform layer	AR	RP	BBS
BBS9	27,241	BB, CC, synapse of the plexiform layer	AR	RP	BBS
C2orf71	388,939	CC (putative)	AR	RP	
C8orf37	157,657	BB, ciliary rootlet	AR	CRD, RP	
CEP290	80,184	CC	AR	LCA	BBS, JBTS, MKS, SLSN
CLRN1	7,401	CC, IS, ribbon synapses	AR	RP	USH3
FAM161A	84,140	Apical part of CC, BB, IS, OPL	AR	RP	
IQCB1	9,657	CC, OS	AR	LCA	SLSN
LCA5	167,691	BB, CC	AR	LCA	
MAK	4,117	CC, OS axoneme	AR	RP	
OFD1	8,481	CC, IS (putative)	XL	RP	JBTS, OFD, SGBS2
RP1	6,101	CC, OS axoneme	AD, AR	RP	
RP1L1	94,137	CC, OS axoneme	AD with reduced penetrance	MD	
RP2	6,102	BB, Golgi, periciliary region, PM	XL	RP	
RPGR	6,103	BB, CC	XL	CD, CRD, MD, RP, RP and sinorespiratory infections, with or without deafness	
RPGRIP1	57,096	CC	AR	CRD, LCA, RP	
TOPORS	10,210	BB, nuclei of ganglion cells, periciliary region	AD	RP	

Table 6.1 (continued) Non-syndromic retinal ciliopathy genes

Entrez gene symbol	Entrez gene identifier	Localisation of the encoded protein in photoreceptors	Pattern of inheritance	Non-syndromic disease association	Syndromic disease association
TTC8	123,016	BB, CC, synapse of the plexiform layer	AR	RP	BBS
TULP1	7,287	CC, IS, OLM	AR	LCA, RP	
USH2A	7,399	CC, IS, OLM, OPL	AR	RP	USH2

AD, autosomal dominant; AR, autosomal recessive; BB, basal body; BBS, Bardet–Biedl syndrome; CC, connecting cilium; CD, cone dystrophy; CRD, cone–rod dystrophy; IS, inner segment; JBTS, Joubert syndrome; LCA, Leber congenital amaurosis; MD, macular dystrophy; MKS, Meckel–Gruber syndrome; OFD, orofaciodigital syndrome; OLM, outer limiting membrane; OMD, occult macular dystrophy; OPL, outer plexiform layer; OS, outer segment; PM, plasma membrane; RP, retinitis pigmentosa; SGBS2, Simpson–Golabi–Behmel syndrome type 2; SLSN, Senior–Løken syndrome; USH2, Usher syndrome type 2; USH3, Usher syndrome type 3; XL, X-linked.

Adapted from Estrada-Cuzcano et al., Non-syndromic retinal ciliopathies: translating gene discovery into therapy. *Human Molecular Genetics*, Volume 21, pp. 111–124, Copyright © 2012 with permission of Oxford University Press.

Clinical management and therapies

For the majority of patients with LCA and non-syndromic retinal ciliopathies the management is currently largely supportive and it is important to ensure that patients have access to the most appropriate advice. For individuals with some remaining vision, they may benefit from the use of low-vision equipment, including raised typeface, electronic, computer-based and optical aids.

Orientation and mobility training and other adaptive training skills are useful in patients with LCA and non-syndromic retinal ciliopathies. As there is no clear association with cognitive disabilities, these patients should access equivalent training to those with non-syndromic blindness.

A number of studies exploring specific disease-based therapies are currently under way. These include gene-replacement therapy, genetic manipulation such as antisense oligonucleotide therapy or pharmacotherapies.

Gene therapy

LCA patients with a mutation in the *RPE65* gene have been treated by gene therapy in recent clinical trials. Therapeutic research has mainly focused on gene replacement by subretinal injection of virus-packaged gene constructs, which appears to be safe and successful in restoring vision in models for non-syndromic retinal dystrophies. Early results of three clinical trials showed recovery of some functional vision and no obvious side effects, thus demonstrating the short-term safety and efficacy of adeno-associated virus in LCA (Bainbridge et al. 2008; Cideciyan et al. 2009; Maguire et al. 2008; Simonelli et al. 2010). In addition, a small phase 1 trial showed some sustained improvement (Maguire et al. 2009). However, while the gene therapy improves vision for at least 3 years, photoreceptor degeneration was reported to continue (Cideciyan et al. 2013).

For the non-syndromic retinal ciliopathy genes, the efficacy of gene replacement therapy has been evaluated in an *Rpgrip1* mutant mouse model (*tm1Tili*) (Zhao et al. 2003). Subretinal delivery of the human replacement construct, packaged in an adeno-associated virus (AAV) serotype 8 vector, results in the expression of functional RPGRIP1 in connecting cilia of photoreceptors, better preservation of rod and cone photoreceptor function and prolongs photoreceptor survival (Pawlyk et al. 2010). AAV-based gene therapies for large genes, such as *CEP290*, which encodes a 2479 amino acid protein and is the most frequently mutated LCA gene, are problematic given the size restriction for the AAV vector (~5 kb) (Wu et al. 2010). In such cases, an alternative mini-gene augmentation therapy, which uses only part of the gene important for gene function in the photoreceptor, would still fulfil the AAV size limit criteria. A mini-gene augmentation approach for *CEP290* has recently been successful in zebrafish. Expression of only the N-terminal 1059 amino acids of CEP290 resulted in a rescue of visual impairment in CEP290 morpholino-injected fish (Baye et al. 2011). Additional studies are required to determine the efficiency of mini-gene replacement therapies in higher vertebrate models.

Genetic manipulation

Promising alternatives for gene augmentation in retinal dystrophies are therapeutic approaches using either modified U1 small nuclear (sn)RNAs or antisense oligonucleotides to correct mutation-induced splicing defects. Lentiviral treatment of fibroblasts derived from individuals with RP who carry a c.479G>A splice donor site mutation in exon 5 of *BBS1* partially corrected aberrant splicing of endogenously expressed *BBS1* transcripts in a dose-dependent manner, indicating that U1 snRNAs can correct pathogenic effects of splice donor site mutations (Schmid et al. 2011). Similar results were obtained for X-linked RP, for which splice defects in *RPGR* have been corrected in patient-derived primary fibroblasts, using U1 snRNAs (Glaus et al. 2011). A recent antisense oligonucleotide-based approach for the most frequent LCA-causative mutation (c.2991 + 1655A.G), which resulted in the inclusion of an aberrant exon in *CEP290* mRNA, was successful in almost completely redirecting normal CEP290 splicing using immortalised lymphoblastoid cells of individuals carrying two intronic *CEP290* mutations (Collin et al. 2012b).

Pharmacological therapies

Pharmacological therapy may offer an early alternative to slow or halt the course of photoreceptor degeneration in ciliary retinopathies. Administration of ciliary neurotrophic factor through an encapsulated cell technology device was effective in slowing photoreceptor degeneration in animal models of RP, and is currently being tested in clinical trials for RP (Wen et al. 2012). Tauroursodeoxycholic acid given subcutaneously was recently shown to preserve photoreceptors in a *Bbs1* mutant knock-in mouse (Drack et al. 2012).

Just a few years ago, most cases of inherited blindness associated with both syndromic and non-syndromic ciliopathies were considered untreatable. In a very short space of time, a number of promising therapies are undergoing evaluation for restoration of visual loss bringing hope to millions worldwide.

References

Abu-Safieh, L., Al-Anazi, S., Al-Abdi, L., Hashem, M., Alkuraya, H., Alamr, M., et al. 2012. In search of triallelism in Bardet–Biedl syndrome. *Eur J Hum Genet* **20**, 420–427.

Aherne, A., Kennan, A., Kenna, P. F., Mcnally, N., Lloyd, D. G., Alberts, I. L., et al. 2004. On the molecular pathology of neurodegeneration in IMPDH1-based retinitis pigmentosa. *Hum Mol Genet* **13**, 641–650.

Aldahmesh, M. A., Al-Owain, M., Alqahtani, F., Hazzaa, S. & Alkuraya F. S. 2010. A null mutation in CABP4 causes Leber's congenital amaurosis-like phenotype. *Mol Vis* **16**, 207–212. PubMed PMID: 20157620; PubMed Central PMCID: PMC2820108.

Andersen, J. S., Wilkinson, C. J., Mayor, T., Mortensen, P., Nigg, E. A. & Mann, M. 2003. Proteomic characterization of the human centrosome by protein correlation profiling. *Nature* **426**, 570–574.

Bainbridge, J. W., Smith, A. J., Barker, S. S., Robbie, S., Henderson, R., Balaggan, K., et al. 2008. Effect of gene therapy on visual function in Leber's congenital amaurosis. *New Engl J Med* **358**, 2231–2239.

Baye, L. M., Patrinostro, X., Swaminathan, S., Beck, J. S., Zhang, Y., Stone, E. M., et al. 2011. The N-terminal region of centrosomal protein 290 (CEP290) restores vision in a zebrafish model of human blindness. *Hum Mol Genet* **20**, 1467–1477.

Boggon, T. J., Shan, W. S., Santagata, S., Myers, S. C. & Shapiro, L. 1999. Implication of tubby proteins as transcription factors by structure-based functional analysis. *Science* **286**, 2119–2125.

Boldt, K., Mans, D. A., Won, J., Van Reeuwijk, J., Vogt, A., Kinkl, N., et al. 2011. Disruption of intraflagellar protein transport in photoreceptor cilia causes Leber congenital amaurosis in humans and mice. *J Clin Invest* **121**, 2169–2180.

Booij, J. C., Florijn, R. J., Ten Brink, J. B., Loves, W., Meire, F., Van Schooneveld, M. J., et al. 2005. Identification of mutations in the AIPL1, CRB1, GUCY2D, RPE65, and RPGRIP1 genes in patients with juvenile retinitis pigmentosa. *J Med Genet* **42**, e67.

Bowne, S. J., Sullivan, L. S., Mortimer, S. E., Hedstrom, L., Zhu, J., Spellicy, C. J., et al. 2006. Spectrum and frequency of mutations in IMPDH1 associated with autosomal dominant retinitis pigmentosa and Leber congenital amaurosis. *Invest. Ophthalmol Vis Sci* **47**, 34–42.

Boylan, J. P. & Wright, A. F. 2000. Identification of a novel protein interacting with RPGR. *Hum Mol Genet* **9**, 2085–2093.

Brecelj, J. & Stirn-Kranjc, B.1999. ERG and VEP follow-up study in children with Leber's congenital amaurosis. *Eye (London)* **13**(Pt 1), 47–54.

Carroll, K., Gomez, C. & Shapiro, L. 2004. Tubby proteins: The plot thickens. *Nat Rev Mol Cell Biol* **5**, 55–63.

Chang, B., Khanna, H., Hawes, N., Jimeno, D., He, S., Lillo, C., et al. 2006. In-frame deletion in a novel centrosomal/ciliary protein CEP290/NPHP6 perturbs its interaction with RPGR and results in early-onset retinal degeneration in the rd16 mouse. *Hum Mol Genet* **15**, 1847–1857.

Chen, S., Wang, Q. L., Nie, Z., Sun, H., Lennon, G., Copeland, N. G., et al. 1997. Crx, a novel Otx-like paired-homeodomain protein, binds to and transactivates photoreceptor cell-specific genes. *Neuron* **19**, 1017–1030.

Chiang, P. W., Wang, J., Chen, Y., Fu, Q., Zhong, J., Chen, Y., et al. 2012. Exome sequencing identifies NMNAT1 mutations as a cause of Leber congenital amaurosis. *Nat Genet* **44**, 972–974.

Chow, R. L. & Lang, R. A. 2001. Early eye development in vertebrates. *Annu Rev Cell Dev Biol* **17**, 255–296.

Cideciyan, A. V., Hauswirth, W. W., Aleman, T. S., Kaushal, S., Schwartz, S. B., Boye, S. L., et al. 2009. Vision 1 year after gene therapy for Leber's congenital amaurosis. *New Engl J Med* **361**, 725–727.

Cideciyan, A. V., Jacobson, S. G., Beltran, W. A., Sumaroka, A., Swider, M., Iwabe, S., et al. 2013. Human retinal gene therapy for Leber congenital amaurosis shows advancing retinal degeneration despite enduring visual improvement. *Proc Natl Acad Sci U.S.A.* **110**, E517–E525.

Collin, G. B., Marshall, J. D., Ikeda, A., So, W. V., Russell-Eggitt, I., Maffei, P., et al. 2002. Mutations in ALMS1 cause obesity, type 2 diabetes and neurosensory degeneration in Alstrom syndrome. *Nat Genet* **31**, 74–78.

Collin, G. B., Marshall, J. D., King, B. L., Milan, G., Maffei, P., Jagger, D. J., et al. 2012a. The Alstrom syndrome protein, ALMS1, interacts with alpha-actinin and components of the endosome recycling pathway. *PLoS One* **7**, e37925.

Collin, R. W., Den Hollander, A. I., Van Der Velde-Visser, S. D., Bennicelli, J., Bennett, J. & Cremers, F. P. 2012b. Antisense oligonucleotide (AON)-based therapy for Leber congenital amaurosis caused by a frequent mutation in CEP290. *Mol Ther, Nucleic Acids* **1**, e14.

Dagi, L. R., Leys, M. J., Hansen, R. M. & Fulton, A. B. 1990. Hyperopia in complicated Leber's congenital amaurosis. *Arch Ophthalmol* **108**, 709–712.

Daiger, S. P., Bowne, S. J. & Sullivan, L. S.2007. Perspective on genes and mutations causing retinitis pigmentosa. *Arch Ophthalmol* **125**, 151–158.

Den Hollander, A. I., Black, A., Bennett, J. & Cremers, F. P. 2010. Lighting a candle in the dark: Advances in genetics and gene therapy of recessive retinal dystrophies. *J Clin Invest* **120**, 3042–3053.

Den Hollander, A. I., Koenekoop, R. K., Mohamed, M. D., Arts, H. H., Boldt, K., Towns, K. V., et al. 2007. Mutations in LCA5, encoding the ciliary protein lebercilin, cause Leber congenital amaurosis. *Nat Genet* **39**, 889–895.

Den Hollander, A. I., Koenekoop, R. K., Yzer, S., Lopez, I., Arends, M. L., Voesenek, K. E., et al. 2006. Mutations in the CEP290 (NPHP6) gene are a frequent cause of Leber congenital amaurosis. *Am J Hum Genet* **79**, 556–561.

Den Hollander, A. I., Roepman, R., Koenekoop, R. K. & Cremers, F. P. 2008. Leber congenital amaurosis: Genes, proteins and disease mechanisms. *Prog Retin Eye Res* **27**, 391–419.

Den Hollander, A. I., Ten Brink, J. B., De Kok, Y. J., Van Soest, S., Van Den Born, L. I., Van Driel, M. A., et al. 1999. Mutations in a human homologue of *Drosophila* crumbs cause retinitis pigmentosa (RP12). *Nat Genet* **23**, 217–221.

Dharmaraj, S., Leroy, B. P., Sohocki, M. M., Koenekoop, R. K., Perrault, I., Anwar, K., et al. 2004. The phenotype of Leber congenital amaurosis in patients with AIPL1 mutations. *Arch Ophthalmol* **122**, 1029–1037.

Dharmaraj, S. R., Silva, E. R., Pina, A. L., Li, Y. Y., Yang, J. M., Carter, C. R., et al. 2000. Mutational analysis and clinical correlation in Leber congenital amaurosis. *Ophthalmic Genet* **21**, 135–150.

Drack, A. V., Dumitrescu, A. V., Bhattarai, S., Gratie, D., Stone, E. M., Mullins, R., et al. 2012. TUDCA slows retinal degeneration in two different mouse models of retinitis pigmentosa and prevents obesity in Bardet–Biedl syndrome type 1 mice. *Invest Ophthalmol Vis Sci* **53**, 100–106.

Estrada-Cuzcano, A., Koenekoop, R. K., Coppieters, F., Kohl, S., Lopez, I., Collin, R. W., et al. 2011. IQCB1 mutations in patients with leber congenital amaurosis. *Invest Ophthalmol Vis Sci* **52**, 834–839.

Estrada-Cuzcano, A., Roepman, R., Cremers, F. P., Den Hollander, A. I. & Mans, D. A. 2012. Non-syndromic retinal ciliopathies: Translating gene discovery into therapy. *Hum Mol Genet* **21**, R111–124.

Falk, M. J., Zhang, Q., Nakamaru-Ogiso, E., Kannabiran, C., Fonseca-Kelly, Z., Chakarova, C., et al. 2012. NMNAT1 mutations cause Leber congenital amaurosis. *Nat Genet* **44**, 1040–1045.

Franceschetti, A. & Dieterle, P.1954. Diagnostic and prognostic importance of the electroretinogram in tapetoretinal degeneration with reduction of the visual field and hemeralopia. *Confin Neurol* **14**, 184–186.

Freund, C. L., Gregory-Evans, C. Y., Furukawa, T., Papaioannou, M., Looser, J., Ploder, L., et al. 1997. Cone–rod dystrophy due to mutations in a novel photoreceptor-specific homeobox gene (CRX) essential for maintenance of the photoreceptor. *Cell* **91**, 543–553.

Freund, C. L., Wang, Q. L., Chen, S., Muskat, B. L., Wiles, C. D., Sheffield, V. C., et al. 1998. *De novo* mutations in the CRX homeobox gene associated with Leber congenital amaurosis. *Nat Genet* **18**, 311–312.

Fulton, A. B., Hansen, R. M. & Mayer, D. L. 1996. Vision in Leber congenital amaurosis. *Arch Ophthalmol* **114**, 698–703.

Furukawa, T., Morrow, E. M. & Cepko, C. L. 1997. Crx, a novel otx-like homeobox gene, shows photoreceptor-specific expression and regulates photoreceptor differentiation. *Cell* **91**, 531–541.

Furukawa, T., Morrow, E. M., Li, T., Davis, F. C. & Cepko, C. L. 1999. Retinopathy and attenuated circadian entrainment in Crx-deficient mice. *Nat Genet* **23**, 466–470.

Gal, A., Li, Y., Thompson, D. A., Weir, J., Orth, U., Jacobson, S. G., et al. 2000. Mutations in MERTK, the human orthologue of the RCS rat retinal dystrophy gene, cause retinitis pigmentosa. *Nat Genet* **26**, 270–271.

Galvin, J. A., Fishman, G. A., Stone, E. M. & Koenekoop, R. K.2005. Evaluation of genotype–phenotype associations in leber congenital amaurosis. *Retina* **25**, 919–929.

Gherman, A., Davis, E. E. & Katsanis, N.2006. The ciliary proteome database: An integrated community resource for the genetic and functional dissection of cilia. *Nat Genet* **38**, 961–962.

Glaus, E., Schmid, F., Da Costa, R., Berger, W. & Neidhardt, J. 2011. Gene therapeutic approach using mutation-adapted U1 snRNA to correct a RPGR splice defect in patient-derived cells. *Mol Ther* **19**, 936–941.

Haeseleer, F., Imanishi, Y., Maeda, T., Possin, D. E., Maeda, A., Lee, A., et al. 2004. Essential role of Ca^{2+} -binding protein 4, a Cav1.4 channel regulator, in photoreceptor synaptic function. *Nat Neurosci* **7**, 1079–1087.

Haeseleer, F., Jang, G. F., Imanishi, Y., Driessen, C. A., Matsumura, M., Nelson, P. S., et al. 2002. Dual-substrate specificity short chain retinol dehydrogenases from the vertebrate retina. *J Biol Chem* **277**, 45537–45546.

Hagstrom, S. A., Adamian, M., Scimeca, M., Pawlyk, B. S., Yue, G. & Li, T. 2001. A role for the Tubby-like protein 1 in rhodopsin transport. *Invest Ophthalmol Vis Sci* **42**, 1955–1962.

Hagstrom, S. A., Duyao, M., North, M. A. & Li, T. 1999. Retinal degeneration in tulp1−/− mice: Vesicular accumulation in the interphotoreceptor matrix. *Invest Ophthalmol Vis Sci* **40**, 2795–2802.

Haim, M. 2002. Epidemiology of retinitis pigmentosa in Denmark. *Acta Ophthalmol Scand, Supplement* **233**, 1–34.

Hamel, C. P. 2007. Cone rod dystrophies. *Orphanet J Rare Dis* **2**, 7.

Hanein, S., Perrault, I., Gerber, S., Tanguy, G., Barbet, F., Ducroq, D., et al. 2004. Leber congenital amaurosis: Comprehensive survey of the genetic heterogeneity, refinement of the clinical definition, and genotype–phenotype correlations as a strategy for molecular diagnosis. *Hum Mutat* **23**, 306–317.

Hanein, S., Perrault, I., Olsen, P., Lopponen, T., Hietala, M., Gerber, S., et al. 2002. Evidence of a founder effect for the RETGC1 (GUCY2D) 2943DelG mutation in Leber congenital amaurosis pedigrees of Finnish origin. *Hum Mutat* **20**, 322–323.

Hasson, T., Heintzelman, M. B., Santos-Sacchi, J., Corey, D. P. & Mooseker, M. S. 1995. Expression in cochlea and retina of myosin VIIa, the gene product defective in Usher syndrome type 1B. *Proc Natl Acad Sci U.S.A.* **92**, 9815–9819.

Heher, K. L., Traboulsi, E. I. & Maumenee, I. H.1992. The natural history of Leber's congenital amaurosis. Age-related findings in 35 patients. *Ophthalmology* **99**, 241–245.

Henderson, R. H., Waseem, N., Searle, R., Van Der Spuy, J., Russell-Eggitt, I., Bhattacharya, S. S., et al. 2007. An assessment of the apex microarray technology in genotyping patients with Leber congenital amaurosis and early-onset severe retinal dystrophy. *Invest Ophthalmol Vis Sci* **48**, 5684–5689.

Henderson, R. H., Williamson, K. A., Kennedy, J. S., Webster, A. R., Holder, G. E., Robson, A. G., et al. 2009. A rare *de novo* nonsense mutation in OTX2 causes early onset retinal dystrophy and pituitary dysfunction. *Mol Vis* **15**, 2442–2447.

Hong, D. H., Yue, G., Adamian, M. & Li, T. 2001. Retinitis pigmentosa GTPase regulator (RPGRr)-interacting protein is stably associated with the photoreceptor ciliary axoneme and anchors RPGR to the connecting cilium. *J Biol Chem* **276**, 12091–12099.

Ikeda, A., Naggert, J. K. & Nishina, P. M. 2002a. Genetic modification of retinal degeneration in tubby mice. *Exp Eye Res* **74**, 455–461.

Ikeda, A., Nishina, P. M. & Naggert, J. K. 2002b. The tubby-like proteins, a family with roles in neuronal development and function. *J Cell Sci* **115**, 9–14.

Ikeda, S., He, W., Ikeda, A., Naggert, J. K., North, M. A. & Nishina, P. M. 1999. Cell-specific expression of tubby gene family members (tub, Tulp1,2, and 3) in the retina. *Invest Ophthalmol Vis Sci* **40**, 2706–2712.

Ikeda, S., Shiva, N., Ikeda, A., Smith, R. S., Nusinowitz, S., Yan, G., et al. 2000. Retinal degeneration but not obesity is observed in null mutants of the tubby-like protein 1 gene. *Hum Mol Genet* **9**, 155–163.

Imanishi, Y., Batten, M. L., Piston, D. W., Baehr, W. & Palczewski, K. 2004a. Noninvasive two-photon imaging reveals retinyl ester storage structures in the eye. *J Cell Biol* **164**, 373–383.

Imanishi, Y., Gerke, V. & Palczewski, K. 2004b. Retinosomes: New insights into intracellular managing of hydrophobic substances in lipid bodies. *J Cell Biol* **166**, 447–453.

Jacobson, S. G., Aleman, T. S., Cideciyan, A. V., Heon, E., Golczak, M., Beltran, W. A., et al. 2007. Human cone photoreceptor dependence on RPE65 isomerase. *Proc Natl Acad Sci U.S.A.* **104**, 15123–15128.

Janecke, A. R., Thompson, D. A., Utermann, G., Becker, C., Hubner, C. A., Schmid, E., et al. 2004. Mutations in RDH12 encoding a photoreceptor cell retinol dehydrogenase cause childhood-onset severe retinal dystrophy. *Nat Genet* **36**, 850–854.

Jin, M., Li, S., Moghrabi, W. N., Sun, H. & Travis, G. H. 2005. Rpe65 is the retinoid isomerase in bovine retinal pigment epithelium. *Cell* **122**, 449–459.

Kantardzhieva, A., Alexeeva, S., Versteeg, I. & Wijnholds, J.2006. MPP3 is recruited to the MPP5 protein scaffold at the retinal outer limiting membrane. *FEBS J* **273**, 1152–1165.

Kantardzhieva, A., Gosens, I., Alexeeva, S., Punte, I. M., Versteeg, I., Krieger, E., et al. 2005. MPP5 recruits MPP4 to the CRB1 complex in photoreceptors. *Invest Ophthalmol Vis Sci* **46**, 2192–2201.

Khanna, H., Hurd, T. W., Lillo, C., Shu, X., Parapuram, S. K., He, S., et al. 2005. RPGR-ORF15, which is mutated in retinitis pigmentosa, associates with SMC1, SMC3, and microtubule transport proteins. *J Biol Chem* **280**, 33580–33587.

Kim, J. C., Badano, J. L., Sibold, S., Esmail, M. A., Hill, J., Hoskins, B. E., et al. 2004. The Bardet–Biedl protein BBS4 targets cargo to the pericentriolar region and is required for microtubule anchoring and cell cycle progression. *Nat Genet* **36**, 462–470.

Kimura, A., Singh, D., Wawrousek, E. F., Kikuchi, M., Nakamura, M. & Shinohara, T. 2000. Both PCE-1/RX and OTX/CRX interactions are necessary for photoreceptor-specific gene expression. *J Biol Chem* **275**, 1152–1160.

Koenekoop, R. K.2004. An overview of Leber congenital amaurosis: A model to understand human retinal development. *Surv Ophthalmol* **49**, 379–398.

Koenekoop, R. K., Lopez, I., Den Hollander, A. I., Allikmets, R. & Cremers, F. P. 2007. Genetic testing for retinal dystrophies and dysfunctions: Benefits, dilemmas and solutions. *Clin Experiment Ophthalmol* **35**, 473–485.

Koenekoop, R. K., Loyer, M., Dembinska, O. & Beneish, R. 2002. Visual improvement in Leber congenital amaurosis and the CRX genotype. *Ophthalmic Genet* **23**, 49–59.

Koenekoop, R. K., Wang, H., Majewski, J., Wang, X., Lopez, I., Ren, H., et al. 2012. Mutations in NMNAT1 cause Leber congenital amaurosis and identify a new disease pathway for retinal degeneration. *Nat Genet* **44**, 1035–1039.

Kohl, S., Marx, T., Giddings, I., Jagle, H., Jacobson, S. G., Apfelstedt-Sylla, E., et al. 1998. Total colourblindness is caused by mutations in the gene encoding the alpha-subunit of the cone photoreceptor cGMP-gated cation channel. *Nat Genet* **19**, 257–259.

Kurth, I., Thompson, D. A., Ruther, K., Feathers, K. L., Chrispell, J. D., Schroth, J., et al. 2007. Targeted disruption of the murine retinal dehydrogenase gene Rdh12 does not limit visual cycle function. *Mol Cell Biol* **27**, 1370–1379.

Leber, T. 1869. Uber Retinitis Pigmentosa und angeborene Amaurose. *von Graefe's Arch Ophthalmol* **15**, 1–25.

Lorenz, B., Gyurus, P., Preising, M., Bremser, D., Gu, S., Andrassi, M., et al. 2000. Early-onset severe rod–cone dystrophy in young children with RPE65 mutations. *Invest Ophthalmol Vis Sci* **41**, 2735–2742.

Lotery, A. J., Namperumalsamy, P., Jacobson, S. G., Weleber, R. G., Fishman, G. A., Musarella, M. A., et al. 2000. Mutation analysis of 3 genes in patients with Leber congenital amaurosis. *Arch Ophthalmol* **118**, 538–543.

Lu, X., Guruju, M., Oswald, J. & Ferreira, P. A. 2005. Limited proteolysis differentially modulates the stability and subcellular localization of domains of RPGRIP1 that are distinctly affected by mutations in Leber's congenital amaurosis. *Hum Mol Genet* **14**, 1327–1340.

Maguire, A. M., High, K. A., Auricchio, A., Wright, J. F., Pierce, E. A., Testa, F., et al. 2009. Age-dependent effects of RPE65 gene therapy for Leber's congenital amaurosis: A phase 1 dose-escalation trial. *Lancet* **374**, 1597–1605.

Maguire, A. M., Simonelli, F., Pierce, E. A., Pugh, E. N. Jr.**Mingozzi, F., Bennicelli, J., et al.** 2008. Safety and efficacy of gene transfer for Leber's congenital amaurosis. *New Engl J Med* **358**, 2240–2248.

Mcbee, J. K., Van Hooser, J. P., Jang, G. F. & Palczewski, K.2001. Isomerization of 11-*cis*-retinoids to all-*trans*-retinoids *in vitro* and *in vivo*. *J Biol Chem* **276**, 48483–48493.

Mchenry, C. L., Liu, Y., Feng, W., Nair, A. R., Feathers, K. L., Ding, X., et al. 2004. MERTK arginine-844-cysteine in a patient with severe rod–cone dystrophy: Loss of mutant protein function in transfected cells. *Invest Ophthalmol Vis Sci* **45**, 1456–1463.

Mclean, J. E., Hamaguchi, N., Belenky, P., Mortimer, S. E., Stanton, M. & Hedstrom, L. 2004. Inosine 5′-monophosphate dehydrogenase binds nucleic acids *in vitro* and *in vivo*. *Biochem J* **379**, 243–251.

Michaelides, M., Hardcastle, A. J., Hunt, D. M. & Moore, A. T. 2006. Progressive cone and cone–rod dystrophies: Phenotypes and underlying molecular genetic basis. *Surv Ophthalmol* **51**, 232–258.

Milam, A. H., Hendrickson, A. E., Xiao, M., Smith, J. E., Possin, D. E., John, S. K., et al. 2000. Localization of tubby-like protein 1 in developing and adult human retinas. *Invest Ophthalmol Vis Sci* **41**, 2352–2356.

Mitton, K. P., Swain, P. K., Chen, S., Xu, S., Zack, D. J. & Swaroop, A. 2000. The leucine zipper of NRL interacts with the CRX homeodomain. A possible mechanism of transcriptional synergy in rhodopsin regulation. *J Biol Chem* **275**, 29794–29799.

Moiseyev, G., Takahashi, Y., Chen, Y., Gentleman, S., Redmond, T. M., Crouch, R. K., et al. 2006. RPE65 is an iron(II)-dependent isomerohydrolase in the retinoid visual cycle. *J Biol Chem* **281**, 2835–2840.

Morimura, H., Fishman, G. A., Grover, S. A., Fulton, A. B., Berson, E. L. & Dryja, T. P. 1998. Mutations in the RPE65 gene in patients with autosomal recessive retinitis pigmentosa or leber congenital amaurosis. *Proc Natl Acad Sci U.S.A.* **95**, 3088–3093.

Mortimer, S. E. & Hedstrom, L.2005. Autosomal dominant retinitis pigmentosa mutations in inosine 5′-monophosphate dehydrogenase type I disrupt nucleic acid binding. *Biochem J* **390**, 41–47.

Mykytyn, K., Braun, T., Carmi, R., Haider, N. B., Searby, C. C., Shastri, M., et al. 2001. Identification of the gene that, when mutated, causes the human obesity syndrome BBS4. *Nat Genet* **28**, 188–191.

Nishida, A., Furukawa, A., Koike, C., Tano, Y., Aizawa, S., Matsuo, I., et al. 2003. Otx2 homeobox gene controls retinal photoreceptor cell fate and pineal gland development. *Nat Neurosci* **6**, 1255–1263.

North, M. A., Naggert, J. K., Yan, Y., Noben-Trauth, K. & Nishina, P. M. 1997. Molecular characterization of TUB, TULP1, and TULP2, members of the novel tubby gene family and their possible relation to ocular diseases. *Proc Natl Acad Sci U.S.A.* **94**, 3128–3133.

Otto, E. A., Loeys, B., Khanna, H., Hellemans, J., Sudbrak, R., Fan, S., et al. 2005. Nephrocystin-5, a ciliary IQ domain protein, is mutated in Senior–Loken syndrome and interacts with RPGR and calmodulin. *Nat Genet* **37**, 282–288.

Pawlyk, B. S., Bulgakov, O. V., Liu, X., Xu, X., Adamian, M., Sun, X., et al. 2010. Replacement gene therapy with a human RPGRIP1 sequence slows photoreceptor degeneration in a murine model of Leber congenital amaurosis. *Hum Gene Ther* **21**, 993–1004.

Peng, G. H. & Chen, S. 2007. Crx activates opsin transcription by recruiting HAT-containing co-activators and promoting histone acetylation. *Hum Mol Genet* **16**, 2433–2452.

Perrault, I., Delphin, N., Hanein, S., Gerber, S., Dufier, J. L., Roche, O., et al. 2007. Spectrum of NPHP6/CEP290 mutations in Leber congenital amaurosis and delineation of the associated phenotype. *Hum Mutat* **28**, 416.

Perrault, I., Hanein, S., Zanlonghi, X., Serre, V., Nicouleau, M., Defoort-Delhemmes, S., et al. 2012. Mutations in NMNAT1 cause Leber congenital amaurosis with early-onset severe macular and optic atrophy. *Nat Genet* **44**, 975–977.

Perrault, I., Rozet, J. M., Ghazi, I., Leowski, C., Bonnemaison, M., Gerber, S., et al. 1999. Different functional outcome of RetGC1 and RPE65 gene mutations in Leber congenital amaurosis. *Am J Hum Genet* **64**, 1225–1228.

Redmond, T. M., Poliakov, E., Yu, S., Tsai, J. Y., Lu, Z. & Gentleman, S. 2005. Mutation of key residues of RPE65 abolishes its enzymatic role as isomerohydrolase in the visual cycle. *Proc Natl Acad Sci U.S.A.* **102**, 13658–13663.

Riazuddin, S. A., Iqbal, M., Wang, Y., Masuda, T., Chen, Y., Bowne, S., et al. 2010. A splice-site mutation in a retina-specific exon of BBS8 causes nonsyndromic retinitis pigmentosa. *Am J Hum Genet* **86**, 805–812.

Richard, M., Roepman, R., Aartsen, W. M., Van Rossum, A. G., Den Hollander, A. I., Knust, E., et al. 2006. Towards understanding CRUMBS function in retinal dystrophies. *Hum Mol Genet* **15 (Spec No 2)**, R235–243.

Rivolta, C., Sweklo, E. A., Berson, E. L. & Dryja, T. P. 2000. Missense mutation in the USH2A gene: Association with recessive retinitis pigmentosa without hearing loss. *Am J Hum Genet* **66**, 1975–1978.

Roepman, R., Bernoud-Hubac, N., Schick, D. E., Maugeri, A., Berger, W., Ropers, H. H., et al. 2000. The retinitis pigmentosa GTPase regulator (RPGR) interacts with novel transport-like proteins in the outer segments of rod photoreceptors. *Hum Mol Genet* **9**, 2095–2105.

Roepman, R., Letteboer, S. J., Arts, H. H., Van Beersum, S. E., Lu, X., Krieger, E., et al. 2005. Interaction of nephrocystin-4 and RPGRIP1 is disrupted by nephronophthisis or Leber congenital amaurosis-associated mutations. *Proc Natl Acad Sci U.S.A.* **102**, 18520–18525.

Rozet, J. M., Perrault, I., Gerber, S., Hanein, S., Barbet, F., Ducroq, D., et al. 2001. Complete abolition of the retinal-specific guanylyl cyclase (retGC-1) catalytic ability consistently leads to leber congenital amaurosis (LCA). *Invest Ophthalmol Vis Sci* **42**, 1190–1192.

Sayer, J. A., Otto, E. A., O'Toole, J. F., Nurnberg, G., Kennedy, M. A., Becker, C., et al. 2006. The centrosomal protein nephrocystin-6 is mutated in Joubert syndrome and activates transcription factor ATF4. *Nat Genet* **38**, 674–681.

Schmid, F., Glaus, E., Barthelmes, D., Fliegauf, M., Gaspar, H., Nurnberg, G., et al. 2011. U1 snRNA-mediated gene therapeutic correction of splice defects caused by an exceptionally mild BBS mutation. *Hum Mutat* **32**, 815–824.

Sergouniotis, P. I., Davidson, A. E., Mackay, D. S., Li, Z., Yang, X., Plagnol, V., et al. 2011. Recessive mutations in KCNJ13, encoding an inwardly rectifying potassium channel subunit, cause leber congenital amaurosis. *Am J Hum Genet* **89**, 183–190.

Simonelli, F., Maguire, A. M., Testa, F., Pierce, E. A., Mingozzi, F., Bennicelli, J. L., et al. 2010. Gene therapy for Leber's congenital amaurosis is safe and effective through 1.5 years after vector administration. *Mol Ther* **18**, 643–650.

Simonelli, F., Ziviello, C., Testa, F., Rossi, S., Fazzi, E., Bianchi, P. E., et al. 2007. Clinical and molecular genetics of Leber's congenital amaurosis: A multicenter study of Italian patients. *Invest Ophthalmol Vis Sci* **48**, 4284–4290.

Sitorus, R. S., Lorenz, B. & Preising, M. N. 2003. Analysis of three genes in Leber congenital amaurosis in Indonesian patients. *Vision Res* **43**, 3087–3093.

Sohocki, M. M., Bowne, S. J., Sullivan, L. S., Blackshaw, S., Cepko, C. L., Payne, A. M., et al. 2000a. Mutations in a new photoreceptor–pineal gene on 17p cause Leber congenital amaurosis. *Nat Genet* **24**, 79–83.

Sohocki, M. M., Perrault, I., Leroy, B. P., Payne, A. M., Dharmaraj, S., Bhattacharya, S. S., et al. 2000b. Prevalence of AIPL1 mutations in inherited retinal degenerative disease. *Mol Genet Metab* **70**, 142–150.

Sohocki, M. M., Sullivan, L. S., Mintz-Hittner, H. A., Birch, D., Heckenlively, J. R., Freund, C. L., et al. 1998. A range of clinical phenotypes associated with mutations in CRX, a photoreceptor transcription-factor gene. *Am J Hum Genet* **63**, 1307–1315.

Stone, E. M. 2007. Leber congenital amaurosis—a model for efficient genetic testing of heterogeneous disorders: LXIV Edward Jackson Memorial Lecture. *Am J Ophthalmol* **144**, 791–811.

Thiadens, A. A., Soerjoesing, G. G., Florijn, R. J., Tjiam, A. G., Den Hollander, A. I., Van Den Born, L. I., et al. 2011. Clinical course of cone dystrophy caused by mutations in the RPGR gene. *Graefe's Archive for Clinical and Experimental Ophthalmology* **249**, 1527–1535.

Thompson, D. A., Gyurus, P., Fleischer, L. L., Bingham, E. L., Mchenry, C. L., Apfelstedt-Sylla, E., et al. 2000. Genetics and phenotypes of RPE65 mutations in inherited retinal degeneration. *Invest Ophthalmol Vis Sci* **41**, 4293–4299.

Tschernutter, M., Jenkins, S. A., Waseem, N. H., Saihan, Z., Holder, G. E., Bird, A. C., et al. 2006. Clinical characterisation of a family with retinal dystrophy caused by mutation in the Mertk gene. *Br J Ophthalmol* **90**, 718–723.

Valente, E. M., Silhavy, J. L., Brancati, F., Barrano, G., Krishnaswami, S. R., Castori, M., et al. 2006. Mutations in CEP290, which encodes a centrosomal protein, cause pleiotropic forms of Joubert syndrome. *Nat Genet* **38**, 623–625.

Vallespin, E., Lopez-Martinez, M. A., Cantalapiedra, D., Riveiro-Alvarez, R., Aguirre-Lamban, J., Avila-Fernandez, A., Villaverde, C., et al. 2007. Frequency of CEP290 c.2991_1655A>G mutation in 175 Spanish families affected with Leber congenital amaurosis and early-onset retinitis pigmentosa. *Mol Vis* **13**, 2160–2162.

Van De Pavert, S. A., Kantardzhieva, A., Malysheva, A., Meuleman, J., Versteeg, I., Levelt, C., et al. 2004. Crumbs homologue 1 is required for maintenance of photoreceptor cell polarization and adhesion during light exposure. *J Cell Sci* **117**, 4169–4177.

Van Der Spuy, G. D., Warren, R. M., Richardson, M., Beyers, N., Behr, M. A. & Van Helden, P. D. 2003. Use of genetic distance as a measure of ongoing transmission of Mycobacterium tuberculosis. *J Clin Microbiol* **41**, 5640–5644.

Van Der Spuy, J., Chapple, J. P., Clark, B. J., Luthert, P. J., Sethi, C. S. & Cheetham, M. E. 2002. The Leber congenital amaurosis gene product AIPL1 is localized exclusively in rod photoreceptors of the adult human retina. *Hum Mol Genet* **11**, 823–831.

Wagner, R. S., Caputo, A. R., Nelson, L. B. & Zanoni, D. 1985. High hyperopia in Leber's congenital amaurosis. *Arch Ophthalmol* **103**, 1507–1509.

Wang, X., Wang, H., Cao, M., Li, Z., Chen, X., Patenia, C., et al. 2011. Whole-exome sequencing identifies ALMS1, IQCB1, CNGA3, and MYO7A mutations in patients with Leber congenital amaurosis. *Hum Mutat* **32**, 1450–1459.

Webb, T. R., Parfitt, D. A., Gardner, J. C., Martinez, A., Bevilacqua, D., Davidson, A. E., et al. 2012. Deep intronic mutation in OFD1, identified by targeted genomic next-generation sequencing, causes a severe form of X-linked retinitis pigmentosa (RP23). *Hum Mol Genet* **21**, 3647–3654.

Wen, R., Tao, W., Li, Y. & Sieving, P. A. 2012. CNTF and retina. *Prog Retin Eye Res* **31**, 136–151.

Wu, Z., Yang, H. & Colosi, P. 2010. Effect of genome size on AAV vector packaging. *Mol Ther* **18**, 80–86.

Xi, Q., Pauer, G. J., Marmorstein, A. D., Crabb, J. W. & Hagstrom, S. A. 2005. Tubby-like protein 1 (TULP1) interacts with F-actin in photoreceptor cells. *Invest Ophthalmol Vis Sci* **46**, 4754–4761.

Yzer, S., Leroy, B. P., De Baere, E., De Ravel, T. J., Zonneveld, M. N., Voesenek, K., et al. 2006. Microarray-based mutation detection and phenotypic characterization of patients with Leber congenital amaurosis. *Invest Ophthalmol Vis Sci* **47**, 1167–1176.

Yzer, S., Van Den Born, L. I., Schuil, J., Kroes, H. Y., Van Genderen, M. M., Boonstra, F. N., et al. 2003. A Tyr368His RPE65 founder mutation is associated with variable expression and progression of early onset retinal dystrophy in 10 families of a genetically isolated population. *J Med Genet* **40**, 709–713.

Zernant, J., Kulm, M., Dharmaraj, S., Den Hollander, A. I., Perrault, I., Preising, M. N., et al. 2005. Genotyping microarray (disease chip) for Leber congenital amaurosis: Detection of modifier alleles. *Invest Ophthalmol Vis Sci* **46**, 3052–3059.

Zhao, Y., Hong, D. H., Pawlyk, B., Yue, G., Adamian, M., Grynberg, M., et al. 2003. The retinitis pigmentosa GTPase regulator (RPGR)- interacting protein: Subserving RPGR function and participating in disk morphogenesis. *Proc Natl Acad Sci U.S.A.* **100**, 3965–3970.

Chapter 7

Meckel–Gruber syndrome

Gabrielle Wheway and Colin A. Johnson

History of the eponym

Meckel–Gruber syndrome (MKS, OMIM number #249000) was first described by Johann Friedrich Meckel in 1822, who described two siblings with occipital encephalocele, polycystic kidneys, polydactyly, microcephaly, cleft palate and club foot, both of whom died within minutes of birth (Meckel, 1882). The same condition was later described in seven individuals by Georg Gruber who termed it dysencephalia splanchnocystica (Gruber, 1934). The condition was further delineated by Opitz and Howe who described one case and reviewed 43 cases in the literature, many of which had been published as 'trisomy 13 with normal karyoptye', and suggested the name 'Meckel syndrome' for the condition (Opitz & Howe, 1969). The term 'Meckel syndrome' is now interchangeable with 'Meckel–Gruber syndrome'. Subsequently, there has been further refinement of the diagnostic criteria for MKS (Fraser & Lytwyn, 1981; Hsia et al. 1971; Mecke & Passarge, 1971; Salonen, 1984). Since the eponym was established, descriptions of neonates with MKS-like phenotypes have been discovered in the literature as early as 1684 (Kompanje, 2003), with a description of a child with facial dysmorphia, polydactyly, microphthalmia and occipital encephalocele, who died within a few minutes of birth (Kompanje, 2003; Krahe, 1684).

Epidemiology

Autosomal recessive inheritance for MKS is confirmed by numerous examples of affected siblings, concordance in monozygotic twins, equal occurrence in males and females, and parental consanguinity in some cases.

The reported frequency of incidence varies between populations and ethnic groups. Worldwide, the incidence has been estimated at 1 in every 135,000 live births (Auber, 2007). The overall incidence in the Spanish population has been estimated to be about 1 in 78,000, accounting for 1 in 1600 cases of congenital birth defect recorded in the country between 1976 and 1988. MKS accounted for 12.3% of all cases of well-recognised autosomal recessive conditions in Spain (Martinezfrias et al. 1991). A study of 21,477 pregnancies in London between 1992 and 1996 found one case of MKS in a non-consanguineous partnership, suggesting a birth incidence in the overall British population of 1 in 20,000.

Populations and ethnic groups with increased levels of endogamy, such as in Gujarati Indians (Young et al. 1985), Tatars (Lurie et al. 1984) and Hutterites (Schurig et al. 1980) have higher incidences of MKS. It is relatively common in Finland (1 in every 9000 live births), due to a probable population bottleneck and subsequent endogamy in the Finnish population (Salonen & Norio, 1984), and it comprises one of the Finnish heritage genetic diseases. In a Belgian study of 10,224 births, three were diagnosed with MKS over a 5-year period, suggesting an incidence rate of about 1 in 3400 (Moerman et al. 1982). A study of Israeli Jews estimated the incidence of MKS to be 1 in every 50,000 births (Fried, 1973).

MKS incidence is also higher in consanguineous populations, such as Bedouin tribes in Kuwait (1 in 3530 births) (Teebi et al. 1992). The rate of first-cousin consanguinity in Kuwait is estimated at between 37.8% (Alawadi et al. 1985) and 54.3% (Al-Nassar et al. 1989), and is particularly high amongst Bedouins. There are similar findings in Saudi Arabia, with an incidence of 1 in 3500 (Teebi & Teebi, 2005). In Palestinian Arabs, MKS accounts for 1 in every 200 genetic birth defects, but the incidence in the general population is unknown (Zlotogora, 1997).

Clinical features

MKS is considered the most severe ciliopathy, since it is almost invariably lethal within the first year of life. The majority of affected individuals die *in utero* or within a few hours of birth, but there are a few examples of survival beyond birth, including one example of survival to 43 months (Genuardi et al. 1993).

MKS has a broad, multi-organ phenotype with considerable variation, but it is generally characterised by a classic 'triad' of polycystic kidneys, occipital encephalocele, and polydactyly. These clinical features are frequently accompanied by ductal proliferation in the portal area of the liver and liver fibrosis which is considered by some to be an additional diagnostic criterion of MKS (Salonen, 1984; Sergi et al. 2000). Some also consider microcephaly, genital malformations and cleft palate as major features of MKS (Hsia et al. 1971). Features of Potter's sequence (clubbed feet, pulmonary hypoplasia and cranial anomalies) are frequent, secondary to oligohydramnios or anhydramnios during pregnancy. Pulmonary hypoplasia is thought to be the main cause of death.

Kidney defects

Cystic dysplasia of the kidneys is the most constant and characteristic feature of MKS. The cystic changes seen in MKS differ from typical polycystic kidney disease. The degree of cyst formation varies between individuals with MKS, but the kidneys will often be grossly enlarged, causing massive swelling of the abdomen (Figure 7.1a & b). Cystic dysplasia is usually bilateral, but unilateral dysplasia has been reported. Large, fluid-filled cysts are visible by eye in most affected individuals, but in others small cysts and cystic swelling of the proximal tubules can be seen microscopically, with little normal renal parenchyma (Figure 7.1c). Cysts develop first in the glomeruli in the cortex, and cystogenesis progresses along the tubules and collecting duct in the medulla. There may be thinning of the cortex in some cases. Abnormal foetal renal function is a frequent cause of oligohydramnios or anhydramnios, a common complication of an MKS pregnancy (Hsia et al. 1971; Majewski et al. 1983).

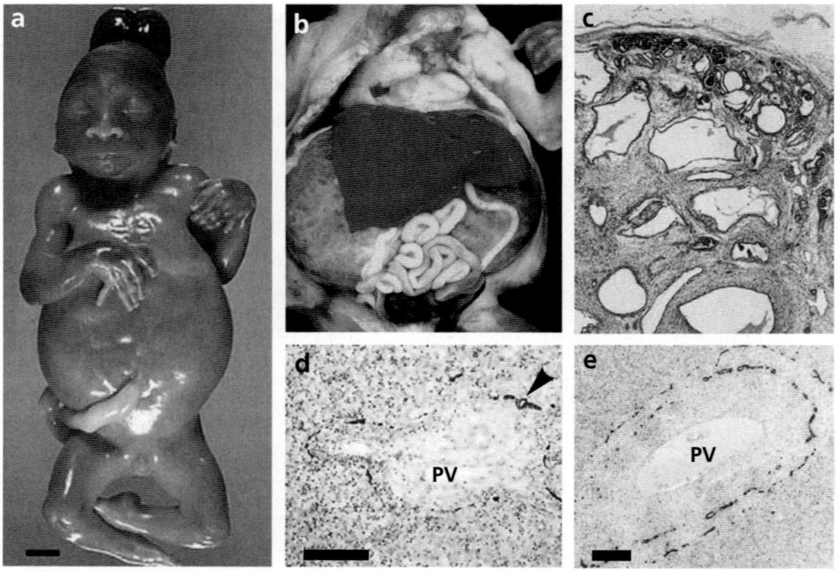

Figure 7.1 Typical external and histopathological features of Meckel–Gruber syndrome. (a) External features of a foetus at gestation age 18 +/40 with Meckel–Gruber syndrome (MKS), showing typical clinical features comprising occipital encephalocele, massive flank masses due to cystic kidneys, post-axial hexadactyly of the hands and a typical Potter's sequence facies with a slanting forehead and flattened nose. Scale bar = 1 cm. (b) Massive swelling of the abdomen of a foetus at gestation age 18 +/40 with MKS due to grossly enlarged, cystic kidneys. (c) Cystic dysplasia of the kidneys comprising large, fluid-filled cysts, small cysts and cystic swelling of the proximal tubules and glomeruli, with absence of normal renal parenchyma. (d) Normal bile duct development in an undiseased 18 +/40 foetal liver visualised by immunohistochemical staining for cytokeratins showing a normal, patent bile duct (arrowhead) and remnants of the embryonic ductal plate surrounding the portal vein (PV). Scale bar = 50 µm. (e) Retention of embryonic structures (ductal plate malformation) in the liver from a foetus at gestation age 18 +/40 with Meckel–Gruber syndrome. Scale bar = 50 µm.

Central nervous system (CNS) defects

MKS has been suggested to be the most frequent syndromic cause of neural tube defects (Simpson, 1991). The phenotype is consistent with a failure of the neural tube to close at the rostral hindbrain (closure point 4) (Vanallen et al. 1993). The common central nervous system (CNS) defects seen in MKS are occipital encephalocele (Figure 7.1a), rhombic roof dysgenesis and prosencephalic dysgenesis which may include olfactory bulb dysgenesis, optic nerve hypoplasia and agenesis of the corpus callosum. In addition to these, other CNS features frequently observed in MKS include microcephaly or total anencephaly and midline defects such as cerebellar vermis hypoplasia and absence of the lateral ventricles. Midline defects may extend to clefting of the cerebellum, and even clefting of the brain stem (Majewski et al. 1983). The Dandy–Walker malformation

(Dandy & Blackfan, 1914; Taggart & Walker, 1942), which comprises cystic swelling of the fourth ventricle to fill the posterior fossa and dysgenesis/agenesis of the cerebellar vermis, is sometimes noted. The Dandy–Walker malformation is often associated with hydrocephaly, another occasional feature of MKS (Lowry et al. 1983). Hypoplastic or ectopic pituitary gland and malformation of the sella turcica have also been recorded in MKS. Micropolygyria and heterotopias are also sometimes present, consistent with neuronal migration defects.

Liver defects

Bile duct proliferation and dilatation, and congenital hepatic fibrosis around the hepatic portal vein appear to be obligate features in MKS (Salonen & Norio, 1984). Fibrosis may affect the architecture of the liver, and can be visualised microscopically in some cases (Figure 7.1d & e). The liver may also be enlarged and have occasional cysts. The gall bladder may be absent. The presence of a hepatic lesion, detected by histopathological methods, is considered by some to be an essential diagnostic feature of MKS.

Craniofacial abnormalities

A Potter's sequence facies is frequent in MKS. The forehead is often slanting, the nose flattened, and the face round and broad with thick, full lips (Figure 7.1a). The ears are often reported as being rudimentary and low set. The neck is often short, and the skin of the neck may be webbed. Micrognathia is a frequent feature. Cleft palate is also frequent, and can be sub-mucousal or incomplete and accompanied by cleft lip. The clefting may be medial or bilateral. Some cases of medial cleft palate and lip are associated with absence of the nasal septum. The tongue may be attached to the floor of the mouth, or lobulation of the tongue may be present. Total absence of the tongue, papillomatous tongue tumours, abnormal epiglottis and larynx, and notches in the alveolar ridges has also been reported.

Skeletal defects

Polydactyly, when present, usually manifests as post-axial hexadactyly (Figure 7.1a), although pre-axial polydactyly has been reported with up to seven digits. Polydactyly is usually seen for all the limbs, but may appear only on the hands, only on the feet, or on hands and limbs of one side only. Polydactyly of the hands is more common than polydactyly of the feet in MKS. Syndactyly may also be present. The polydactyly may manifest as additional true digits, or due to bifurcations of metatarsals or metacarpals. The additional digits usually sit parallel to the other fingers or toes, but sometimes sit at an angle, or even perpendicular to the other digits. Skull bone defects associated with occipital encephalocele often comprise a deep posterior fossa with an occipital bone defect which may extend to the foramen magnum. Large anterior and occipital fontanelles are often seen, often accompanied by hydrocephalus. The anterior and middle cranial fossae are often shallow, housing a reduced size brain. Shortening and bowing of the long bones is an occasional feature of MKS, in addition to infrequent absence of the sternum and shortening of the ribs. The neck is often short, and the first and second cervical vertebrae may not be fully formed, in association with occipital bone defects.

Genitourinary tract defects

The most common genital malformation is external male genital hypoplasia, but may also include absence of testes, ambiguous genitalia, rudimentary vagina, uterus bicornis, rudimentary intra-abdominal testicular anlagen and doubled anlagen of the uterus and vagina. Hermaphroditism has been reported in several cases Defects in the urinary tract seen in MKS include urinary bladder agenesis or hypoplasia (Pettersen, 1983), agenesis or hypoplasia or duplication of the ureter, or ureter disconnected from the bladder.

Eye defects

Microphthalmia is the most common eye anomaly noted in MKS, but other occasional ocular defects include anophthalmia, iris coloboma and optic nerve coloboma or hypoplasia. Aniridia, lens defects, hypertelorism, and slanting and shortening of the palpebral fissures have also been reported.

Defects of other visceral organs

Congenital heart defects seen in MKS patients include patent ductus arteriosus, atrial/ventricular septal defects, dextrocardia, persistent left superior vena cava, aortic valve stenosis, aorta hypoplasia and ventricle hypoplasia. The intestines may be malrotated, and the caecum and appendix may be displaced. Anal atresia is also sometimes seen. Accessory spleens are frequent, but the spleen may also be enlarged or totally absent. Cystic and fibrotic changes of the pancreas are occasional features.

Range of the phenotype

There is extensive intrafamilial and even intra-individual variation in the MKS phenotype. Cystic kidney dysplasia appears to be the only near-obligate feature for MKS, although this could be due to ascertainment bias since a renal defect is necessary for this particular diagnosis. To exclude this potential bias, one study determined that in a cohort of the affected siblings of MKS probands 100% of the siblings had cystic kidney dysplasia (Fraser & Lytwyn, 1981). A separate study of 67 cases of MKS in Finland also found polycystic kidneys in all cases (Salonen, 1984). A third study of 141 MKS cases found polycystic kidneys in 93% of cases (Majewski et al. 1983), and one earlier study of 51 MKS cases reported kidney cysts in 80% of individuals (Hsia et al. 1971). On average, cystic renal dysplasia therefore occurs in about 93% of all MKS cases.

In contrast, occipital encephalocele and post-axial polydactyly have a more variable presentation. Averaging all the data from the Hsia, Fraser, Majewski and Salonen papers, occipital encephalocele is observed in about 84% of MKS cases, and polydactyly in 78% of cases. Some of this variation in phenotype can be correlated to the genotype of the affected individual. For example, occipital encephalocele and post-axial polydactyly are nearly obligatory in MKS caused by mutations in *MKS1*, but are less frequent in patients with *MKS3* mutations (Consugar et al. 2007).

Several studies have reported a high incidence of congenital hepatic fibrosis with bile duct proliferation in MKS (Moerman et al. 1982; Salonen, 1984), with one report of liver fibrosis in 100% of cases studied (Salonen, 1984). However, other literature reviews reported lower rates of hepatic fibrosis of between 30 and 40% (Fraser & Lytwyn, 1981; Majewski et al. 1983).

The most commonly reported MKS anomalies (beyond the main hallmark features of cystic kidney dysplasia, encephalocele, polydactyly and hepatic fibrosis) are genital malformations and cleft palate and/or lip, both seen in around 40% of cases, and microcephaly or anencephaly, seen in 43% of cases. Micrognathia is reported in 34% of MKS cases, and eye anomalies in 31% of cases. Urinary tract anomalies are seen in 33% of cases and the rate of congenital heart defects in MKS is 26% (Fraser & Lytwyn, 1981; Majewski et al. 1983; Salonen, 1984). All other features are seen at an incidence of less than 20%.

Diagnosis

The risk of MKS can be ascertained by taking a thorough family history. Evidence of consanguinity increases the likelihood of autosomal recessive conditions such as MKS. Any incidence of MKS or MKS-like disease in the family also greatly increases risk. As MKS is an autosomal recessive condition, if the condition occurred in previous pregnancies then the recurrence risk is 25%. High-risk pregnancies should be investigated as early as possible, as a positive diagnosis will usually lead to parents opting for induced termination which carries less risk of complications when carried out earlier in pregnancy.

In general, a diagnosis of MKS can be made if cystic kidney dysplasia is present, along with at least one other hallmark feature of the disease which comprises occipital encephalocele, polydactyly or ductal proliferation in the portal area of the liver (Hsia et al. 1971; Salonen, 1984). However, there has been debate about the minimum diagnostic criteria for MKS, as the phenotype varies so widely between and within families. Whilst some argue that cystic kidney dysplasia must be present for a diagnosis of MKS, there have been examples of MKS siblings with only microcysts in the kidneys, which cannot be detected prenatally (Wright et al. 1994). With this in mind, some in the field believe that MKS can be diagnosed if any one of the hallmark features of MKS is present, plus two other 'relevant' MKS features. Hepatic lesions are considered by some to be an essential diagnostic criterion, but are visible only during post-mortem examination. Definitive diagnosis may be possible by DNA testing to screen for mutations in the known MKS genes.

Prenatal diagnosis is possible using a combination of imaging techniques, α-fetoprotein (AFP) testing of amniotic fluid, and DNA testing of foetus and parents.

Ultrasonography

In the first instance, transabdominal ultrasound should be used to identify foetal anomalies. These anomalies can be further investigated by transvaginal scanning. Visualisation can be compromised by oligohydramnios (Verjaal & Meyer, 1980), although this is less problematic when scanning in the first trimester of pregnancy (Braithwaite & Economides, 1995). Transabdominal ultrasonography, performed at 10–14 weeks gestation, has been shown to successfully detect MKS in both high-risk and low-risk pregnancies (Braithwaite & Economides, 1995; Sepulveda et al. 1997). Occipital bone defects can clearly be seen by ultrasound from 10 weeks (Braithwaite & Economides, 1995) and encephalocele from 13 weeks gestation (Pachi et al. 1989). If an encephalocele is suspected, this can be confirmed by taking fluid from the suspected encephalocele and testing for absence of cells and α-glucosidase, and by measuring for IgG and AFP levels that are consistent with cerebrospinal fluid. Other CNS defects can also be detected in the first trimester of pregnancy using ultrasound, including Dandy–Walker malformation (Nizard et al. 2005) but it is important to confirm this diagnosis later in pregnancy and/or using prenatal magnetic resonance imaging (MRI). Foetal kidneys can be visualised from 9 weeks, and adopt the adult form and position at about 11 weeks. Measurements of the foetal trunk can give an early indication of presence of polycystic kidneys if the trunk is enlarged (Kaffe et al. 1976). Renal dysplasia is also suggested by unsually heterogeneous corticomedullary differentiation, reduced echogenicity of the medulla, increased echogenicity of the cortex and the visualisation of small medullary cysts (Ickowicz et al. 2006). The foetal bladder can also be visualised by ultrasonography from 11 weeks. It is useful to look for the absence of a visible foetal bladder, as this is indicative of renal dysfunction. Foetal polydactyly can be detected from 11 weeks, and diagnosis of polydactyly is easier in the first trimester as the hands tend to be clenched from the second trimester onwards (Souka & Nicolaides, 1997). Ultrasonography can be useful for detecting dilated lateral ventricles, indicative of the hydrocephalus often seen in MKS (Wapner et al. 1981).

Magnetic resonance imaging (MRI)

MRI can be a very useful tool for diagnosis of MKS, but there are few published examples. MRI has been successfully used for prenatal diagnosis in pregnancies at risk for Joubert syndrome (JBTS), a related ciliopathy that can be allelic to MKS (Doherty et al. 2005). MRI is mainly used when ultrasonography findings are inconclusive, or provides an alternative to ultrasonography if lack of amniotic fluid prevents clear ultrasound imaging.

It is useful in detecting origin and extent of an abnormality but must be performed after 18 weeks' gestation. MRI will offer better soft-tissue resolution than an ultrasonograph, and can provide clearer images of intracranial structures to enable an accurate diagnosis of CNS malformations. Although foetal movement and maternal aortic pulsation do not necessarily prevent successful diagnosis of MKS using MRI (Chao et al. 2005), imaging artefacts caused by foetal movement can be prevented by treatment with a foetal neuromuscular blockade (Williamson et al. 1989) or by general anaesthesia of the mother.

Embryofetoscopy

Visualisation of the embryo is possible in the first trimester of pregnancy, from as early as 7 weeks, using transabdominal or vaginal endoscopy. Placement of the endoscope into the uterus through a thin-gauge needle through the abdomen can be guided by ultrasound. This method can provide clear visualisation of the surface anatomy of the embryo, allowing observation of polydactyly and occipital encephalocele from 11 weeks gestational age (Quintero et al. 1993). Alternatively, fetoscopy can be carried out by introducing the fetoscope to the surface of the amniotic sac through the cervix. This has enabled a diagnosis of MKS in an at-risk embryo from 10 weeks gestational age (Dumez et al. 1994). However, these invasive techniques are only recommended for investigating high-risk pregnancies.

Testing for α-fetoprotein (AFP)

α-Fetoprotein (AFP) testing of amniotic fluid can also give an indication of MKS (Chemke et al. 1977). AFP can be measured in amniotic fluid from 12 weeks gestation and in maternal blood from 15 weeks. Elevated AFP levels may suggest an open neural tube defect, but it is important to note that most encephaloceles are closed and do not elevate amniotic AFP levels, or an encephalocele may not be present. Testing for AFP levels can also be complicated by oligohydramnios (Verjaal et al. 1980) and/or presence of foetal blood in the amniotic fluid.

DNA testing

DNA testing of both parents and foetus can be an additional useful diagnostic tool. Testing can be by direct DNA sequencing of known MKS genes or by microsatellite linkage analysis of known MKS loci. If the identity of the mutated MKS gene in a multiplex or consanguineous family is not known, microsatellite linkage analysis is often more rapid and straight-forward method of genotyping the parents and foetus. If both parents are heterozygous at a known MKS locus and the foetus is homozygous or compound heterozygous, this can suggest a diagnosis of MKS. However, it is essential that DNA testing is accompanied by thorough ultrasonography imaging to confirm the diagnosis. It is important to note that biallelic mutations in several MKS genes can cause other related ciliopathies that are compatible with life, principally Joubert syndrome (JBTS).

Differential diagnoses

Due to the complex multi-organ phenotype and wide phenotypic variability in MKS, diagnosis can be difficult and MKS can be confused with other conditions. MKS is sometimes misdiagnosed as trisomy 13. Cystic kidneys, microcephaly and polydactyly are features of trisomy 13, but neither cystic and fibrotic changes of the liver nor occipital encephalocele occur in trisomy 13. Trisomy 13 is associated with more severe muscular defects, whereas skeletal defects are more common in MKS (Pettersen, 1983). Chromosome analysis of foetal cells can be used to exclude trisomy 13.

Prenatally, the features of Bardet–Biedl syndrome (BBS) can be confused with the features of MKS, leading to misdiagnosis. Although mental retardation and obesity are features of BBS, these do not present until later in life. The other common features of BBS (polydactyly, renal defects, hepatic anomalies, genital hypoplasia and heart malformations) are present in most BBS foetuses, which may lead to a misdiagnosis of MKS (Karmous-Benailly et al. 2005). With this in mind, a prenatal diagnosis of MKS should be made with caution in the absence of any family history of MKS or occipital encephalocele.

The many overlapping features of Smith–Lemli–Opitz syndrome (SLO) and MKS can lead to misdiagnosis (Lowry, 1983). Important phenotypic distinctions between MKS and SLO include differences in the kidney phenotype. Kidneys tend to be hypoplastic with small cysts in SLO, whereas in MKS the kidneys are cystic but not hypoplastic and are often greatly enlarged. Genital ambiguity is seen in SLO, whereas the defect seen in MKS is male genital hypoplasia rather than true male pseudohermaphroditism. Occipital encephalocele is not a feature of SLO.

Accurate diagnosis is also important because there have been several incidences of MKS patients surviving beyond birth, particularly if encephalocele is absent or treated immediately, and some normal kidney function remains (Genuardi et al. 1993; Kaplan et al. 1993; Lowry et al. 1983; Schurig et al. 1980). If kidney cystic dysplasia is unilateral, or mild and bilateral, there is a small possibility of viability of the foetus. Continuation of the pregnancy and/or administration of life-extending drugs should be considered.

Interesting features of the condition

As a disorder of neural tube closure, it is perhaps surprising that there have been no cases of spina bifida reported in MKS patients. It is also interesting that the expressivity of the MKS phenotype is so variable, with even intra-individual variation within families for individuals that carry the same disease-causing mutation (Valente et al. 2010). The effect of genetic modifiers can explain some of this variation, but the report of significant differences in the phenotype of two monozygotic twins with MKS suggests that other non-genetic factors are involved (Hsia et al. 1971).

Genetics

MKS is a genetically heterogeneous condition. Ten genes have been identified which, when mutated, cause MKS. Mutations in these genes only account for around 50% of known cases of MKS in the Leeds MKS patient cohort, so it is clear that there are several more unidentified disease genes (Szymanska et al. 2012). It may prove a challenge to find the remaining causative genes, as 'private mutations' in a gene in a single family may account for many of them. The remaining unidentified genetic causes of MKS may also be difficult to elucidate because of complex digenic and trigenic inheritance. The known MKS disease genes are:

- *MKS1*—MKS1 (Kyttala et al. 2006)
- *MKS2*—TMEM216 (Valente et al. 2010)
- *MKS3*—TMEM67/meckelin (Smith et al. 2006)
- *MKS4*—CEP290 (Baala et al. 2007a)
- *MKS5*—RPGRIP1-like (Delous et al. 2007)

- *MKS6*—CC2D2A (Tallila et al. 2008)
- *MKS7*—nephrocystin 3 (Bergmann et al. 2008)
- *MKS8*—TCTN2 (Shaheen et al. 2011)
- *MKS9*—B9D1 (Hopp et al. 2011)
- *MKS10*—B9D2 (Dowdle et al. 2011).

Mutations in *MKS1* account for around 7% of all MKS cases and around 70% of Finnish MKS cases. The vast majority of Finnish *MKS1* patients share a common so-called 'Finn major' mutation (*MKS1* IVS15-7_35del) which is thought to have arisen from a common founder of the Finnish population (Khaddour et al. 2007). The Finn major mutation is associated with the 'campomelic' form of Meckel–Gruber syndrome, which has a significant involvement of the skeleton in the disease phenotype. In one study, six out of eight patients with the Finn major mutation had shortening and bowing of the long bones, which is far less common in cases of MKS caused by other mutations (Auber, 2007). Mutations in *MKS1* are almost always associated with polydactyly and occipital encephalocele. In the study by Auber et al. (2007), all MKS1 patients had the classic triad of MKS symptoms with the addition of hepatic ductal dysplasia. This can be contrasted with *TMEM67/MKS3*-mutated patients, which have a lower incidence of polydactyly and CNS malformations (Consugar et al. 2007). In another study, five out of 12 *MKS1*-mutated cases were found to have bone dysplasia and cleft palate and two of the 12 showed *situs inversus*, which tends to be less frequent in other cases of MKS (Khaddour et al. 2007). Similar genotype–phenotype correlations were found in the Leeds MKS patient cohort (Szymanska et al. 2012).

Biallelic mutations in the known MKS genes can also cause other related and allelic ciliopathies. Mutations in *TMEM216* (*MKS2*) are also associated with JBTS (Valente et al. 2010). Mutations in *TMEM67* (*MKS3*) are associated with JBTS (Baala et al. 2007b), cerebellar vermis hypo/aplasia, oligophrenia, congenital ataxia, ocular coloboma and hepatic fibrosis (COACH) syndrome (Brancati et al. 2009), nephronophthisis (NPHP) (Seeman et al. 2010) and a disorder with features of polycystic kidney disease, JBTS and NPHP (Gunay-Aygun et al. 2009). Homozygous mutations in *CEP290* (*MKS4*) occur in patients with NPHP (Chang et al. 2006), Senior–Løken syndrome (SLS), JBTS (Sayer, 2006), BBS (Leitch et al. 2008) and Leber congenital amaurosis (LCA) (den Hollander et al. 2006). Homozygous mutations in *RPGRIP1L* (*MKS5*) are found in patients with JBTS (Arts et al. 2007; Delous et al. 2007) and COACH syndromes (Doherty et al. 2010). Homozygous mutations in *MKS7* can also cause NPHP (Olbrich et al. 2003) and mutations in *CC2D2A* (*MKS6*) can lead to JBTS (Gorden et al. 2008) and COACH syndromes (Doherty et al. 2010). Mutations in *TCTN2* (*MKS8*) have also been found in JBTS patients (Sang et al. 2011).

Underlying ciliary defect

The MKS1, TMEM216 (MKS2) and meckelin/TMEM67 (MKS3) proteins play a role in centrosome migration. Cells lacking MKS1, TMEM216 or meckelin either do not develop primary cilia, as the centrosome and associated centrioles do not migrate to the apical cell surface to establish the basal body of the cilium, (Brancati et al. 2009; Dawe et al. 2007) or have multiple centrioles, multiple cilia and longer cilia (Tammachote et al. 2009).

MKS1 interacts with B9D1 (MKS9) (Bialas et al. 2009), TCTN2 (MKS8) and CC2D2A (MKS6) at the basal body of the cilium to regulate hedgehog (Hh) signalling, which patterns the neural tube and developing limb (Sang et al. 2011). If any of these proteins are mutated or lost, Hh signalling is perturbed.

Physiological effect of the ciliary defect

The loss of a functional cilium due to mutations in MKS genes leads to devastating effects during development. This is because the cilium is a centre for cell signalling, which is crucial in organising cell division, migration, tissue patterning and organogenesis. Hedgehog signalling patterns the neural tube and limb, and therefore the occurrence of polydactyly and neural tube closure defects (occipital encephalocele, midline defects such as cerebellar vermis hypoplasia) in the MKS phenotype is directly attributable to abnormal hedgehog signalling due to a loss of cilia function in the developing neural tube and limb.

The primary cilium is also believed to act as a flow sensor, to measure rates of flow of urine in the kidney and cerebrospinal fluid (CSF) in the CNS, to allow regulation of cell proliferation in response to fluid flow. With defects in primary cilia, epithelial cells in the renal tubules are unable to regulate appropriate tissue growth in response to fluid flow, resulting in the development of renal cysts (Nauli et al. 2003). This is attributed to defects in planar cell polarity division, causing retarded longitudinal growth and predominantly transverse growth of cells, leading to epithelial cyst formation (Simons et al. 2005). Defects in mechanosensory and motile cilia at the embryonic node, and defects in the surrounding immotile cilia that detect fluid flow and transduce an appropriate developmental signal are thought to cause situs and laterality defects in MKS and other ciliopathies. Defects in motile cilia on ependymal cells is thought to reduce flow of cerebrospinal fluid, leading to hydrocephaly and other CNS defects (McGrath et al. 2003; Nonaka et al. 1998).

Cilia on cholangiocytes in the bile duct also act as chemosensors to organise bile secretion and proliferation in response to hypotonicity or to biliary nucleotides through Ca^{2+} and cAMP signalling (Gradilone et al. 2007; Masyuk et al. 2006, 2008). Loss of cilia therefore leads to abnormal proliferation in the portal area of the liver, as seen in MKS.

Clinical management

Individuals with MKS almost always die *in utero* or within a few hours of birth. Death after birth is normally attributed to respiratory insufficiency due to pulmonary hypoplasia as a result of severe prenatal renal dysfunction. Survival beyond birth is normally only seen in patients with mild renal dysplasia, which has not severely compromised kidney function, or had secondary effects on lung development. Extended survival tends to be associated with relatively mild CNS defects (absence of occipital encephalocele, or encephalocele containing no brain material and/or mild hydrocephalus correctable with a shunt).

Encephaloceles are not incompatible with life, especially if little brain tissue is externalised (Budorick et al. 1995).

The longest recorded survival of an MKS patient was of a boy with all of the hallmark signs of MKS but no occipital encephalocele (although Dandy–Walker malformation was present), who survived to 43 months. This boy had mild cystic kidney dysplasia, which only began to affect renal function at 37 months, progressively deteriorating until death due to pneumonia and loss of kidney function at 43 months (Genuardi et al. 1993). In a separate report, two sibs were described with MKS, one of which died within 3 hours of birth due to severe bilateral cystic renal dysplasia and severe hydrocephalus, and the other of which survived to 28 months. The prolonged survival of this second sibling was attributed to the relatively mild renal dysplasia, absence of occipital encephalocele (although hydrocephalus and cerebellar vermis hypoplasia were present) and treatment of hydrocephalus with a ventriculo-atrial shunt. The baby was resuscitated at birth, given respiratory assistance and had metabolic acidosis corrected. Respiratory distress was a recurrent problem and this baby died at 28 months due to pneumonia and progressive loss of renal function (Lowry et al. 1983).

Another case of MKS survival beyond birth was a boy with post-axial polydactyly of the hands, enlarged, cystic kidneys, mild microcephaly and large occipital fluid-filled sac. The baby survived to 5 months. Survival was attributed to a lack of brain tissue in the occipital sac, and early intervention to close the encephalocele. The child suffered from repeated spells of apnoea, as did his sibling, whose breathing problems were successfully treated with three doses of aminophylline each day (Schurig et al. 1980). Similarly, Kaplan et al. (1993) described a case of occipital encephalocele and unilateral polycystic kidney in an infant. The encephalocele was treated early, and the function of the unaffected kidney permitted life to 5 months (Kaplan et al. 1993). Paavola and colleagues reported a number of cases of MKS, all of which died prenatally or at birth, and one which survived to 18 months. This surviving child had all the main features of MKS, including occipital encephalocele which was successfully corrected, and cystic kidneys which did not immediately severely impact on renal function. He suffered from severe apnoeas but survived to 18 months with no other medical interventions. The cause of death was not established (Paavola et al. 1997).

Many of these cases may be classified as 'atypical' MKS, due to the mild renal phenotype. In most cases, survival was attributed to a renal defect which did not inhibit renal function or have secondary effects on prenatal lung development. Although these features may not be so severe as to cause death immediately after birth, they will inevitably deteriorate until death due to renal failure and/or bronchopneumonia, irrespective of medical intervention. Increased lifespan in these cases is also often associated with a lack of encephalocele, with early surgical intervention to close the encephalocele and/or treatment of hydrocephalus with a shunt. The treatment of any CNS defect is the most valuable medical treatment which could be offered to a patient with MKS to prolong life.

In all cases of MKS, whether terminating before term, surviving to term or surviving beyond birth, genetic counselling should be offered to parents to inform them of the basis of the foetal anomaly, and the likelihood of recurrence of the problem. If the condition has been diagnosed prenatally, this counselling should be offered promptly, to inform parents of the risks associated with MKS and allow parents to make a timely decision about the future of the pregnancy. Many parents will opt for an induced termination, although induced labour should be offered if termination is not possible on legal, ethical or religious grounds.

After the loss of a pregnancy or child, further counselling should be offered to help parents make a decision about whether to permit a post-mortem examination and donate blood samples for genetic testing. A genetic counsellor should explain the possible benefits of post-mortem examination in providing valuable information that could aid diagnosis and genetic counselling for future pregnancies. The use of DNA samples for genetic testing from affected individuals, parents and any unaffected siblings can be used to test for mutations in any of the ten known MKS genes. If causative mutations are found, this can enable rapid diagnosis of future pregnancies and identify the carrier status of any unaffected siblings who wish to consider genetic counselling when they reach reproductive age. If no mutations are found in the known MKS genes, DNA samples are a valuable resource for the identification of novel MKS genes in future research. When taking DNA samples which may be used for research, local guidelines from a research ethics committee, institutional review board or equivalent must be followed, in accordance with the ethical principles stated in the Code of Ethics of the World Medical Association (Declaration of Helsinki) for medical research involving human subjects.

References

Al-Nassar, K., Kelly, C. & El-Kazimi, A. 1989. Patterns of consanguinity in the population of Kuwait. *Am J Hum Genet* **45**(suppl 4), A0915.

Alawadi, S. A., Moussa, M. A., Naguib, K. K., Farag, T. I., Teebi, A. S., Elkhalifa, M. et al. 1985. Consanguinity among the Kuwaiti population. *Clin Genet* **27**, 483–486.

Arts, H. H., Doherty, D., Van Beersum, S. E. C., Parisi, M. A., Letteboer, S. J. F., Gorden, N. T., et al. 2007. Mutations in the gene encoding the basal body protein RPGRIP1L, a nephrocystin-4 interactor, cause Joubert syndrome. *Nat Genet* **39**, 882–888.

Auber, B., Burfeind, P., Herold, S., Schoner, K., Simson, G., Rauskolb, R., et al. 2007. A disease causing deletion of 29 base pairs in intron 15 in the *MKS1* gene is highly associated with the campomelic variant of the Meckel–Gruber syndrome. *Clin Genet* **72**, 454–459.

Baala, L., Audollent, S., Martinovic, J., Ozilou, C., Babron, M.-C., Sivanandamoorthy, S., et al. 2007a. Pleiotropic effects of CEP290 (NPHP6) mutations extend to Meckel syndrome. *Am J Hum Genet* **81**, 170–179.

Baala, L., Romano, S., Khaddour, R., Saunier, S., Smith, U. M., Audollent, S., et al. 2007b. The Meckel–Gruber syndrome gene, MKS3, is mutated in Joubert syndrome. *Am J Hum Genet* **80**, 186–194.

Bergmann, C., Fliegauf, M., Bruchle, N. O., Frank, V., Olbrich, H., Kirschner, J., et al. 2008. Loss of nephrocystin-3 function can cause embryonic lethality, Meckel–Gruber-like syndrome, situs inversus, and renal–hepatic–pancreatic dysplasia. *Am J Hum Genet* **82**, 959–970.

Bialas, N. J., Inglis, P. N., Li, C. M., Robinson, J. F., Parker, J. D. K., Healey, M. P., et al. 2009. Functional interactions between the ciliopathy-associated Meckel syndrome 1 (MKS1) protein and two novel MKS1-related (MKSR) proteins. *J Cell Sci* **122**, 611–624.

Braithwaite, J. M. & Economides, D. L. 1995. First-trimester diagnosis of Meckel–Gruber syndrome by transabdominal sonography in a low-risk case. *Prenat Diagn* **15**, 1168–1170.

Brancati, F., Iannicelli, M., Travaglini, L., Mazzotta, A., Bertini, E., Boltshauser, E., et al. 2009. MKS3/TMEM67 mutations are a major cause of COACH syndrome, a Joubert syndrome related disorder with liver involvement. *Hum Mutat* **30**, E432–E442.

Budorick, N. E., Pretorius, D. H., Mcgahan, J. P., Grafe, M. R., James, H. E. & Slivka, J. 1995. Cephalocele detection in utero—sonographic and clinical features. *Ultrasound Obstet Gynecol* **5**, 77–85.

Chang, B., Khanna, H., Hawes, N., Jimeno, D., He, S., Lillo, C., et al. 2006. In-frame deletion in a novel centrosomal/ciliary protein CEP290/NPHP6 perturbs its interaction with RPGR and results in early-onset retinal degeneration in the rd16 mouse. *Hum Mol Genet* **15**, 1847–1857.

Chao, A.,Wong, A. M., Hsueh, C., Chang, Y-L., Wang, T-H. 2005. Integration of imaging and pathological studies in Meckel–Gruber syndrome. *Prenat Diagn* **25**, 267–268.

Chemke, J., Miskin, A., Ravacha, Z., Porath, A., Sagiv, M. & Katz, Z. 1977. Prenatal diagnosis of Meckel syndrome—alpha-feto protein and beta-trace protein in amniotic fluid. *Clin Genet* **11**, 285–289.

Consugar, M., Kubly, V., Lager, D., Hommerding, C., Wong, W., Bakker, E., et al. 2007. Molecular diagnostics of Meckel–Gruber syndrome highlights phenotypic differences between MKS1 and MKS3. *Hum Genet* **121**, 591–599.

Dandy, W. E. & Blackfan, K. D. 1914. Internal hydrocephalus—An experimental, clinical and pathological study. *Am J Dis Child***8**, 406–482.

Dawe, H. R., Smith, U. M., Cullinane, A. R., Gerrelli, D., Cox, P., Badano, J. L., et al. 2007. The Meckel–Gruber syndrome proteins MKS1 and meckelin interact and are required for primary cilium formation. *Hum Mol Genet* **16**, 173–186.

Delous, M., Baala, L., Salomon, R., Laclef, C., Vierkotten, J., Tory, K., et al. 2007. The ciliary gene RPGRIP1L is mutated in cerebello-oculo-renal syndrome (Joubert syndrome type B) and Meckel syndrome. *Nat Genet* **39**, 875–881.

Den Hollander, A. I., Koenekoop, R. K., Yzer, S., Lopez, I., Arends, M. L., Voesenek, K. E. J., et al. 2006. Mutations in the CEP290 (NPHP6) gene are a frequent cause of Leber congenital amaurosis. *Am J Hum Genet* **79**, 556–561.

Doherty, D., Glass, I. A., Siebert, J. R., Strouse, P. J., Parisi, M. A., Shaw, D. W. W., et al. 2005. Prenatal diagnosis in pregnancies at risk for Joubert syndrome by ultrasound and MRI. *Prenat Diagn* **25**, 442–447.

Doherty, D., Parisi, M. A., Finn, L. S., Gunay-Aygun, M., Al-Mateen, M., Bates, D., et al. 2010. Mutations in 3 genes (MKS3, CC2D2A and RPGRIP1L) cause COACH syndrome (Joubert syndrome with congenital hepatic fibrosis). *J Med Genet* **47**, 8–21.

Dowdle, W. E., Robinson, J. F., Kneist, A., Sirerol-Piquer, M. S., Frints, S. G. M., Corbit, K. C., et al. 2011. Disruption of a ciliary B9 protein complex causes Meckel syndrome. *Am J Hum Genet* **89**, 94–110.

Dumez, Y., Dommergues, M., Gubler, M. C., Bunduki, V., Narcy, F., Lemerrer, M., et al. 1994. Meckel–Gruber syndrome—prenatal diagnosis at 10 menstrual weeks using embryoscopy. *Prenat Diagn* **14**, 141–144.

Fraser, F. C. & Lytwyn, A. 1981. Spectrum of anomalies in the Meckel syndrome, or maybe there is a malformation syndrome with at least one constant anomaly. *Am J Med Genet* **9**, 67–73.

Fried, K. 1973. Relatively high prevalence of Meckel syndrome among Jews. *Isr J Med Sci* **9**, 1399–1399.

Genuardi, M., Dionisivici, C., Sabetta, G., Mignozzi, M., Rizzoni, G., Cotugno, G., et al. 1993. Cerebro-reno-digital (Meckel-like) syndrome with Dandy–Walker malformation, cystic kidneys, hepatic fibrosis and polydactyly. *Am J Med Genet* **47**, 50–53.

Gorden, N. T., Arts, H. H., Parisi, M. A., Coene, K. L. M., Letteboer, S. J. F., Van Beersum, S. E. C., et al. 2008. CC2D2A is mutated in Joubert syndrome and interacts with the ciliopathy-associated basal body protein CEP290. *Am J Hum Genet* **83**, 559–571.

Gradilone, S. A., Masyuk, A. I., Splinter, P. L., Banales, J. M., Huang, B. Q., Tietz, P. S., et al. 2007. Cholangiocyte cilia express TRPV4 and detect changes in luminal tonicity inducing bicarbonate secretion. *Proc Natl Acad Sci U.S.A.* **104**, 19138–19143.

Gruber, G. 1934. Beitrage zur frage "gekoppelter" Miszbildungen (Akrocephalosyndactylie und Dysencephalia splanchnocystica). *Beitrage Zur Pathologischen Anatomie Und Zur Allgemeinen Pathologie* **93**, 459–476.

Gunay-Aygun, M., Parisi, M. A., Doherty, D., Tuchman, M., Tsilou, E., Kleiner, D. E., et al. 2009. MKS3-related ciliopathy with features of autosomal recessive polycystic kidney disease, nephronophthisis, and Joubert syndrome. *J Pediatr* **155**, 386–392.

Hopp, K., Heyer, C. M., Hommerding, C. J., Henke, S. A., Sundsbak, J. L., Patel, S., et al. 2011. B9D1 is revealed as a novel Meckel syndrome (MKS) gene by targeted exon-enriched next-generation sequencing and deletion analysis. *Hum Mol Genet* **20**, 2524–2534.

Hsia, Y. E., Bratu, M. & Herbordt, A. 1971. Genetics of Meckel syndrome (dysencephalia–splanchnocystica). *Pediatrics* **48**, 237–247.

Ickowicz, V., Eurin, D., Maugey-Laulom, B., Didier, F., Garel, C., Gubler, M. C., et al. 2006. Meckel–Gruber syndrome: Sonography and pathology. *Ultrasound Obstet Gynecol* **27**, 296–300.

Kaffe, S., Godmilow, L., Walker, B., Kerenyi, T. D., Beratis, N. G., Ainbender, E., et al. 1976. Prenatal diagnosis of Meckel syndrome. *Pediatr Res* **10**, 367–367.

Kaplan, M., Benneriah, Z. & Achiron, R. 1993. Survival in an infant with a prenatally diagnosed Meckel syndrome variant. *Am J Perinatol* **10**, 172–174.

Karmous-Benailly, H., Martinovic, J., Gubler, M. C., Sirot, Y., Clech, L., Ozilou, C., et al. 2005. Antenatal presentation of Bardet–Biedl syndrome may mimic Meckel syndrome. *Am J Hum Genet* **76**, 493–504.

Khaddour, R., Smith, U., Baala, L., Martinovic, J., Clavering, D., Shaffiq, R., et al. 2007. Spectrum of MKS1 and MKS3 mutations in Meckel syndrome: A genotype–phenotype correlation. Mutation in brief #960. Online. *Hum Mutat* **28**, 523–524.

Kompanje, E. J. O. 2003. Features described and illustrated in 1684 suggesting Meckel–Gruber syndrome. *Pediatr Dev Pathol* **6**, 595–598.

Krahe, C. 1684. The description of a monstrous child. *Phil Trans* **14**, 599.

Kyttala, M., Tallila, J., Salonen, R., Kopra, O., Kohlschmidt, N., Paavola-Sakki, P., et al. 2006. MKS1, encoding a component of the flagellar apparatus basal body proteome, is mutated in Meckel syndrome. *Nat Genet* **38**, 155–157.

Leitch, C. C., Zaghloul, N. A., Davis, E. E., Stoetzel, C., Diaz-Font, A., Rix, S., et al. 2008. Hypomorphic mutations in syndromic encephalocele genes are associated with Bardet–Biedl syndrome. *Nat Genet* **40**, 443–448.

Lowry, R. B. 1983. Variability in the Smith–Lemli–Opitz syndrome—overlap with the Meckel syndrome. *Am J Med Genet* **14**, 429–433.

Lowry, R. B., Hill, R. H. & Tischler, B. 1983. Survival and spectrum of anomalies in the Meckel syndrome. *Am J Med Genet* **14**, 417–421.

Lurie, I. W., Prytkov, A. N. & Meldere, L. V. 1984. Meckel syndrome in different populations. *Am J Med Genet* **18**, 661–669.

Majewski, F., Stoss, H., Goecke, T. & Kemperdick, H. 1983. Are bowing of long tubular bones and preaxial polydactyly signs of the Meckel syndrome? *Hum Genet* **65**, 125–133.

Martinezfrias, M. L., Bermejo, E., Cereijo, A., Sanchez, M., Lopez, M. & Gonzalo, C. 1991. Epidemiological aspects of Mendelian syndromes in a Spanish population sample. II—autosomal recessive malformation syndromes. *Am J Med Genet* **38**, 626–629.

Masyuk, A. I., Gradilone, S. A., Banales, J. M., Huang, B. Q., Masyuk, T. V., Lee, S. O., et al. 2008. Cholangiocyte primary cilia are chemosensory organelles that detect biliary nucleotides via P.2Y(12) purinergic receptors. *Am J Physiol, Gastrointestinal Liver Physiol* **295**, G725–G734.

Masyuk, A. I., Masyuk, T. V., Splinter, P. L., Huang, B. Q., Stroope, A. J. & Larusso, N. F. 2006. Cholangiocyte cilia detect changes in luminal fluid flow and transmit them into intracellular Ca^{2+} and cAMP signaling. *Gastroenterology* **131**, 911–920.

Mcgrath, J., Somlo, S., Makova, S., Tian, X. & Brueckner, M. 2003. Two populations of node monocilia initiate left–right asymmetry in the mouse. *Cell* **114**, 61–73.

Mecke, S. & Passarge, E. 1971. Encephalocele, polycystic kidneys and polydactyly as an autosomal recessive trait simulating certain other disorders—Meckel syndrome. *Annales De Genetique* **14**, 97–103.

Meckel, J. 1882. Beschreibung zweier, durch sehr ähnliche Bildungsabweichungen entstellter Geschwister. *Deutsch Arch Physiol* **7**, 99–172.

Moerman, P., Verbeken, E., Fryns, J. P., Goddeeris, P. & Lauweryns, J. M. 1982. The Meckel syndrome—pathological and cytogenetic observations in 8 cases. *Hum Genet* **62**, 240–245.

Nauli, S. M., Alenghat, F. J., Luo, Y., Williams, E., Vassilev, P., Lil, X. G., et al. 2003. Polycystins 1 and 2 mediate mechanosensation in the primary cilium of kidney cells. *Nat Genet* **33**, 129–137.

Nizard, J., Bernard, J. P. & Ville, Y. 2005. Fetal cystic malformations of the posterior fossa in the first trimester of pregnancy. *Fetal Diagn Ther* **20**, 146–151.

Nonaka, S., Tanaka, Y., Okada, Y., Takeda, S., Harada, A., Kanai, Y., et al. 1998. Randomization of left-right asymmetry due to loss of nodal cilia generating leftward flow of extraembryonic fluid in mice lacking KIF3B motor protein. *Cell* **95**, 829–837.

Olbrich, H., Fliegauf, M., Hoefele, J., Kispert, A., Otto, E., Volz, A., et al. 2003. Mutations in a novel gene, NPHP3, cause adolescent nephronophthisis, tapeto-retinal degeneration and hepatic fibrosis. *Nat Genet* **34**, 455–459.

Opitz, J. & Howe, J. 1969. The Meckel syndrome (dysencephalia splanchnocystica, the Gruber syndrome), in *Malformation Syndromes*, ed. by **Bergsma D. Elsevier**: New York, pp 167–179.

Paavola, P., Salonen, R., Baumer, A., Schinzel, A., Boyd, P. A., Gould, S., et al. 1997. Clinical and genetic heterogeneity in Meckel syndrome. *Hum Genet* **101**, 88–92.

Pachi, A., Giancotti, A., Torcia, F., Deprosperi, V. & Maggi, E. 1989. Meckel–Gruber syndrome—ultrasonographic diagnosis at 13 weeks gestational age in an at-risk case. *Prenat Diagn* **9**, 187–190.

Pettersen, J. C. 1983. Gross anatomical studies of a newborn with the Meckel syndrome. *Teratology* **28**, 157–164.

Quintero, R. A., Abuhamad, A., Hobbins, J. C. & Mahoney, M. J. 1993. Transabdominal thin-gauge embryofetoscopy—a technique for early prenatal diagnosis and its use in the diagnosis of a case of Meckel-Gruber syndrome. *Am J Obstet Gynecol* **168**, 1552–1557.

Salonen, R. 1984. The Meckel syndrome—clinicopathological findings in 67 patients. *Am J Med Genet* **18**, 671–689.

Salonen, R. & Norio, R. 1984. The Meckel syndrome in Finland: Epidemiologic and genetic aspects. *Am J Med Genet* **18**, 8.

Sang, L. Y., Miller, J. J., Corbit, K. C., Giles, R. H., Brauer, M. J., Otto, E. A., et al. 2011. Mapping the NPHP-JBTS-MKS protein network reveals ciliopathy disease genes and pathways. *Cell* **145**, 513–528.

Sayer, J., Otto, EA., O'Toole JF, Nurnberg G, Kennedy MA, Becker C, et al. 2006. The centrosomal protein nephrocystin-6 is mutated in Joubert syndrome and activates transcription factor ATF4. *Nat Genet* **38**, 674–681.

Schurig, V., Bowen, P., Harley, F. & Schiff, D. 1980. The Meckel syndrome in the Hutterites. *Am J Med Genet* **5**, 373–381.

Seeman, T., Seemanova, E., Nuernberg, G., Nuernberg, P., Janssen, S. & Otto, E. A. 2010. Polycystic kidney and hepatic disease with mental retardation is nephronophthisis 11 caused by MKS3/TMEM67 mutations. *Pediatr Nephrol* **25**, 2375–2376.

Sepulveda, W., Sebire, N. J., Souka, A., Snijders, R. J. M. & Nicolaides, K. H. 1997. Diagnosis of the Meckel–Gruber syndrome at eleven to fourteen weeks' gestation. *Am J Obstet Gynecol* **176**, 316–319.

Sergi, C., Adam, S., Kahl, P. & Otto, H. F. 2000. Study of the malformation of ductal plate of the liver in Meckel syndrome and review of ether syndromes presenting with this anomaly. *Pediatr Dev Pathol* **3**, 568–583.

Shaheen, R., Faqeih, E., Seidahmed, M. Z., Sunker, A., Alali, F. E., Alqahtani, K., et al. 2011. A TCTN2 mutation defines a novel Meckel Gruber syndrome locus. *Hum Mutat* **32**, 573–578.

Simons, M., Gloy, J., Ganner, A., Bullerkotte, A., Bashkurov, M., Kronig, C., et al. 2005. Inversin, the gene product mutated in nephronophthisis type II, functions as a molecular switch between Wnt signaling pathways. *Nat Genet* **37**, 537–543.

Simpson, J., Mills, J., Rhoads, G., Cunningham, G., Conley, M., Hoffman, H. 1991. Genetic heterogeneity in neural tube defects. *Ann Genet* **34**, 279–286.

Smith, U. M., Consugar, M., Tee, L. J., Mckee, B. M., Maina, E. N., Whelan, S., et al. 2006. The transmembrane protein meckelin (MKS3) is mutated in Meckel–Gruber syndrome and the wpk rat. *Nat Genet* **38**, 191–196.

Souka, A. P. & Nicolaides, K. H. 1997. Diagnosis of fetal abnormalities at the 10–14-week scan. *Ultrasound Obstet Gynecol* **10**, 429–442.

Szymanska, K., Berry, I., Logan, C., Cousins, S., Lindsay, H., Jafri, H., et al. 2012. Founder mutations and genotype–phenotype correlations in Meckel–Gruber syndrome and associated ciliopathies. *Cilia* **1**, 18.

Taggart, J. K. & Walker, A. E. 1942. Congenital atresia of the foramens of Luschka and Magendie. *Arch Neurol Psychiatry* **48**, 583–612.

Tallila, J., Jakkula, E., Peltonen, L., Salonen, R. & Kestilä, M. 2008. Identification of CC2D2A as a Meckel syndrome gene adds an important piece to the ciliopathy puzzle. *Am J Hum Genet* **82**, 1361–1367.

Tammachote, R., Hommerding, C. J., Sinders, R. M., Miller, C. A., Czarnecki, P. G., Leightner, A. C., et al. 2009. Ciliary and centrosomal defects associated with mutation and depletion of the Meckel syndrome genes MKS1 and MKS3. *Hum Mol Genet* **18**, 3311–3323.

Teebi, A. S., Alsaleh, Q. A. & Odeh, H. 1992. Meckel syndrome and neural-tube defects in Kuwait. *J Med Genet* **29**, 140–140.

Teebi, A. S. & Teebi, S. A. 2005. Genetic diversity among the Arabs. *Community Genet* **8**, 21–26.

Valente, E. M., Logan, C. V., Mougou-Zerelli, S., Lee, J. H., Silhavy, J. L., Brancati, F., et al. 2010. Mutations in TMEM216 perturb ciliogenesis and cause Joubert, Meckel and related syndromes. *Nat Genet* **42**, 619–625.

Vanallen, M. I., Kalousek, D. K., Chernoff, G. F., Juriloff, D., Harris, M., Mcgillivray, B. C., et al. 1993. Evidence for multisite closure of the neural-tube in humans. *Am J Med Genet* **47**, 723–743.

Verjaal, M., Meyer, A. H., Beckerbloemkolk, M. J., Leschot, N. J., Weduwen, J. J. D. & Gras, J. 1980. Oligohydramnios hampering prenatal diagnosis of Meckel syndrome. *Am J Med Genet* **7**, 85–86.

Wapner, R. J., Kurtz, A. B., Ross, R. D. & Jackson, L. G. 1981. Ultrasonographic parameters in the prenatal diagnosis of Meckel syndrome. *Obstet Gynecol* **57**, 388–392.

Williamson, R. A., Weiner, C. P., Yuh, W. T. C. & Abuyousef, M. M. 1989. Magnetic-resonance imaging of anomalous fetuses. *Obstet Gynecol* **73**, 952–956.

Wright, C., Healicon, R., English, C. & Burn, J. 1994. Meckel syndrome—what are the minimum diagnostic criteria? *J Med Genet* **31**, 482–485.

Young, I. D., Rickett, A. B. & Clarke, M. 1985. High-incidence of Meckels syndrome in Gujarati Indians. *J Med Genet* **22**, 301–304.

Zlotogora, J. 1997. Genetic disorders among Palestinian Arabs. 2. Hydrocephalus and neural tube defects. *Am J Med Genet* **71**, 33–35.

Chapter 8

Nephronophthisis

Shalabh Srivastava and John A. Sayer

History of the eponym

Nephronophthisis (NPHP) may also be known as juvenile nephronophthisis or familial juvenile nephronophthisis. NPHP literally means 'disintegration of nephrons'. The histological defect is characterised by a chronic tubulo-interstitial nephritis, leading to nephron drop-out and multiple cysts of varying sizes forming at the corticomedullary junction and medulla. It is inherited as an autosomal recessive condition. Affected patients develop end-stage renal failure (ESRF) within the first three decades of life. The genes involved encode protein products, termed 'nephrocystins'. Nephrocystins localise to the primary cilium and basal body; thus NPHP may be termed a ciliopathy.

In 1951, Fanconi first coined the phrase 'familial juvenile nephronophthisis'. However, in these early descriptions of nephronophthisis, the disease was also descriptively called 'idiopathic parenchymal contracted kidney' (Fanconi et al. 1951). Subsequently, a series of cases were also reported descriptively with the label 'medullary cystic disease of the kidney'. An article in the *New England Journal of Medicine* in 1967 described in detail the virtually identical presentations of juvenile nephronophthisis and medullary cystic kidney disease (Strauss & Sommers, 1967). However, even though the histopathological changes are identical this does not mean these names can be used interchangeably. To distinguish between these conditions it must be remembered that medullary cystic kidney disease is inherited as an autosomal dominant trait, whereas NPHP is autosomal recessive.

Epidemiology

The incidence of NPHP has been estimated to be 9/8.3 million live births in United States and 1/50,000 live births in Canada (Potter et al. 1980). A Finnish study calculated an incidence of ~1 in 60,000 live births (Ala-Mello & Al, 1998). NPHP represents the most frequent genetic cause of ESRF in the first three decades of life (Hildebrandt et al. 2006) accounting for 10–15% of children with ESRF.

Clinical features

A typical presentation of childhood NPHP occurs at around 6 years of age and includes symptoms of polyuria, nocturia or secondary enuresis, polydipsia and lethargy

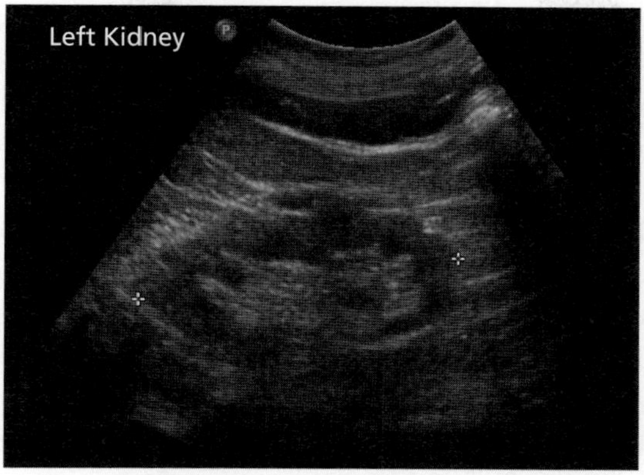

Figure 8.1 Ultrasound scan features of nephronophthisis.Renal ultrasound scan demonstrating corticomedullary cysts.

Reprinted by permission from Macmillan Publishers Ltd: *European Journal of Human Genetics*, Volume 17, Issue 4, pp. 406–416, advanced online publication, 10 December 2008, doi:10.1038/ejhg.2008.238

(secondary to anaemia) (Ala-Mello et al. 1996). These features are secondary to renal salt wasting and an inability to concentrate urine. These features are not exclusive to NPHP but do implicate dysfunction of the renal cortical collecting duct (Krishnan et al. 2008). Renal ultrasound scanning may identify kidneys of normal or reduced size, with increased echogenicity and corticomedullary cysts formation (see Figure 8.1) (Hildebrandt et al. 2009).

There is a rarer infantile variant of NPHP in which children reach ESRF by 3 years of age and have enlarged cystic kidneys on renal ultrasound (Salomon et al. 2009). Also, in contrast, it is possible for a diagnosis of NPHP to be delayed into adulthood (Bollee et al. 2006; Hoefele et al. 2011).

Range of the phenotype

Traditionally, and given the wide range of presentations, NPHP has been divided into infantile, juvenile and adolescent forms.

Infantile NPHP may be defined as an early onset of ESRF (less than 5 years of age). The presentation may be apparent antenatally with features of foetal oliguria and oligohydramnios (Haider et al. 1998). Associated extra-renal features peculiar to infantile NPHP include severe early onset hypertension, situs inversus and ventricular septal defect (Simms et al. 2009).

Juvenile NPHP has a median onset of ESRF of 12 years of age. Polyuria and polydipsia (and salt wasting) are apparent in early childhood (4–6 years of age) and are associated with a urinary concentration defect (<400 mosm/kg in early morning urine) that is not responsive to desmopressin. There may also be evidence of growth retardation (secondary to salt wasting, dehydration and renal insufficiency). Typically there is an absence (or minimal features) of haematuria and proteinuria.

Adolescent NPHP, as the name implies, gives a slightly later mean onset of ESRF, typically 15 years (Olbrich et al. 2003). Molecular genetic advances make these clinical distinctions less relevant. Infantile NPHP may mimic other severe childhood cystic kidney diseases including autosomal recessive polycystic kidney disease (ARPD) and oral–facial–digital syndrome.

A key feature of NPHP is the association of extra-renal manifestations. These occur in 10–15% of patients (Hildebrandt & Zhou, 2007). The most frequent extra-renal feature is retinal degeneration, but a wide variety of other associated features may be evident including cerebellar vermis hypoplasia (known as Joubert syndrome, JS), occipital encephalocele (known as Meckel–Gruber syndrome, MKS), hepatic fibrosis, situs inversus, bronchiectasis and skeletal defects (Hildebrandt & Zhou, 2007). Some of the rare syndromes associated with NPHP are listed in Table 8.1.

Table 8.1 Syndromes associated with nephronophthisis

Syndrome	Extra-renal features	Other variable features
Alström syndrome	Progressive cone–rod dystrophy leading to blindness, sensorineural hearing loss, childhood obesity associated with hyperinsulinaemia, and type 2 diabetes mellitus	Dilated cardiomyopathy
Arima syndrome	Agenesis of the cerebellar vermis, ocular abnormalities	Liver disease
COACH syndrome (a sub-type of Joubert syndrome)	Cerebellar vermis hypo/aplasia, oligophrenia (mental retardation), ataxia, ocular coloboma, and he patic fibrosis	
Cogan's oculomotor apraxia	Defective or absent horizontal voluntary eye movements	
Ellis–van Creveld syndrome	Skeletal dysplasia characterised by short limbs, short ribs, postaxial polydactyly, and dysplastic nails and teeth	
Jeune's syndrome (asphyxiating thoracic dystrophy)	Chondrodysplasia characterized by a severely constricted thoracic cage, short-limbed short stature, and polydactyly	Hepatic disease, retinal degeneration, pancreatic cysts
Joubert syndrome	Hypoplasia of the cerebellar vermis; neurologic symptoms including dysregulation of breathing pattern and developmental delay	Retinal dystrophy, renal anomalies
Meckel–Gruber syndrome	CNS malformation (including occipital encephalocele) and hepatic abnormalities, including portal fibrosis or ductal proliferation	Polydactyly, most often post-axial

Table 8.1 (continued) Syndromes associated with nephronophthisis

Syndrome	Extra-renal features	Other variable features
RHYNS syndrome	Retinitis pigmentosa, hypopituitarism and mild skeletal dysplasia	
Senior–Løken syndrome	Leber congenital amaurosis	
Sensenbrenner syndrome (cranioectodermal dysplasia)	Skeletal abnormalities, including craniosynostosis, narrow rib cage, short limbs, and brachydactyly, and ectodermal defects	Hepatic fibrosis, heart defects, and retinitis pigmentosa

Diagnosis

The diagnosis of NPHP relies on a clinical suspicion of the disorder (Figure 8.2). Patients may present to a variety of clinicians, not just paediatricians or nephrologists. NPHP should initially be investigated non-invasively. The key features required in order make a clinical diagnosis would include a history of polyuria and polydipsia, and secondary enuresis. Presenting features at any point during childhood or later life may be complications of renal insufficiency/renal failure such as nausea, vomiting, itch, fatigue (anaemia) and growth retardation.

In order to make a diagnosis of any inherited disease a detailed family history is required. The may be a family history of renal disease (typically in an autosomal recessive pattern) and a history of consanguinity. Some caution regarding pedigrees is required, especially in consanguineous families, where a recessive trait such as NPHP may appear to run in an autosomal dominant pattern (Hoefele et al. 2011).

Clinical examination should include an assessment of blood pressure and a search for extra-renal manifestations such as retinal pigmentation, abnormal eye movements and polydactyly.

Baseline clinic investigations should include a urine dipstick. Typically, low levels of urinary protein (<0.5g/l) and minimal haematuria are noted. An early-morning urine is useful to assess urinary concentration.

Imaging investigations should include an ultrasound scan of the abdomen and kidneys to assess renal size, to look for corticomedullary cysts, corticomedullary differentiation, to exclude renal tract dilatation and to examine for liver fibrosis/splenomegaly.

Additional imaging may include brain magnetic resonance imaging scanning and full neurological evaluation to assess cerebellar function if there are any neurological symptoms. In addition, a baseline ophthalmological examination is essential to look for minor degrees of coloboma, retinopathy and oculomotor apraxia. Visual evoked potential studies may be performed in newborn children and electro-retinogram studies may be performed from 8 months of age.

Baseline blood tests should include renal function (urea, creatinine), liver function (albumin, transaminases, bilirubin), full blood count (to look for renal anaemia) and clotting

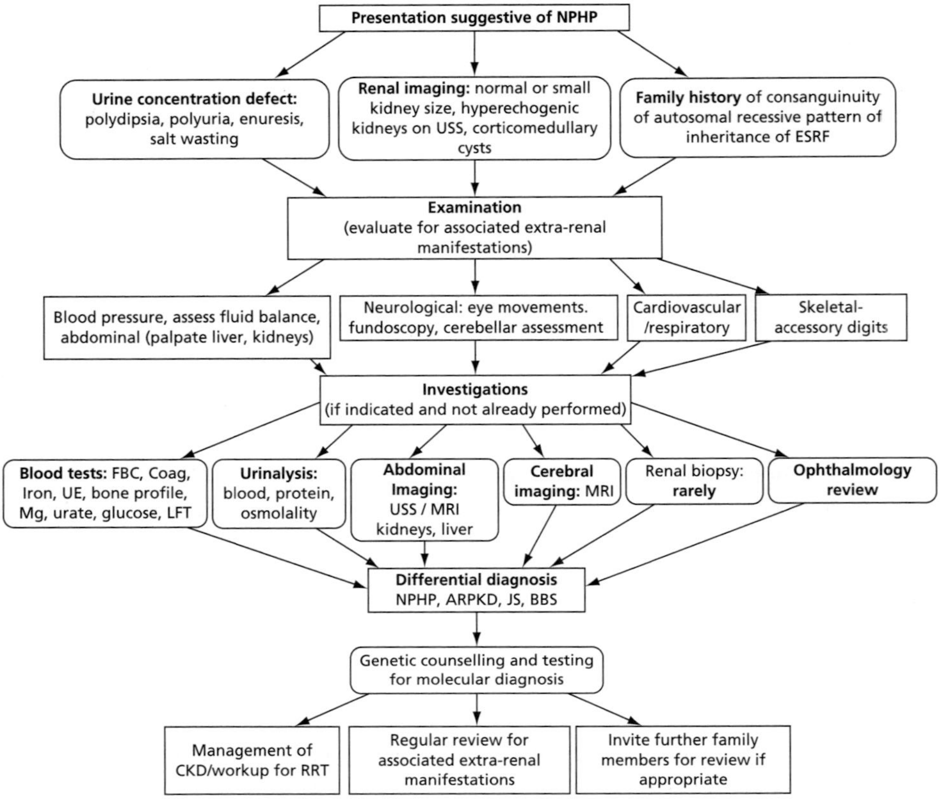

Figure 8.2 Diagnostic algorithm for suspected cases of NPHP. ARPKD, autosomal recessive polycystic kidney disease; BBS, Bardet–Biedl syndrome; CKD, chronic kidney disease; Coag, coagulation; ESRF, end-stage renal failure; FBC, full blood count; JS, Joubert syndrome; LFT, liver function tests; Mg, magnesium; MRI, magnetic resonance imaging; NPHP, nephronophthisis; RRT, renal replacement therapy; UE, urea and electrolytes; USS, ultrasound scanning.

Adapted from Roslyn J. Simms, AnnMarie Hynes, Lorraine Eley, and John A. Sayer, Nephronophthisis: a genetically diverse ciliopathy, *International Journal of Nephrology*, Volume 2011.

studies (prothrombin time as a marker of liver function and before renal biopsy, if necessary). If renal failure is advanced, screening for renal osteodystrophy, hyperparathyroidism and metabolic acidosis should be performed.

Genetics tests are available for the most common form of NPHP, namely *NPHP1* where mutations account for ~25% of cases. Following informed consent/genetic counselling DNA should be obtained from index cases, parents, siblings and any other available family members. It is our own practise to discuss with the whole family the implications of a molecular genetic diagnosis within a special family clinic setting. Here a multi-specialty team of adult nephrologist, paediatric nephrologist and clinical geneticist provide information and guidance regarding the utility, limitations and implications of a genetic diagnosis.

Additional genes may be tested but given the number of genes involved this has traditionally been arduous. Modern genetics techniques may help to deliver a more streamlined approach.

Renal biopsy findings include tubular atrophy, interstitial fibrosis and tubular basement membrane defects, including abrupt transition between thickening and attenuation or disintegration (Hildebrandt et al. 1992; Krishnan et al. 2008; Zollinger et al. 1980). Infantile NPHP differs from typical NPHP in that there is moderate renal enlargement, histological changes that include cortical microcysts, cystic dilatation of Bowman's spaces and lack of tubular basement membrane disruption (Gagnadoux et al. 1989). In advance stages of disease the histological findings demonstrate diffuse sclerosing interstitial nephropathy. The histological appearance of NPHP may be mimicked exactly by medullary cystic kidney disease (Hildebrandt et al. 2006). Recently, mutations in *FAN1* were shown to be a cause of karyomegalic interstitial nephritis (Zhou et al. 2012). It is noteworthy that the renal histology in karyomegalic interstitial nephritis is indistinguishable from that of nephronophthisis, except for the presence of karyomegaly (Mihatsch et al. 1979; Palmer et al. 2007).

Genetics

NPHP, as previously stated, is an autosomal recessively inherited disorder, with many gene defects now associated with this disorder. To date, mutations have been identified in 16 genes (Table 8.2) which collectively account for approximately 30% of patients (Hurd & Hildebrandt, 2011). There is also a number of examples of oligogenicity in NPHP. Here a third allele (often a heterozygous mutation in a second NPHP gene) is identified in addition to a homozygous or compound heterozygous mutation in the first NPHP gene. Such additional alleles may modify the phenotype. Patients with combinations of *NPHP1* and *AHI1*, *NPHP6* and *AHI1* and *NPHP1* and *NPHP6* have been reported in a cohort of JS patients (Tory et al. 2007). *AHI1* may be a particular modifier of brain (Tory et al. 2007) and retinal (Coppieters et al. 2010a) phenotypes.

The other noteworthy feature of some of the NPHP genes is the hugely wide phenotypic variability (Valente et al. 2008). Thus, for example, mutations in *CEP290* (also known as *NPHP6*) may cause isolated LCA (den Hollander et al., 2006), isolated NPHP (Coppieters et al. 2010b), a JS phenotype (Sayer et al. 2006) and even a MKS phenotype (Baala et al. 2007), resulting in perinatal lethality. Exactly how a single gene disorder gives this wide spectrum of disease phenotypes is not clear, but modifier alleles may be playing an important role.

Traditional genetic testing based on polymerase chain reaction analysis and direct Sanger sequencing will detect homozygous or heterozygous *NPHP1* deletions and mutations (found in around 25% of cases).Other genes may be tested using emerging high throughput technologies such as DNA capture and automated sequencing. A renal biopsy should not be necessary if a molecular genetic diagnosis can be made. If a molecular diagnosis is not available, a renal biopsy may be required to confirm or exclude NPHP.

Table 8.2 Mutated genes in nephronophthisis and associated extra-renal manifestations

Gene (alias)	Chromosome	Protein	Mutation frequency (Wolf & Hildebrandt, 2011)	Extra-renal features	Reference
NPHP1	2q13	nephrocystin-1	23%	SLS, JS,	(Hildebrandt et al. 1997; Saunier et al. 1997)
INV (*NPHP2*)	9q31	inversin	1–2%	SLS, HF VSD, situs inversus	(Otto et al. 2003)
NPHP3	3q22.1	nephrocystin-3	<1%	SLS, HF, MKS, situs inversus	(Olbrich et al. 2003)
NPHP4	1p36.22	nephrocystin-4 or nephroretinin	2–3%	SLS	(Mollet et al. 2002; Otto et al. 2002)
IQCB1 (*NPHP5*)	3q21.1	nephrocystin-5 or IQ motif containing B1	3–4%	SLS	(Otto et al. 2005)
CEP290 (*NPHP6*)	12q21.32	centrosomal protein 290	1%	LCA, SLS, JS, MKS, BBS	(Sayer et al. 2006; Valente et al. 2006)
GLIS2 (*NPHP7*)	16p13.3	GLI similar 2	<0.5%		(Attanasio et al. 2007)
RPGRIP1L (*NPHP8*)	16q12.2	RPGRIP1-like	0.5%	SLS, JS, MKS	(Wolf et al. 2007)
NEK8 (*NPHP9*)	17q11.1	NIMA related kinase 8	<0.5%	SLS	(Otto et al. 2008)
SDCCAG8 (*NPHP10*)	1q44	serologically defined colon cancer antigen 8	<0.5%	SLS, BBS-like	(Otto et al. 2010)
TMEM67 (*NPHP11*)	8q22.1	transmembrane protein 67	<0.5%	JS, HF, MKS	(Otto et al. 2009)
XPNPEP3 (*NPHPL1*)	22q13	X-prolyl aminopeptidase 3	<0.5%	cardiomyopathy, seizures	(O'Toole et al. 2010)
TTC21B (*NPHP12*)	2q24.3	intraflagellar transport protein 139	<1%	JS, MKS, BBS, JATD	(Davis et al. 2011)
WDR19 (*NPHP13*)	4p14	WD repeat domain 19	<1%	cranioectodermal dysplasia, Jeune syndrome	(Bredrup et al. 2011)

Table 8.2 (continued) Mutated genes in nephronophthisis and associated extra-renal manifestations

Gene (alias)	Chromosome	Protein	Mutation frequency (Wolf & Hildebrandt, 2011)	Extra-renal features	Reference
ZNF423 (*NPHP14*)	16q12.1	Zinc finger protein 423	<1%	JS, situs inversus	(Chaki et al. 2012)
CEP164 (*NPHP15*)	11q23.3	centrosomal protein of 164 kDa	<1%	SLS	(Chaki et al. 2012)

BBS, Bardet–Biedl syndrome; HF, hepatic fibrosis; JATD, Jeune asphyxiating thoracic dystrophy; JS, Joubert syndrome; LCA, Leber's congenital amaurosis; MKS, Meckel–Gruber syndrome; SLS, Senior–Løken syndrome; VSD, ventricular septal defect.

Underlying ciliary defect

The protein products of all NPHP genes localise to primary cilia and related structures (basal bodies, centrosomes), resulting in a unifying hypothesis that nephrophthisis is a ciliopathy (Hildebrandt & Otto, 2005).

The ciliary expression of each of the nephrocystin proteins may explain the pleiotropic effects of multi-organ involvement. Within the kidney, cilia are considered to be involved in mechanosensation of urinary flow in the renal tubules (Nauli et al. 2003) but the precise role of the nephrocystin proteins remains unclear. Within photoreceptor cells, a role for nephrocystin proteins in maintenance of connecting cilium function seems likely, and defects lead to retinal dystrophy and degeneration (Hildebrandt & Otto, 2005).

As nephrocystin proteins are multi-domain proteins, often rich in coiled-coil domains and other protein–protein interaction domains (Simms et al. 2011) they are thought of as adapter molecules. There localisation is dynamic (Morgan et al. 2002; Sayer et al. 2006) and may not be limited to ciliary structures. However, there is strong evidence for Cep290 having a role as 'gatekeeper' of the primary cilium (Betleja & Cole, 2010; Craige et al. 2010), regulating the entry of proteins into the ciliary axoneme. Evidence for nephrocystins forming discrete functional complexes was recently obtained using elegant protein-protein interaction studies (Sang et al. 2011). These studies revealed that nephrocystin proteins 1, 4 and 8 were functioning as a complex at the apical surface whilst nephrocystins 5 and 6 were at a centrosomal location (Sang et al. 2011). This data implies that loss of function of each gene may contribute a unique mechanism to the disease pathogenesis.

Clinical management

Once a diagnosis of NPHP has been made, regular clinical reviews are required to appropriately manage chronic kidney disease (CKD)/ESRF and individuals with extra-renal manifestations should be referred to appropriate colleagues and are ideally best managed in specialist clinics (Figure 8.2).

Genetic counselling should include a discussion of the recurrence risk, which is 25% in most families, although X-linked inheritance should also be considered. In addition, as we have mentioned, examples of oligogenicity for NPHP exist, where the phenotype (especially extra-renal manifestations such as CNS and eye involvement) may be variable and unpredictable. The identification of the molecular defect(s) in couples at risk allows early prenatal genetic testing. The Ciliopathy Alliance (http://www.ciliopathyalliance.org/) brings together patient support groups, researchers, doctors and allied health professionals representing patients and families living with and affected by ciliopathies such as NPHP and is a valuable resource for families and their physicians.

Presently, there is no cure for NPHP and related ciliopathies. Clinicians must focus on optimising the delivery of renal replacement therapy, ideally with renal transplantation where possible. However, with a growing understanding of the pathophysiology of NPHP, the future is more hopeful. In recent years various drugs including vasopressin receptor antagonists (Gattone et al. 2003), mTOR inhibitors (mammalian target of rapamycin,) (Shillingford et al. 2006), triptolide (Leuenroth et al. 2008) and roscovitine (cyclin-dependent kinase inhibitor) (Bukanov et al. 2006) have been shown to be effective in reducing renal cysts in animal models of NPHP and ADPKD. Many of these drugs are currently or have recently been involved in clinical trials in adult patients.

Summary

The discovery of genetic causes of NPHP and its associated syndromes has led to important insights into this disease. To date, at least 16 genes are implicated in the pathogenesis of NPHP, with variable associated phenotypes.

The ciliary hypothesis underlying cystic kidney disease allows disease pathways to be explored and clinical trials to be undertaken, targeting cyst formation.

The clinical management of NPHP centres on the management of progressive CKD and preparation for renal replacement therapy. Genetic advances should allow a molecular genetic diagnosis to be made in around a third of cases and allows screening of at risk relatives and informed decisions regarding future pregnancies to be made.

References

Ala-Mello, S. & Al, E. 1998. Molecular studies in Finnish patients with familial juvenile nephronophthisis exclude a founder effect and support a common mutation causing mechanism. *J Med Genet* **35**, 279–283.

Ala-Mello, S., Kivivuori, S. M., Ronnholm, K. A., Koskimies, O. & Siimes, M. A. 1996. Mechanism underlying early anaemia in children with familial juvenile nephronophthisis. *Pediatr Nephrol* **10**, 578–581.

Attanasio, M., Uhlenhaut, N. H., Sousa, V. H., O'Toole J. F., Otto, E., Anlag, K., et al. 2007. Loss of GLIS2 causes nephronophthisis in humans and mice by increased apoptosis and fibrosis. *Nat Genet* **39**, 1018–1024.

Baala, L., Audollent, S., Martinovic, J., Ozilou, C., Babron, M. C., Sivanandamoorthy, S., et al. 2007. Pleiotropic effects of CEP290 (NPHP6) mutations extend to Meckel syndrome. *Am J Hum Genet* **81**, 170–179.

Betleja, E. & Cole, D. G. 2010. Ciliary trafficking: CEP290 guards a gated community. *Curr Biol* **20**, R928–R931.

Bollee, G., Fakhouri, F., Karras, A., Noel, L. H., Salomon, R., Servais, A., et al. 2006. Nephronophthisis related to homozygous NPHP1 gene deletion as a cause of chronic renal failure in adults. *Nephrol Dial Transplant* **21**, 2660–2663.

Bredrup, C., Saunier, S., Oud, M. M., Fiskerstrand, T., Hoischen, A., Brackman, D., et al. 2011. Ciliopathies with skeletal anomalies and renal insufficiency due to mutations in the IFT-A gene WDR19. *Am J Hum Genet* **89**, 634–643.

Bukanov, N. O., Smith, L. A., Klinger, K. W., Ledbetter, S. R. & Ibraghimov-Beskrovnaya, O. 2006. Long-lasting arrest of murine polycystic kidney disease with CDK inhibitor roscovitine. *Nature* **444**, 949–952.

Chaki, M., Airik, R., Ghosh, A., Giles, R., Chen, R., Slaats, G., et al. 2012. Exome capture reveals ZNF423 and CEP164 mutations, linking renal ciliopathies to DNA damage response signaling. *Cell*. **150**, 533–548.

Coppieters, F., Casteels, I., Meire, F., De Jaegere, S., Hooghe, S., Van Regemorter, N., et al. 2010a. Genetic screening of LCA in Belgium: Predominance of CEP290 and identification of potential modifier alleles in AHI1 of CEP290-related phenotypes. *Hum Mutat* **31**, E1709–1766.

Coppieters, F., Lefever, S., Leroy, B. P. & De Baere, E. 2010b. CEP290, a gene with many faces: Mutation overview and presentation of CEP290base. *Hum Mutat* **31**, 1097–1108.

Craige, B., Tsao, C. C., Diener, D. R., Hou, Y., Lechtreck, K. F., Rosenbaum, J. L. & Witman, G. B. 2010. CEP290 tethers flagellar transition zone microtubules to the membrane and regulates flagellar protein content. *J Cell Biol* **190**, 927–940.

Davis, E. E., Zhang, Q., Liu, Q., Diplas, B. H., Davey, L. M., Hartley, J., et al. 2011. TTC21B contributes both causal and modifying alleles across the ciliopathy spectrum. *Nat Genet* **43**, 189–196.

Den Hollander, A. I., Koenekoop, R. K., Yzer, S., Lopez, I., Arends, M. L., Voesenek, K. E., et al. 2006. Mutations in the CEP290 (NPHP6) gene are a frequent cause of Leber congenital amaurosis. *Am J Hum Genet* **79**, 556–561.

Fanconi, G., Hanhart, E., Von, A. A., Uhlinger, E., Dolivo, G. & Prader, A. 1951. [Familial, juvenile nephronophthisis (idiopathic parenchymal contracted kidney)]. *Helv Paediatr Acta*, **6**, 1–49.

Gagnadoux, M. F., Bacri, J. L., Broyer, M. & Habib, R. 1989. Infantile chronic tubulo-interstitial nephritis with cortical microcysts: Variant of nephronophthisis or new disease entity? *Pediatr Nephrol* **3**, 50–55.

Gattone, V. H. 2nd, Wang, X., Harris, P. C. & Torres, V. E. 2003. Inhibition of renal cystic disease development and progression by a vasopressin V2 receptor antagonist. *Nat Med* **9**, 1323–1326.

Haider, N. B., Carmi, R., Shalev, H., Sheffield, V. C. & Landau, D. 1998. A Bedouin kindred with infantile nephronophthisis demonstrates linkage to chromosome 9 by homozygosity mapping. *Am J Hum Genet* **63**, 1404–1410.

Hildebrandt, F., Attanasio, M. & Otto, E. 2009. Nephronophthisis: Disease mechanisms of a ciliopathy. *J Am Soc Nephrol* **20**, 23–35.

Hildebrandt, F. & Otto, E. 2005. Cilia and centrosomes: A unifying pathogenic concept for cystic kidney disease? *Nat Rev Genet* **6**, 928–940.

Hildebrandt, F., Otto, E., Rensing, C., Nothwang, H. G., Vollmer, M., Adolphs, J., et al. 1997. A novel gene encoding an SH3 domain protein is mutated in nephronophthisis type 1. *Nat Genet* **17**, 149–153.

Hildebrandt, F., Sayer, J. A., Jungers, P. & Grunfeld, J.-P. 2006. Nephronophthisis–medullary cystic and medullary sponge kidney disease, in *Diseases of the Kidney and Urinary Tract*, ed. by **Schrier, R. W.** Philadelphia: Lippincott Williams & Wilkins.pp. 478–501.

Hildebrandt, F., Waldherr, R., Kutt, R. & Brandis, M. 1992. The nephronophthisis complex: Clinical and genetic aspects. *Clin Invest* **70**, 802–808.

Hildebrandt, F. & Zhou, W. 2007. Nephronophthisis-associated ciliopathies. *J Am Soc Nephrol* **18**, 1855–1871.

Hoefele, J., Nayir, A., Chaki, M., Imm, A., Allen, S. J., Otto, E. A., et al. 2011. Pseudodominant inheritance of nephronophthisis caused by a homozygous NPHP1 deletion. *Pediatr Nephrol* **26**, 967–971.

Hurd, T. W. & Hildebrandt, F. 2011. Mechanisms of nephronophthisis and related ciliopathies. *Nephron Exp Nephrol* **118**, e9–14.

Krishnan, R., Eley, L. & Sayer, J. A. 2008. Urinary concentration defects and mechanisms underlying nephronophthisis. *Kidney Blood Press Res* **31**, 152–162.

Leuenroth, S. J., Bencivenga, N., Igarashi, P., Somlo, S. & Crews, C. M. 2008. Triptolide reduces cystogenesis in a model of ADPKD. *J Am Soc Nephrol* **19**, 1659–1662.

Mihatsch, M. J., Gudat, F., Zollinger, H. U., Heierli, C., Tholen, H. & Reutter, F. W. 1979. Systemic karyomegaly associated with chronic interstitial nephritis. A new disease entity? *Clin Nephrol* **12**, 54–62.

Mollet, G., Salomon, R., Gribouval, O., Silbermann, F., Bacq, D., Landthaler, G., et al. 2002. The gene mutated in juvenile nephronophthisis type 4 encodes a novel protein that interacts with nephrocystin. *Nat Genet* **32**, 300–305.

Morgan, D., Eley, L., Sayer, J., Strachan, T., Yates, L. M., Craighead, A. S., et al. 2002. Expression analyses and interaction with the anaphase promoting complex protein Apc2 suggest a role for inversin in primary cilia and involvement in the cell cycle. *Hum Mol Genet* **11**, 3345–3350.

Nauli, S. M., Alenghat, F. J., Luo, Y., Williams, E., Vassilev, P., Li, X., et al. 2003. Polycystins 1 and 2 mediate mechanosensation in the primary cilium of kidney cells. *Nat Genet* **33**, 129–137.

O'Toole, J. F., Liu, Y., Davis, E. E., Westlake, C. J., Attanasio, M., Otto, E. A., et al. 2010. Individuals with mutations in XPNPEP3, which encodes a mitochondrial protein, develop a nephronophthisis-like nephropathy. *J Clin Invest* **120**, 791–802.

Olbrich, H., Fliegauf, M., Hoefele, J., Kispert, A., Otto, E., Volz, A., et al. 2003. Mutations in a novel gene, *NPHP3*, cause adolescent nephronophthisis, tapeto-retinal degeneration and hepatic fibrosis. *Nat Genet* **34**, 455–459.

Otto, E., Hoefele, J., Ruf, R., Mueller, A. M., Hiller, K. S., Wolf, M. T., et al. 2002. A gene mutated in nephronophthisis and retinitis pigmentosa encodes a novel protein, nephroretinin, conserved in evolution. *Am J Hum Genet* **71**, 1167–1171.

Otto, E. A., Hurd, T. W., Airik, R., Chaki, M., Zhou, W., Stoetzel, C., et al. 2010. Candidate exome capture identifies mutation of SDCCAG8 as the cause of a retinal–renal ciliopathy. *Nat Genet* **42**, 840–850.

Otto, E. A., Loeys, B., Khanna, H., Hellemans, J., Sudbrak, R., Fan, S., et al. 2005. Nephrocystin-5, a ciliary IQ domain protein, is mutated in Senior–Loken syndrome and interacts with RPGR and calmodulin. *Nat Genet* **37**, 282–288.

Otto, E. A., Schermer, B., Obara, T., O'Toole, J. F., Hiller, K. S., Mueller, A. M., et al. 2003. Mutations in INVS encoding inversin cause nephronophthisis type 2, linking renal cystic disease to the function of primary cilia and left–right axis determination. *Nat Genet* **34**, 413–420.

Otto, E. A., Tory, K., Attanasio, M., Zhou, W., Chaki, M., Paruchuri, Y., et al. 2009. Hypomorphic mutations in meckelin (MKS3/TMEM67) cause nephronophthisis with liver fibrosis (NPHP11). *J Med Genet* **46**, 663–670.

Otto, E. A., Trapp, M. L., Schultheiss, U. T., Helou, J., Quarmby, L. M. & Hildebrandt, F. 2008. NEK8 mutations affect ciliary and centrosomal localization and may cause nephronophthisis. *J Am Soc Nephrol* **19**, 587–592.

Palmer, D., Lallu, S., Matheson, P., Bethwaite, P. & Tompson, K. 2007. Karyomegalic interstitial nephritis: A pitfall in urine cytology. *Diagn Cytopathol* **35**, 179–182.

Potter, D. E., Holliday, M. A., Piel, C. F., Feduska, N. J., Belzer, F. O. & Salvatierra Jr, **O.** 1980. Treatment of end-stage renal disease in children: A 15-year experience. *Kidney Int* **18**, 103–109.

Salomon, R., Saunier, S. & Niaudet, P. 2009. Nephronophthisis. *Pediatr Nephrol* **24**, 2333–2344.

Sang, L., Miller, J. J., Corbit, K. C., Giles, R. H., Brauer, M. J., Otto, E. A., et al. 2011. Mapping the NPHP-JBTS-MKS protein network reveals ciliopathy disease genes and pathways. *Cell*, **145**, 513–528.

Saunier, S., Calado, J., Heilig, R., Silbermann, F., Benessy, F., Morin, G., et al. 1997. A novel gene that encodes a protein with a putative src homology 3 domain is a candidate gene for familial juvenile nephronophthisis. *Hum Mol Genet* **6**, 2317–2323.

Sayer, J. A., Otto, E. A., O'Toole, J. F., Nurnberg, G., Kennedy, M. A., Becker, C., et al. 2006. The centrosomal protein nephrocystin-6 is mutated in Joubert syndrome and activates transcription factor ATF4. *Nat Genet* **38**, 674–681.

Shillingford, J. M., Murcia, N. S., Larson, C. H., Low, S. H., Hedgepeth, R., Brown, N., et al. 2006. The mTOR pathway is regulated by polycystin-1, and its inhibition reverses renal cystogenesis in polycystic kidney disease. *Proc Natl Acad Sci U.S.A.* **103**, 5466–5471.

Simms, R. J., Eley, L. & Sayer, J. A. 2009. Nephronophthisis. *Eur J Hum Genet* **17**, 406–416.

Simms, R. J., Hynes, A. M., Eley, L. & Sayer, J. A. 2011. Nephronophthisis: A genetically diverse ciliopathy. *Int J Nephrol* **2011**, 527137.

Strauss, M. B. & Sommers, S. C. 1967. Medullary cystic disease and familial juvenile nephronophthisis. *New Engl J Med* **277**, 863–864.

Tory, K., Lacoste, T., Burglen, L., Moriniere, V., Boddaert, N., Macher, M. A., et al. 2007. High NPHP1 and NPHP6 mutation rate in patients with Joubert syndrome and nephronophthisis: Potential epistatic effect of NPHP6 and AHI1 mutations in patients with NPHP1 mutations. *J Am Soc Nephrol* **18**, 1566–1575.

Valente, E. M., Brancati, F. & Dallapiccola, B. 2008. Genotypes and phenotypes of Joubert syndrome and related disorders. *Eur J Med Genet* **51**, 1–23.

Valente, E. M., Silhavy, J. L., Brancati, F., Barrano, G., Krishnaswami, S. R., Castori, M., et al. 2006. Mutations in CEP290, which encodes a centrosomal protein, cause pleiotropic forms of Joubert syndrome. *Nat Genet* **38**, 623–625.

Wolf, M. T. & Hildebrandt, F. 2011. Nephronophthisis. *Pediatr Nephrol* **26**, 181–194.

Wolf, M. T., Saunier, S., O'Toole, J. F., Wanner, N., Groshong, T., Attanasio, M., et al. 2007. Mutational analysis of the RPGRIP1L gene in patients with Joubert syndrome and nephronophthisis. *Kidney Int* **72**, 1520–1526.

Zhou, W., Otto, E. A., Cluckey, A., Airik, R., Hurd, T. W., Chaki, M., et al. 2012. FAN1 mutations cause karyomegalic interstitial nephritis, linking chronic kidney failure to defective DNA damage repair. *Nat Genet* **44**, 910–915.

Zollinger, H. U., Mihatsch, M. J., Edefonti, A., Gaboardi, F., Imbasciati, E. & Lennert, T. 1980. Nephronophthisis (medullary cystic disease of the kidney). A study using electron microscopy, immunofluorescence, and a review of the morphological findings. *Helv Paediatr Acta* **35**, 509–530.

Chapter 9

Oral–facial–digital type I syndrome

Brunella Franco

History of the eponym

In 1941 Mohr described a family with a distinct pattern of oral and digital anomalies. Four male siblings out of a family of seven were affected (Mohr, 1941). Papillon-Leage and Psaume in 1954 reported a group of female patients with similar findings and first described this rare hereditary condition as a clear defined entity. The condition was further defined by Gorlin and Psaume in 1962 (Gorlin & Psaume, 1962). The disease consisting of congenital anomalies of oral, facial and digital structures was initially called and known as orofaciodigital dysostosis. Over the years polycystic kidney disease was recognised as a clinical sign commonly observed in the disease. In 1967 Rimoin and Edgerton suggested that Mohr, as well as Papillon-Leage and Psaume, had identified two genetically distinct syndromes that could be distinguished on the basis of clinical features and genetic inheritance and proposed the terms oral–facial–digital type I syndrome for the X-linked dominant form and oral–facial–digital type II syndrome for the autosomal recessive trait (Mohr syndrome) (Rimoin & Edgerton, 1967). Abbreviations for the disease include: OFDI, OFD1, OFDSI, OFDS1. The OMIM entry is MIM 311200 and can be retrieved at http://omim.org/entry/311200.

Epidemiology

Oral–facial–digital (OFD) type I is part of the heterogeneous group of OFD syndromes of which 13 different forms have been described to date (Gurrieri et al. 2007). This hereditary condition is quite rare. Wahrman and colleagues suggested that the frequency of the syndrome is approximately 1/50,000 live births (Wahrman et al. 1966). Melnick and collegues indicated an estimated incidence of 1/250,000 live births in 1975 (Melnick & Shields, 1975). To date, over 130 molecularly diagnosed cases (including affected relatives of familial cases) have been described worldwide from Europe, North America, Australia, Asia and the Middle East (Macca & Franco, 2009). The disease occurs in different ethnic backgrounds (Prattichizzo et al. 2008; Salinas et al. 1991). The vast majority of cases (>75%) has an apparently sporadic presentation and represents *de novo* mutations (Prattichizzo et al. 2008; Thauvin-Robinet et al. 2006).

Clinical features

OFD type I is an X-linked dominant male lethal disorder. Male lethality usually occurs within the second trimester of pregnancy (Doege et al. 1964; Wettke Schäfer & Kantner, 1983). The clinical spectrum of the disease includes malformations of the face, oral cavity and digits (Figure 9.1). Involvement of the central nervous system (CNS) and renal cystic disease are frequently reported in this condition. Affected individuals may display a mild phenotype especially since some of the defects are often surgically corrected in childhood.

Oral abnormalities

Malformation of the oral cavity are present in almost 97% of cases and include tongue abnormalities (lobulated tongue, hamartomas, clefts); hyperplastic and aberrant oral frenula; cleft and/or high arched palate; alveolar ridge clefting and abnormalities of the teeth (malposition of the maxillary canine teeth, infra-occlusions, agenesis of lower lateral incisor teeth and supernumerary teeth).

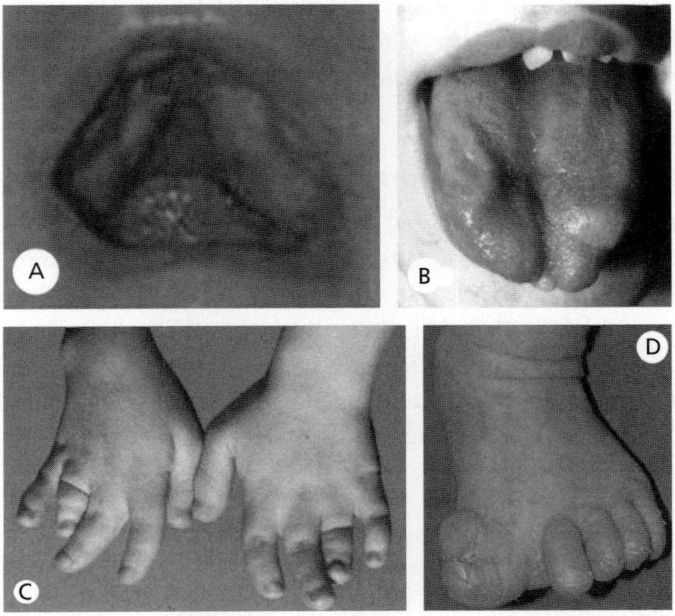

Figure 9.1 Examples of the oral–facial–digital findings observed in OFDI patients. Cleft palate (A). Bifid and lobulated tongue (B). Limb abnormalities are also a frequent finding and include brachydactyly and clinodactyly (C) and duplication of the hallux (D).

Reproduced with permission from Prattichizzo et al., Mutational spectrum of the oral–facial–digital type I syndrome: a study on a large collection of patients, *Human Mutation*, Volume 29, Issue 10, pp. 1237–1246, Copyright © 2008 Wiley-Liss, Inc.

Craniofacial abnormalities

Craniofacial abnormalities are observed in >87% of cases and include facial dysmorphism (down-slanting palpebral fissures, hypoplasia alae nasi, hypertelorism, telecanthus, microretrognatia, flat face, facial asymmetry, frontal bossing) and cleft/pseudocleft of the lips; evanescent milia of the face and ears, which tend to disappear within the third year of life; abnormal (dry or brittle) hair and/or alopecia, particularly evident over skull prominences.

Skeletal abnormalities

Skeletal abnormalities are observed in >88% of cases and include brachydactyly (>60%), clinodactyly (>45%), syndactyly (>49%) and pre-axial (19%) and more rarely post-axial polydactyly (>3%). A broad thumb and duplication of the hallux can be often observed. Interestingly, forelimbs seem to be more affected than hind limbs (83% versus 44%). Short stature is present in 7% of cases. Other skeletal manifestations reported in few cases include increasing of the nasion-sella-basion (cranial base) angle (Aduss & Pruzansky, 1954; Gorlin & Psaume, 1962; Stahl & Furhmann, 1968) and abnormalities of the long and short bones. A few studies report that the tubular bones of the hands and feet are irregularly short. In addition, thick and irregular reticular pattern of radiolucency and/or spicule like formation in metacarpals and, especially phalanges, have been described (Thauvin-Robinet et al. 2006; Wood et al. 1975). Cone-shaped epiphyses of the fingers and irregularities of the long bones have also been observed (Stapleton et al. 1982).

Central nervous system malformations

Involvement of the CNS is reported in about 50% of cases and includes mental retardation/selective cognitive impairment and/or CNS malformations. The malformations most commonly observed comprise agenesis of the corpus callosum, intracerebral single or multiple epithelial or arachnoidal cysts and porencephaly, heterotopia grey matter, cerebellar abnormalities, abnormal gyration, microcephaly (Bisschoff et al. 2013; Coll et al. 1997; Connacher et al. 1987; Gorlin & Psaume, 1962; Holub et al. 2005; Odent et al. 1998; Towfighi et al. 1985). In a few cases precocious puberty due to hypothalamic hamartoma has been reported (Rakkolainen et al. 2002; Somer et al. 1986).

Renal abnormalities

Renal cystic disease is observed in over 60% of adult OFD type I patients (Prattichizzo et al. 2008; Saal et al. 2010). Renal cystic disease usually starts in the second and third decade of life with few reports of occurrence in the first decade of life and examples of patients in which the renal involvement completely dominates the clinical course of the disease (Coll et al. 1997; Feather et al. 1997a). Histological analysis demonstrated that the majority of renal cysts are of glomerular origin. In contrast to what is observed in autosomal dominant polycystic kidney disease (ADPKD), the cystic kidneys in OFD type I are usually of normal size or moderately increased and the renal cysts do not alter the contour of the kidneys (see Table 9.1).

Table 9.1 The clinical spectrum in oral–facial–digital type I syndrome

Clinical features	**Published cases with mutations in % (*n*)**
CRANIOFACIAL	87.30% (110/126)
Abnormal hair/alopecia	21.5% (29/135)
Milia	29.4% (37/126)
Facial dysmorphism	69.1% (87/126)
Cleft lip/pseudocleft of the upper lip	32.6% (44/135)
ORAL	96.8% (122/126)
Tongue anomalies	84.1% (106/126)
Oral frenula	63.7% (86/135)
Alveolar ridge clefting	22.2% (28/126)
Cleft palate/high arched palate	49.6% (67/135)
Teeth abnormalities	43.3% (58/134)
SKELETAL	88.1% (111/126)
Forelimb	83.3% (105/126)
Hind limb	44.4% (56/126)
Brachydactyly	63.7% (86/135)
Clinodactyly	47.4% (64/135)
Syndactyly	49.6% (67/135)
Polydactyly pre-axial	19.3% (26/135)
Polydactyly post-axial	3.7% (5/135)
KIDNEYS	
Cystic kidney disease	37.3% (50/134)
NEUROLOGIC	
CNS involvement*	48.4% (61/126)
Mental retardation (MR)	28.9% (39/135)

The total number of cases for each category was determined according to the available information. #135 includes all patients described in Ferrante et al. 2001a; Morisawa et al. 2004; Prattichizzo et al. 2008; Rakkolainen et al. 2002; Romio et al. 2003; Thauvin-Robinet et al. 2006, 2009. #126 includes the same patients with the exception of those described in Thauvin-Robinet et al. 2009. #134 include all cases but one described in Thauvin-Robinet et al. 2009 for which no information was available for that specific clinical feature. Index cases as well as affected relatives carrying the mutation were considered. *Includes mental retardation (MR)/selective cognitive impairment and/or CNS malformations.

Table 9.1 summarises the clinical spectrum observed in OFD type I syndrome. Additional details have been given by other authors (Franco, 2008; Macca & Franco, 2009). A few cases (>5%) were reported with conductive/sensorineural/central hearing loss while pancreatic, hepatic, and/or ovarian cysts were reported in about 5% of cases (Macca & Franco, 2009; Prattichizzo et al. 2008). A recent report further enhanced the possible occurrence of fibrocystic disease of liver and pancreas in this condition (Chetty-John et al. 2010). Phenotypic variability, both intrafamilial and interfamilial, is often seen in affected females possibly due to X-inactivation (Franco & Ballabio, 2006; Morleo & Franco, 2008). Few exceptional OFDI male cases have been described to date: a patient with Klinefelter syndrome (Wahrman et al. 1966); a 34-week live-born male from a family displaying a clear X-linked dominant inheritance pattern of the disease, this patient, however, developed cardiac failure and died 21 hours after delivery (Goodship et al. 1991); a newborn male born at term who died 4 hours after birth with typical signs of OFD type I including cystic kidneys (Gillerot et al. 1993). A recent report described a family with three affected male neonates with an 'unclassified' X-linked lethal congenital malformation syndrome. Exome sequencing revealed a splicing mutation of *OFD1*. The affected males manifested severe multisystem complications in addition to the cardinal features of OFD I and the carrier female showed only subtle features of OFD I (Tsurusaki et al. 2012).

Range of the phenotype

A large body of evidences indicates that mutations of the same gene may cause different ciliopathy phenotypes. To date, the gene mutated in OFD type I, named *OFD1*, has been found mutated in other disorders, which represent different aspects of ciliopathies. Budny and colleagues reported mutation in the OFD1 transcript in a novel X-linked recessive mental retardation syndrome comprising microcephaly, dysmorphic facies, broad thumbs, short fingers and obesity and ciliary dysfunction of respiratory cilia resulting in recurrent respiratory tract infections (Budny et al. 2006). Phenotypic similarities with a family diagnosed with a severe form of Simpson–Golabi–Behmel syndrome type 2 (SGBS2) have been reported (Terespolsky et al. 1995) and subsequently mapped to Xp22 in a region comprising the *OFD1* gene (Brzustowicz et al. 1999). The *OFD1* transcript has been associated also to mutations in X-linked recessive Joubert syndrome (JBTS10) (Coene et al. 2009; Field et al. 2012; Juric-Sekhar et al. 2012) and more recently in X-linked recessive retinitis pigmentosa (RP23) (Webb et al. 2012). In all these disorders a ciliary dysfunction has been demonstrated and further studies will be needed to fully understand the molecular mechanisms underlying the different phenotypic manifestations occurring when the *OFD1* transcript is mutated.

Diagnosis

The diagnosis of OFD type I is usually established at birth on the basis of characteristic oral, facial, and digital anomalies observed in the majority of affected infants; in other instances, and especially in mildly affected cases in which malformations are surgically corrected in early childhood, the diagnosis is suspected only at more advanced stages, in later childhood or adulthood, after the appearance of polycystic kidney disease. Molecular genetic testing is available for this condition and information on participating laboratories can be obtained at the Gene Tests Laboratory Directory (http://www.genetests.org/), at the Orphanet site (http://www.orpha.net/consor/cgi-bin/index.php, and the European Directory of DNA Laboratories (EDDNAL) (http://www.eddnal.com). Molecular testing is recommended in all female cases when the diagnosis of OFD type I is suspected.

Molecular testing relies on DNA direct sequencing of all coding exons and dosage analysis of negative cases to detect genomic rearrangements not identifiable by direct sequencing due to the presence of the wild-type allele in affected female patients. The association of the two approaches combined allows the identification of mutation in about 85% of cases. The remaining patients may represent conditions in which the available techniques did not allow the detection of the mutations, other type of OFD syndromes or different genetic diseases. Differential diagnosis should be considered with the additional forms of OFD syndromes. Recent studies demonstrated that TCTN3 is mutated in OFD IV, Mohr–Majewski syndrome, an extreme form of OFD associated with bone dysplasia, tibial defect, cystic kidneys, and brain anomalies (Thomas et al. 2012). In 2010 Valente and co-authors reported two unrelated Ashkenazi Jewish patients with Joubert syndrome type 2 (JBTS2) caused by the same homozygous mutation in the *TMEM216* gene. These patients had tongue tumours or multiple oral frenula, reminiscent of OFD VI, in addition to the molar tooth sign on brain imaging and polydactyly (Valente et al. 2010). Subsequent studies on a larger cohort of cases established that oral–facial–digital syndrome type VI represents a rare phenotypic subtype of Joubert syndrome and related disorders (JSRD) (Poretti et al. 2012).

Interesting features of the condition

OFD type I syndrome shares with other ciliopathies common features such as renal cystic disease, involvement of the central nervous system and skeletal abnormalities. On the other hand there are specific features such as teeth malformations, and involvement of the skin, which are not usually observed in disorders associated to ciliary dysfunction. In addition while OFD I cases do not display retinitis pigmentosa recent reports indicate that reduced dose of correctly spliced *OFD1* transcript may lead to isolated retinal degeneration (Webb et al. 2012). These observations suggest that much more needs to be understood on the pathological processes underlying ciliopathies and it is possible that a deeper understanding of these processes will impact on our understanding of more common pathological processes.

Genetics

The locus for OFD type I syndrome was mapped to the Xp22 region (Feather et al. 1997b). A systematic mutation analysis of Xp22 transcripts identified CXORF5 (alias 71-7A) as the causative gene and the transcript was subsequently named *OFD1* (Ferrante et al. 2001b).

To date, 130 different mutations (nine genomic deletions and 121 point mutations, mostly represented by frameshifts) have been identified and over 70% of patients represent sporadic cases (reviewed in Macca & Franco, 2009 and Bisschoff et al. 2013). Frameshifts account for 51,5% of mutations. The point mutations include missense mutations (11.5%), nonsense mutations (11.5%) splice site mutations (16.9%) and two in-frame deletions. Mutations are distributed along the first 17 exons of the protein and the exons most frequently involved are exons 3, 8, 9, 13, 16 and 12. Interestingly, some of the mutations occurred in more than one case and in particular a stretch of nine A at the 5′ end of exon 8 host 12 of the mutations identified to date suggesting the possibility that DNA replication errors may occur. Genomic deletions accounting for 23% of OFD I cases negative for DNA sequencing have also been described (Thauvin-Robinet et al. 2009). The majority of mutations identified to date in OFD I-affected cases result in premature truncation of the protein in its N-terminal region and are therefore predicted to act with a loss of function mechanism. Several studies have made attempts but to date no convincing genotype–phenotype correlation could be identified.

Underlying ciliary defect

Cilia are specialised organelles protruding from the cell surface and can be recognised on almost all mammalian cells. The primary cilium acts as an antenna for the cell, and several signalling pathways, which play important roles in development, such as hedgehog, Wnt and planar cell polarity are transduced through it. In addition, studies in animal models have revealed that during embryogenesis the primary cilium has an essential role in defining the correct patterning of the body. The *OFD1* transcript codifies for a protein localised at the centrosome and basal body (Giorgio et al. 2007; Romio et al. 2004) which play an important role in the formation of primary cilia as demonstrated by functional studies in animal models (Bimonte et al. 2010; D'Angelo et al. 2012; Ferrante et al. 2006, 2009; Zullo et al. 2010) and in *in vitro* systems (Singla et al. 2010).

Physiological effect of the ciliary defect

Mutations in ciliary proteins result in defective cilia and cause disorders called 'ciliopathies'. Many of the ciliopathies, including OFD type I display a clearly developmental phenotype given that defective primary cilia cannot exert their role in transducing signals throughout development. Functional studies have clearly demonstrated that impairment in cilia formation and function in the absence of *OFD1* causes defective sonic hedgehog (Shh) signalling which is the most probable cause of the skeletal abnormalities observed in OFD type I (Bimonte et al. 2010; Ferrante et al. 2006). Renal cystic disease is a common feature of ciliopathies and a large body of experimental evidences indicates that the vast majority of transcripts responsible for inherited forms of renal cystic disease codify for proteins localised to cilia (Yoder, 2007). Cilia are usually absent from epithelial cells lining renal cysts; however, recent data demonstrate that in a conditional murine model with kidney-specific inactivation of the *OFD1* gene, primary cilia initially form and then disappear after the development of cysts, suggesting that the absence of primary cilia is a consequence rather than the primary cause of renal cystic disease (Zullo et al. 2010). Future studies directed at understanding the relationship between the cilium and the renal cystic disease will hopefully provide important insights into the mechanisms of renal cyst pathogenesis and lead to better approaches for therapeutic intervention. CNS involvement is also frequently observed in OFD type I and includes mental retardation, cognitive defects and a wide range of malformations (Macca & Franco, 2009). Recent studies have provided experimental evidences on the role of Ofd1 transcript in dorsal–ventral patterning and axoneme elongation during embryonic brain development in the mouse (D'Angelo et al. 2012). Interestingly, recent lines of investigation have provided data on the role of primary cilia in neuronal signalling, development and in the maintenance of adult brain (Louvi & Grove, 2011). Cilia are critical regulators of Shh signalling on postnatal precursor cells and participate in orchestrating postnatal forebrain development and stem/precursor cell maintenance (Breunig et al. 2008). Impaired Shh signalling due to cilia dysfuction leads to defective hippocampal morphogenesis and dysregulation of mitotic activity in the mice's postnatal brain (Breunig et al. 2008), stressing the hippocampus as a target organ for CNS-related ciliopathic manifestations. Further research will help in clarifying the role of *OFD1* and primary cilia in brain development and higher brain functions.

Clinical management

OFD type I is an inherited developmental disorder and the clinical management should include treatment, surveillance and genetic counseling.

Treatment

Plastic surgery is indicated for correctable defects such as cleft lip/palate, tongue nodules, accessory frenulas, and digital abnormalities (syndactyly, polydactyly); removal of accessory teeth, and orthodontia for malocclusion. Specific learning disabilities should be identified and eventually speech therapy and special education may be warranted. Medical treatment should be considered for specific neurological condition (e.g. seizures). Treatment is routine for renal disease, when necessary patients require renal replacement therapy.

Surveillance

Periodic assessment for early diagnosis of the visceral manifestations of OFD I represent an important step for proper care and prevention of hepatic, pancreatic, and renal complications. A timely diagnosis and understanding of the fibrocystic nature of the hepatorenal and pancreatic disease in this condition may also prevent unnecessary and invasive diagnostic and therapeutic interventions. Regular speech and hearing assessment should be performed if cleft lip is present.

Genetic counselling

Appropriate genetic counselling should be provided to the patient and his/her family. *OFD1* is inherited as an X-linked dominant trait. Approximately 75% of affected individuals are sporadic cases with no family history. A female proband with *OFD1* may have the disorder as the result of a *de novo* gene mutation and recent studies indicate that the majority of cases (>77%) are indeed *de novo* mutations (Prattichizzo et al. 2008). The risk that the unaffected mother of an affected female who is a sporadic case will have another female with *OFD1* is less than 1%. At conception, the risk to the offspring of females with *OFD1* of inheriting the disease-causing *OFD1* allele is 50%; however, most male conceptuses with the disease-causing allele will be aborted. In few familial cases the mutation could not be detected from DNA extracted from peripheral blood samples of the affected mothers suggesting the possibility that mosaicism could occur and should thus be considered in OFD type I syndrome (Prattichizzo et al. 2008). Prenatal diagnosis for pregnancies at increased risk is possible in cases where the mutation underlying the genetic disease has been determined. Prenatal ultrasound examination may detect structural brain malformations and/or other defects (e.g. digital).

References

Aduss, H. & Pruzansky, S. 1954. Postnatal craniofacial development in children with the OFD syndrome. *Arch Oral Biol*, **9**, 193–203.

Bimonte, S., De Angelis, A., Quagliata, L., Giusti, F., Tammaro, R., Dallai, R., et al. 2010. Ofd1 is required in limb bud patterning and endochondral bone development. *Dev Biol*, **349**, 179–191.

Bisschoff, I. J., Zeschnigk, C., Horn, D. et al. (2013). Novel mutations including deletions of the entire OFD1 gene in 30 families with type 1 orofaciodigital syndrome: A study of the extensive clinical variability. *Hum Mut* **34**, 237–247.

Breunig, J. J., Sarkisian, M. R., Arellano, J. I., Morozov, Y. M., Ayoub, A. E., Sojitra, S., et al. 2008. Primary cilia regulate hippocampal neurogenesis by mediating sonic hedgehog signaling. *Proc Natl Acad Sci U.S.A.* **105**, 13127–13132.

Brzustowicz, L. M., Farrell, S., Khan, M. B. & Weksberg, R. 1999. Mapping of a new SGBS locus to chromosome Xp22 in a family with a severe form of Simpson–Golabi–Behmel syndrome. *Am J Hum Genet* **65**, 779–783.

Budny, B., Chen, W., Omran, H., Fliegauf, M., Tzschach, A., Wisniewska, M., et al. 2006. A novel X-linked recessive mental retardation syndrome comprising macrocephaly and ciliary dysfunction is allelic to oral–facial–digital type I syndrome. *Hum Genet* **120**, 171–178.

Chetty-John, S., Piwnica-Worms, K., Bryant, J., Bernardini, I., Fischer, R. E., Heller, T., et al. 2010. Fibrocystic disease of liver and pancreas; under-recognized features of the X-linked ciliopathy oral–facial–digital syndrome type 1 (OFD I). *Am J Med Genet, Part A*, **152A**, 2640–2645.

Coene, K. L., Roepman, R., Doherty, D., Afroze, B., Kroes, H. Y., Letteboer, S. J., Ngu, L. H., et al. 2009. OFD1 is mutated in X-linked Joubert syndrome and interacts with LCA5-encoded lebercilin. *Am J Hum Genet* **85**, 465–481.

Coll, E., Torra, R., Pascual, J., Botey, A., Ara, J., Perez, L., et al. 1997. Sporadic orofaciodigital syndrome type 1 presenting as end-stage renal disease. *Nephrol Dial Transplant* **12**, 1040–1042.

Connacher, A. A., Forsyth, C. C. & Stewart, W. K. 1987. Orofaciodigital syndrome type 1 associated with polycystic kidneys and agenesis of the corpus callosum. *J Med Genet* **24**, 116–118.

D'Angelo, A., De Angelis, A., Avallone, B., Piscopo, I., Tammaro, R., Studer, M., et al. 2012. Ofd1 controls dorso-ventral patterning and axoneme elongation during embryonic brain development. *PLoS One* **7**, e52937.

Doege, T., Thuline, H., Priest, J., Norby, D. & Bryant, J. 1964. Studies of a family with the oral–facial–digital syndrome. *New Engl J Med* **271**, 1073–1080.

Feather, S. A., Winyard, P. J., Dodd, S. & Woolf, A. S. 1997a. Oral–facial–digital syndrome type 1 is another dominant polycystic kidney disease: Clinical, radiological and histopathological features of a new kindred. *Nephrol-Dial-Transplant* **12**, 1354–1361.

Feather, S. A., Woolf, A. S., Donnai, D., Malcolm, S. & Winter, R. M. 1997b. The oral–facial–digital syndrome type 1 (OFD1), a cause of polycystic kidney disease and associated malformations, maps to Xp22.2–Xp22.3. *Hum Mol Genet* **6**, 1163–1167.

Ferrante, M. I., Ghiani, M., Bulfone, A. & Franco, B. 2001a. IL1RAPL2 maps to Xq22 and is specifically expressed in the central nervous system. *Gene* **275**, 217–221.

Ferrante, M. I., Giorgio, G., Feather, S. A., Bulfone, A., Wright, V., Ghiani, M., et al. 2001b. Identification of the gene for oral–facial–digital type I syndrome. *Am J Hum Genet* **68**, 569–576.

Ferrante, M. I., Romio, L., Castro, S., Collins, J. E., Goulding, D. A., Stemple, D. L., et al. 2009. Convergent extension movements and ciliary function are mediated by ofd1, a zebrafish orthologue of the human oral–facial–digital type 1 syndrome gene. *Hum Mol Genet* **18**, 289–303.

Ferrante, M. I., Zullo, A., Barra, A., Bimonte, S., Messaddeq, N., Studer, M., et al. 2006. Oral–facial–digital type I protein is required for primary cilia formation and left–right axis specification. *Nat Genet* **38**, 112–117.

Field, M., Scheffer, I. E., Gill, D., Wilson, M., Christie, L., Shaw, M., et al. 2012. Expanding the molecular basis and phenotypic spectrum of X-linked Joubert syndrome associated with OFD1 mutations. *Eur J Hum Genet* **20**, 806–809.

Franco, B. 2008. The molecular basis of oral–facial–digital type I (OFDI) syndrome, in *Inborn Errors of Development*, second edition, ed. by **Epstein, J. C. Erickson, R. P. & Wynshaw-Boris, A.** New York: Oxford University Press.

Franco, B. & Ballabio, A. 2006. X-inactivation and human disease: X-linked dominant male-lethal disorders. *Curr Opin Genet Dev* **16**, 254–259.

Gillerot, Y., Heimann, M., Fourneau, C., Verellen-Dumoulin, C. & Van Maldergem, L. 1993. Oral–facial–digital syndrome type I in a newborn male. *Am J Med Genet* **46**, 335–338.

Giorgio, G., Alfieri, M., Prattichizzo, C., Zullo, A., Cairo, S. & Franco, B. 2007. Functional characterization of the OFD1 protein reveals a nuclear localization and physical interaction with subunits of a chromatin remodeling complex. *Mol Biol Cell,* **18**, 4397–4404.

Goodship, J., Platt, J., Smith, R. & Burn, J. 1991. A male with type I orofaciodigital syndrome. *J Med Genet* **28**, 691–694.

Gorlin, R. J. & Psaume, J. 1962. Orofaciodigital dysotosis: A new syndrome. A study of 22 cases. *J Pediatr* **61**, 520–530.

Gurrieri, F., Franco, B., Toriello, H. & Neri, G. 2007. Oral–facial–digital syndromes: A review and diagnostic guidelines. *Am J Med Genet* **143A**, 3314–3323.

Holub, M., Potocki, L. & Bodamer, O. A. 2005. Central nervous system malformations in oral–facial–digital syndrome, type 1. *Am J Med Genet A*, **136**, 218.

Juric-Sekhar, G., Adkins, J., Doherty, D. & Hevner, R. F. 2012. Joubert syndrome: Brain and spinal cord malformations in genotyped cases and implications for neurodevelopmental functions of primary cilia. *Acta Neuropathol* **123**, 695–709.

Louvi, A. & Grove, E. A. 2011. Cilia in the CNS: The quiet organelle claims center stage. *Neuron* **69**, 1046–1060.

Macca, M. & Franco, B. 2009. The molecular basis of oral–facial–digital syndrome, type 1. *Am J Med Genet, C, Semin Med Genet* **151C**, 318–325.

Melnick, M. & Shields, E. 1975. Oral–facial–digital syndrome type 1: A phenotypic and genetic analysis. *Oral Surg* **40**, 599–610.

Mohr, O. 1941. A hereditary lethal syndrome in man. *Avh Norske Videnskad, Oslo* **14**, 1–18.

Morisawa, T., Yagi, M., Surono, A., Yokoyama, N., Ohmori, M., Terashi, H. et al. 2004. Novel double-deletion mutations of the OFD1 gene creating multiple novel transcripts. *Hum Genet* **115**, 97–103.

Morleo, M. & Franco, B. 2008. Dosage compensation of the mammalian X chromosome influences the phenotypic variability of X-linked dominant male-lethal disorders. *J Med Genet* **45**, 401–408.

Odent, S., Le Marec, B., Toutain, A., David, A., Vigneron, J., Treguier, C., et al. 1998. Central nervous system malformations and early end-stage renal disease in Oro-facio-digital syndrome type 1: A review. *Am J Med Genet* **75**, 389–394.

Poretti, A., Vitiello, G., Hennekam, R. C., Arrigoni, F., Bertini, E., Borgatti, R., et al. 2012. Delineation and diagnostic criteria of oral–facial–digital syndrome type VI. *Orphanet J Rare Dis* **7**, 4.

Prattichizzo, C., Macca, M., Novelli, V., Giorgio, G., Barra, A. & Franco, B. 2008. Mutational spectrum of the oral–facial–digital type I syndrome: A study on a large collection of patients. *Hum Mutat* **29**, 1237–1246.

Rakkolainen, A., Ala-Mello, S., Kristo, P., Orpana, A. & Jarvela, I. 2002. Four novel mutations in the OFD1 (Cxorf5) gene in Finnish patients with oral–facial–digital syndrome 1. *J Med Genet* **39**, 292–296.

Rimoin, D. & Edgerton, M. 1967. Genetic and clinical heterogeneity in the oral–facial–digital syndromes. *J Pediatr* **71**, 94–102.

Romio, L., Fry, A. M., Winyard, P. J., Malcolm, S., Woolf, A. S. & Feather, S. A. 2004. OFD1 is a centrosomal/basal body protein expressed during mesenchymal–epithelial transition in human nephrogenesis. *J Am Soc Nephrol* **15**, 2556–2568.

Romio, L., Wright, V., Price, K., Winyard, P. J., Donnai, D., Porteous, M. E., et al. 2003. OFD1, the gene mutated in oral–facial–digital syndrome type 1, is expressed in the metanephros and in human embryonic renal mesenchymal cells. *J Am Soc Nephrol* **14**, 680–689.

Saal, S., Faivre, L., Aral, B., Gigot, N., Toutain, A., Van Maldergem, L., et al. 2010. Renal insufficiency, a frequent complication with age in oral–facial–digital syndrome type I. *Clin Genet* **77**, 258–265.

Salinas, C., Pai, G., Vera, C., Milutinovich, J., Hagerty, R., Cooper, J., et al. 1991. Variability of expression of the orofaciodigital syndrome type 1 in black females: Six cases. *Am J Med Genet* **38**, 574–582.

Singla, V., Romaguera-Ros, M., Garcia-Verdugo, J. M. & Reiter, J. F. 2010. Ofd1, a human disease gene, regulates the length and distal structure of centrioles. *Dev Cell* **18**, 410–424.

Somer, M., Lindahl, E. & Perheentupa, J. 1986. Precocious puberty associated with oral–facial–digital syndrome type I. *Acta Paediatr Scand* **75**, 672–675.

Stahl, A. & Furhmann, W. 1968. Oro-facio-digitales Syndrom. Klinik und Genetik. *Deutsche medizinische Wochenschrift* **93**, 1224–1228.

Stapleton, F. B., Bernstein, J., Koh, G., Roy, S.3rd & Wilroy, R. S. 1982. Cystic kidneys in a patient with oral–facial–digital syndrome type I. *Am J Kidney Dis* **1**, 288–293.

Terespolsky, D., Farrell, S. A., Siegel-Bartelt, J. & Weksberg, R. 1995. Infantile lethal variant of Simpson–Golabi–Behmel syndrome associated with hydrops fetalis. *Am J Med Genet* **59**, 329–333.

Thauvin-Robinet, C., Cossee, M., Cormier-Daire, V., Van Maldergem, L., Toutain, A., Alembik, Y., et al. 2006. Clinical, molecular, and genotype–phenotype correlation studies from 25 cases of oral–facial–digital syndrome type 1: A French and Belgian collaborative study. *J Med Genet* **43**, 54–61.

Thauvin-Robinet, C., Franco, B., Saugier-Veber, P., Aral, B., Gigot, N., Donzel, A., et al. 2009. Genomic deletions of OFD1 account for 23% of oral–facial–digital type 1 syndrome after negative DNA sequencing. *Hum Mutat* **30**, E320–329.

Thomas, S., Legendre, M., Saunier, S., Bessieres, B., Alby, C., Bonniere, M., et al. 2012. TCTN3 mutations cause Mohr–Majewski syndrome. *Am J Hum Genet* **91**, 372–378.

Towfighi, J., Berlin, C. M. Jr. Ladda, R. L., Frauenhoffer, E. E. & Lehman, R. A. 1985. Neuropathology of oral–facial–digital syndromes. *Arch Pathol Lab Med* **109**, 642–646.

Tsurusaki, Y., Kosho, T., Hatasaki, K., Narumi, Y., Wakui, K., Fukushima, Y., et al. 2012. Exome sequencing in a family with an X-linked lethal malformation syndrome: Clinical consequences of hemizygous truncating OFD1 mutations in male patients. *Clin Genet* **83**, 135–144.

Valente, E. M., Logan, C. V., Mougou-Zerelli, S., Lee, J. H., Silhavy, J. L., Brancati, F., et al. 2010. Mutations in TMEM216 perturb ciliogenesis and cause Joubert, Meckel and related syndromes. *Nat Genet* **42**, 619–625.

Wahrman, J., Berant, M., Jacobs, J., Aviad, I. & Ben-Hur, N. 1966. The oral–facial–digital syndrome: A male-lethal condition in a boy with 47/xxy chromosomes. *Pediatrics* **37**, 812–821.

Webb, T. R., Parfitt, D. A., Gardner, J. C., Martinez, A., Bevilacqua, D., Davidson, A. E., et al. 2012. Deep intronic mutation in OFD1, identified by targeted genomic next-generation sequencing, causes a severe form of X-linked retinitis pigmentosa (RP23). *Hum Mol Genet* **21**, 3647–3654.

Wettke Schäfer, R. & Kantner, G. 1983. X-linked dominant inherited diseases with lethality in hemizygous males. *Hum Genet* **64**, 1–23.

Wood, B. P., Young, L. W. & Townes, P. L. 1975. Cerebral abnormalities in the oral–facial–digital syndrome. *Pediatr Radiol* **3**, 130–136.

Yoder, B. K. 2007. Role of primary cilia in the pathogenesis of polycystic kidney disease. *JASN, J Am Soc Nephrol* **18**, 1381–1388.

Zullo, A., Iaconis, D., Barra, A., Cantone, A., Messaddeq, N., Capasso, G., et al. 2010. Kidney-specific inactivation of Ofd1 leads to renal cystic disease associated with upregulation of the mTOR pathway. *Hum Mol Genet* **19**, 2792–2803.

Chapter 10

Autosomal dominant polycystic kidney disease

Richard Sandford

History of the eponym

Historically, polycystic kidney disease (PKD) remained a poorly described clinical entity until the 1800s despite early autopsy descriptions in the 16th and 17th centuries (Torres & Watson, 1998). Perhaps the first reported medical case was that of the Polish king Stefan Bathory, who died in 1585 at the age of 53 from symptoms suggestive of uraemia. His autopsy almost certainly revealed PKD. A description in 1820 by Adelon of what was likely to be extensive polycystic kidney and polycystic liver disease and a further description by Reyer in 1841 of 'cystic degeneration of the kidneys' formed the link between abnormal organ structure and disease. In the following years other case reports linked cystic disease of the kidneys and liver, and Lejars in 1888 first used the term 'polycystic kidneys'. By the end of the 19th century PKD became an established clinical, pathological and anatomical diagnosis. Its genetic basis was first recognised by Steiner in 1899.

The disease is currently known as autosomal dominant polycystic kidney disease (ADPKD), reflecting its inheritance and pathology, a term that also distinguishes it from other polycystic diseases, specifically autosomal recessive polycystic kidney disease (ARPKD). Further refinement using molecular pathology is also used in the OMIM description of the disease (www.ncbi.nlm.nih.gov/omim). Other terms, such as 'adult polycystic kidney disease', are no longer used.

Epidemiology

Unlike most ciliopathies, ADPKD is not considered a rare disease. Estimates of its population prevalence vary between 1:400 and 1:1000, making it the commonest single gene disorder leading to end-stage renal disease (ESRD) (Iglesias et al. 1983). It has been reported worldwide in all ethnic groups although accurate prevalence figures in different populations are difficult to establish. They vary between 1:400 in the US to 1:4000 in Japan (Higashihara et al. 1998). Approximately 85% of cases in most large series are due to mutations in the *PKD1* gene with the remainder due to mutations in *PKD2* (Rossetti et al. 2007).

These prevalence figures predict over 60,000 people in the UK with and at risk of developing ADPKD and its complications. In the UK the annual incidence rate for

ESRF caused by ADPKD is 6.4 per million, similar to the US and the rest of Europe. ADPKD accounts for 6.7% of incident adult cases of all ages accepted for renal replacement therapy (RRT, 10.2% of those aged <65 years and male:female ratio 0.8). 9.6% of prevalent cases and 12.2% of those with a renal transplant have a diagnosis of ADPKD (UK Renal Registry, Thirteenth Annual Report 2010, www.renalreg.com). In 2009 the average age at start of RRT in the UK was 54.7 years compared to 55.1 years in 1997 (UK Renal Registry, personal communication). However, longer-term studies such as the Danish National Registry on Regular Dialysis and Transplantation (NRDT) show an increase in age at ESRD from 55.9 years (1990–1995) to 60.6 years (2002–2007) (Orskov et al. 2010) In a single-centre UK study age at RRT increased from 51.1 years (1971–1985) to 54.4 years (1986–2000) (Haynes et al. 2011). These and other centre series suggest the natural history of the disease can potentially be modified although do not identify how.

Clinical features

The main clinical features of ADPKD are listed in Table 10.1.

Cystic features

ADPKD is characterised by progressive bilateral enlargement of polycystic kidneys (Figure 10.1). However, there is considerable clinical variability between individuals both within and between families. Disease progression is highly variable with ESRD occurring in about 50% of affected individuals by the sixth decade of life with severe disease in childhood being rare. Renal imaging allows the diagnosis to be made in most affected individuals by 15 years of age although accurate age specific ultrasound diagnostic criteria based on disease genotype (*PKD1* or *PKD2*) have been established. (Mekahli et al. 2010; Pei et al. 2009).

Table 10.1 Common clinical manifestations of autosomal dominant polycystic kidney disease

Feature	Prevalence
Cystic	
Polycystic kidneys	100% by age 40 years
Liver cysts	>90% by age 50 years
Pancreatic cysts	10%
Seminal vesicle cysts	~40% of males
Non-cystic	
Hypertension	80–90%
Mitral valve prolapse	~25%
Intracranial aneurysms	~8%

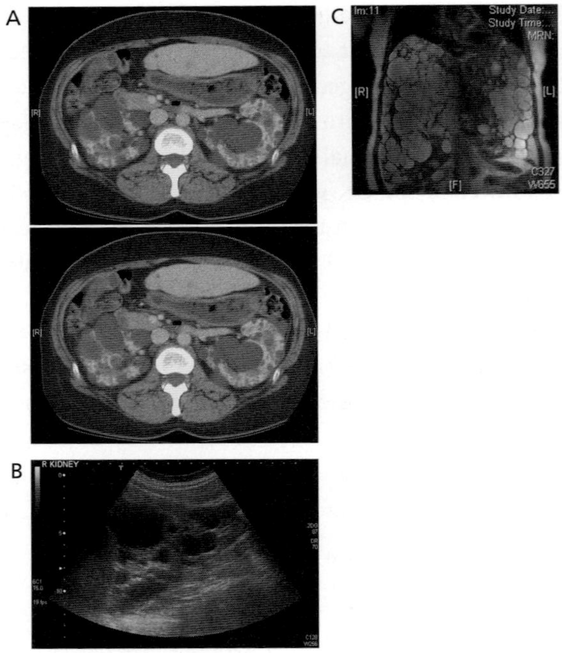

Figure 10.1 Ultrasound, CT and MR imaging in ADPKD. (A) CT imaging of the abdomen can be used to accurately define the architecture of PKD and PLD and estimate renal volumes. (B) Routine trans-abdominal ultrasound is commonly used to screen for ADPKD in at risk individuals. (C) MR is also of value in imaging PKD and PLD especially when repeated images are required such as in the assessment of massive PLD or serial measurements of renal volume.

The renal complications associated with ADPKD include a near universal progressive decline in renal function, loin pain, nephrolithiasis and macroscopic haematuria. Despite the renal enlargement and cyst formation seen in ADPKD, it is not a premalignant condition (Bonsib, 2009). The decline in renal function may occur after many years of apparently stable function measured in the clinic. Many affected individuals are asymptomatic and therefore present late, in their thirties or forties, often with established complications such as hypertension and CKD3 or higher (Dicks et al. 2006). Once the glomerular filtration rate (GFR) has started to decline it progresses at an average rate of ~5 ml/min until ESRD occurs. An elevated serum creatinine therefore occurs relatively late in the course of the disease.

However, functional renal abnormalities are seen in younger patients with a normal GFR, early in the course of their disease, providing a rational basis for pre-symptomatic screening in at-risk individuals (Helal et al. 2012). These abnormalities include a urinary-concentrating defect, reduction in effective renal plasma flow, and increases in filtration fraction and urinary albumin excretion (Meijer et al. 2010). These may be early biomarkers for disease severity and risk of progression to ERSD (Meijer et al. 2010; Torres et al. 2007b). In children, glomerular hyperfiltration (defined as a creatinine clearance ≥140 ml/min/1.73 m^2) may also predict a faster rate of decline of renal function (Helal et al. 2011).

It has also been shown that children with ADPKD and hypertension or blood pressure ≥75th percentile are at particular risk for deterioration in their renal function (Cadnapaphornchai et al. 2008).

Several other factors are also known to be associated with a faster decline in renal function. These include early age of diagnosis, male gender, hypertension, macroscopic haematuria and proteinuria. However, *PKD1* mutation status and total renal volume have been shown to have better predictive value for more rapid disease progression (Chapman et al. 2003; Grantham et al. 2006b; Harris et al. 2006; Hateboer et al. 1999). Currently, their use is mainly confined to research studies although this is likely to change as more accurate prediction of disease severity will be required to identify those patients who may benefit from earlier targeted therapeutic intervention. A mutation in *PKD1* is associated with an average age of ERSD of 53 years compared to 69 years in those with a *PKD2* mutation (Hateboer et al. 1999). The CRISP (Consortium for Radiologic Imaging Study of PKD) and HALT (Halt Progression of Polycystic Kidney Disease) studies have both confirmed a strong association between total renal volume and functional measures such as GFR (Grantham et al. 2006a; Torres et al. 2012b). A baseline total kidney volume (TKV) over 1500 ml predicts a faster rate of decline in GFR of 4.33 ± 8.07 ml/min/year (Grantham et al. 2006b). Using height adjusted TKV (htTKV) further analysis of the same cohort has shown that a baseline htTKV ≥600 cc/m best predicts the risk of developing renal insufficiency (CKD3: GFR 30–59 ml/min/1.73 m^2) within 8 years of follow-up compared to other standard clinical measures (Chapman et al. 2012). These studies have also shown that the exponential growth in renal volume is a defining characteristic of each individual and that *PKD1* kidneys are larger at baseline than those from individuals with *PKD2* due to a greater number of cysts although the rate of cyst growth is similar (Grantham et al. 2008; Harris et al. 2006).

What these studies also show is that disease progression in ADPKD can be monitored in a clinically relevant time scale, 6 months to a few years, and that younger patients with larger renal volumes, hypertension and more rapid renal growth represent a high risk group. Therefore rate of change of renal volume can be used as a surrogate clinical biomarker and outcome measure of disease progression.

Hypertension is very common in ADPKD, affecting 80–90% of individuals with ADPKD and virtually all with ESRD (Dicks et al. 2006). Most adults are hypertensive at the time of diagnosis before any significant reduction in GFR has occurred. It may also be diagnosed in a significant proportion of affected children (Cadnapaphornchai et al. 2008; Mekahli et al. 2010). In children with normal blood pressures, 24-hour monitoring may show a reduction in the normal nocturnal variation and an exaggerated response to exercise.

Pain is common in ADPKD and is a frequent presenting symptom along with family history, hypertension and haematuria (Bajwa et al. 2001, 2004). It may occur early in the disease, often in the presence of normal renal function. It is typically described as back or abdominal pain with a dull or full character (Bajwa et al. 2004). It is most likely related to enlarged kidneys or altered spinal posture. Severe pain may be associated with acute complications such as cyst haemorrhage or cyst infection but may occur in the absence of these in a small number of individuals. However, it is likely that pain is under diagnosed unless specifically sought about during medical evaluation. Its impact is more difficult to assess but pre-dialysis ADPKD patients assess their quality of life as similar to the general population (Rizk et al. 2009).

Cysts are also seen in other organs including the liver and pancreas. The combination of multiple renal and hepatic cysts with or without a positive family history is highly specific for ADPKD. The frequency of hepatic cysts also increases with age. Their prevalence in the CRISP study using MRI is 58% in 15- to 24-year olds, 85% in 25- to 34-year olds, and 94% in 35- to 46-year olds (Bae et al. 2006)., They are rare in children with ADPKD. Liver cysts are usually asymptomatic and detected during diagnostic renal imaging. They are not associated with impaired liver function. Massive polycystic liver disease (PLD) may occur, typically in females, and may be associated with severe pain, abdominal distention and early satiety with gastro-oesophageal reflux (Figure 10.1). Complications associated with liver cysts include infection, haemorrhage and rupture presenting with pain, fever and abdominal tenderness. Pancreatic cysts are usually asymptomatic.

Non-cystic features

The most clinically significant non-cystic manifestation seen in ADPKD is intracranial aneurysm (ICA). Asymptomatic ICA occurs in 8% of ADPKD cases, considerably higher than in the general population (Pirson et al. 2002). It is further increased, ~16%, if there is a family history of ruptured ICA. Indications for screening for ICA in patients include family history of aneurysm or subarachnoid haemorrhage, previous aneurysm rupture, major elective surgery and anxiety. Widespread screening of ADPKD patients is not indicated (Irazabal et al. 2011). Most ICAs detected by screening in ADPKD are small and in the anterior circulation. Their risk of rupture appears to be similar to ICAs in the general population except that it occurs up to 10 years earlier. It may also occur with normal renal function and blood pressure. The only well-defined risk factor for rupture is a family history of rupture. The precise incidence of ICA rupture in ADPKD has been difficult to evaluate but has been estimated to be ~1/2000 person-years in one study (Schievink et al. 1992). After rupture, ICAs in ADPKD are associated with a considerable morbidity and mortality as seen in the general population.

Other vascular abnormalities also occur in ADPKD including mitral valve prolapse and aortic root dilatation and dissection. Screening for these complications is not indicated unless there is strong clinical suspicion or a positive family history.

Range of the phenotype

ADPKD typically presents in the fourth and fifth decades following screening in at risk individuals, or during the investigation of hypertension, loin pain or haematuria. Its clinical course is highly variable. A long asymptomatic period is common. ESRD typically occurs after the sixth decade but may rarely occur as early as the third decade or even in childhood. However, up to one third of individuals will retain good renal function into old age and not require dialysis or transplantation. This group is more likely to have disease caused by a mutation in the *PKD2* gene. At the other extreme of age, ADPKD can be diagnosed *in utero*, in the post-natal period or during early childhood. Recent molecular studies have shown that hypomorphic alleles of *PKD1* exist and produce a wide range of phenotypes. Severe *in utero* disease, seen in a small number of ADPKD families, may be caused by the inheritance *in trans* of a pathogenic germline mutation from an affected parent and a hypomorphic allele from the unaffected parent (Rossetti et al. 2009). This also predicts a high recurrence risk for this presentation.

A severe early onset *in utero* presentation similar to autosomal recessive polycystic kidney disease (ARPKD) has also been seen in children with no family history of ADPKD (Vujic et al. 2010). This is caused by the inheritance of two hypomorphic alleles in *PKD1*, one from each parent. Molecular evaluation is therefore important in cases of atypical PKD.

Diagnosis

The identification of any of the clinical features of ADPKD in an individual with a positive family history is highly suggestive of the diagnosis. However, a definitive diagnosis is typically made using renal imaging (Chapman & Wei, 2011). In an individual with no family history after parental screening (approximately 10% of cases), the typical features on abdominal imaging (multiple renal cysts with or without hepatic cysts) are sought whilst excluding features that may suggest an alternative diagnosis (Table 10.2). Molecular testing is becoming more widely available and can be used to confirm a diagnosis. It is also of value in the assessment of multiple renal cysts where the radiological features are not typical of ADPKD. Genetic counselling should be offered prior to screening or testing to discuss the benefits and unintended consequences of establishing a diagnosis of ADPKD.

New unified diagnostic criteria for individuals at 50% risk of inheriting ADPKD have recently been published (Table 10.3) (Pei et al. 2009). These are based on ultrasound imaging and have not been validated for computed tomography (CT) or magnetic resonance imaging (MRI). However, the widespread availability and low cost of ultrasound suggest they will remain in widespread use for ADPKD. Ultrasound imaging may also provide some information about total renal volume that reflects severity and prognosis although it lacks the precision of CT and MRI required for serial measurements (O'Neill et al. 2005).

In families where the underlying gene mutation is unknown, a total of three or more renal cysts are required to confirm the diagnosis in individuals aged 15–39 years. In individuals aged 40–59 years, two or more cysts in each kidney are required with four or more cysts in each kidney for individuals' ≥60 years. The sensitivity of these criteria is 95% for individuals 30 years or older or those under 30 years with a *PKD1* mutation, but only 67% for those with a *PKD2* mutation under 30 years (Table 10.3).

Table 10.2 The differential diagnosis of autosomal dominant polycystic kidney disease

Disease	Gene(s)	Features	OMIM
Renal cysts and diabetes (RCAD)	*HNF1B*	Maturity onset diabetes of the young (MODY5), PKD, abnormal liver function, genital tract anomalies	137920
von Hippel–Lindau disease (VHL)	*VHL*	Renal cysts, renal cell carcinoma, retinal and CNS haemangioblastomas, phaeochromocytoma, pancreatic cysts	193300
Tuberous sclerosis complex (TSC)	*TSC1* and *TSC2*	Multisystem hamartomatous disease. May occur with ADPKD due to contiguous gene deletion. Co-existence of renal angiomyolipomas and renal cysts diagnostic	TSC1 191100 TSC2 613254
UMOD nephropathy (meduallary cystic kidney disease 2)	*UMOD*	Small kidneys, medullary cysts, hyperuricaemia	603860
Acquired renal cystic disease	–	Usually in end-stage kidneys of any aetiology	
Autosomal recessive polycystic kidney disease	*PKHD1*	Bilateral PKD, CHF	263200
Simple cysts	–	Present in ageing population	
Autosomal dominant polycystic liver disease	*PRKCSH, SEC63*	PLD without PKD	174050
Oral–facial–digital syndrome type I	*OFD1*	PKD, oral frenulae, cleft tongue, cleft palate, dysmorphic features, digital anomalies	311200

Fewer than two cysts in individuals aged ≥40 years are sufficient to exclude the diagnosis (Table 10.3).

These diagnostic criteria can also be used for the assessment of at-risk potential living related kidney donors where disease exclusion is required. In an individual ≥40 years a normal scan or a single cyst has a negative predictive value of 100%. For those aged 30–39 years a normal scan has a false-negative rate of 0.7% whilst for younger individuals a normal scan may not be sufficient to exclude disease. In these instances CT or MRI scanning can be used combined with molecular genetic testing if the familial mutation is known (Figure 10.1).

Table 10.3 (A) Ultrasound criteria for the diagnosis of autosomal dominant polycystic kidney disease (Pei et al. 2009)

Age, years	*PKD1*	*PKD2*	Unknown
15–30	≥3 cysts PPV 100% SEN 94%	PPV 100% SEN 69%	PPV 100% SEN 82%
30–39	≥3 cysts PPV 100% SEN 97%	PPV 100% SEN 95%	PPV 100% SEN 95%
40–59	≥2 cysts bilat PPV 100% SEN 93%	PPV 100% SEN 89%	PPV 100% SEN 90%

(B) Ultrasound criteria for exclusion of the diagnosis of autosomal dominant polycystic kidney disease (Pei et al. 2009)

Age, years	*PKD1*	*PKD2*	Unknown
15–30	≤1 cyst NPV 99% SPEC 98%	NPV 83% SPEC 97%	NPV 91% SPEC 97%
30–39	≤1 cyst NPV 100% SPEC 96%	PV 97% SPEC 94%	NPV 98% SPEC 95%
40–59	≤1 cyst NPV 100% SPEC 94%	NPV 100% SPEC 94%	NPV 100% SPEC 94%

PPV: positive predictive value; SEN: sensitivity.

Source: Data from York Pei et al., Unified criteria for ultrasonographic diagnosis of ADPKD, *Journal of American Society of Nephrology*, Volume **20**, pp. 205–212, Copyright © 2009 by the American Society of Nephrology, DOI: 10.1681/ASN.2008050507.

Molecular testing for ADPKD is now widely available. In a research setting a mutation detection rate of up to 88% can be achieved (Rossetti et al. 2007). The majority are sequence variants with deletions and duplications being much frequent (Figure 10.2). Linkage analysis can also be carried out if multiple affected and unaffected family members are available for analysis. This method is therefore suitable for only a minority of families but is accurate and suitable for diagnosis or disease exclusion if mutation testing is not available or a mutation has not been identified.

A patient's genotype can also be predicted from the family history and this can be used to target *PKD1* or *PKD2* testing first (Barua et al. 2009; Robinson et al. 2012). The presence of an affected family member with ESRD at ≤55 years is highly predictive of a *PKD1* mutation. However, the presence an affected family member with preserved renal function or onset of ESRD at >70 years is highly predictive of a *PKD2* mutation.

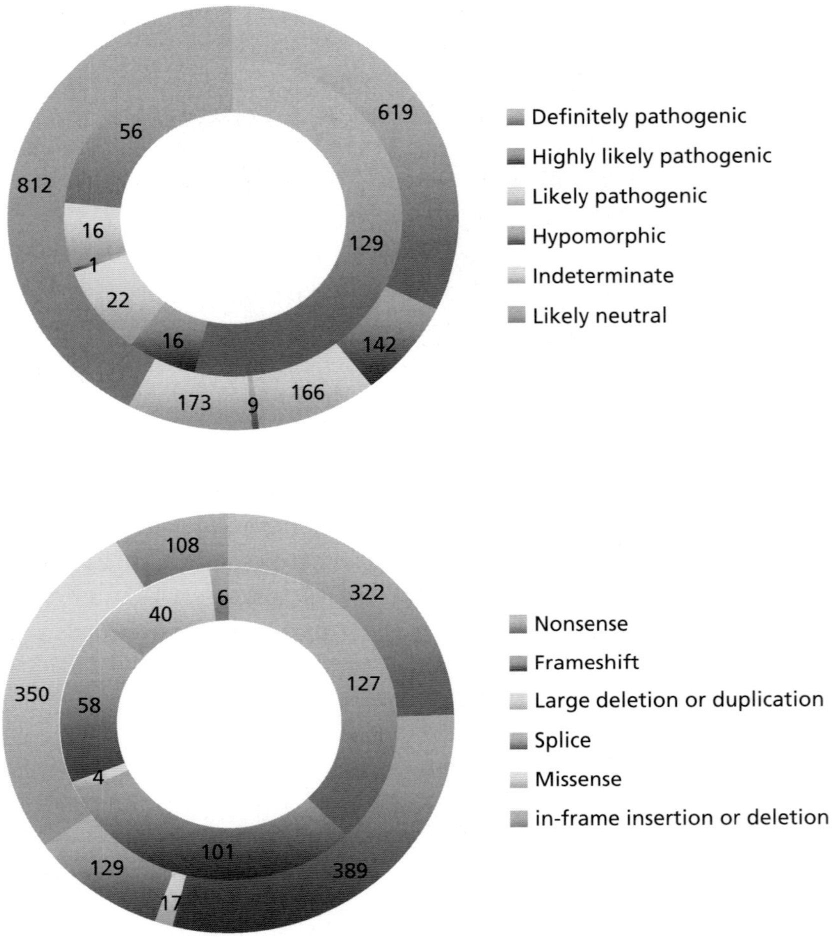

Figure 10.2 Number and type of mutation identified in *PKD1* (outer circle) and *PKD2* (inner circle).
Source: data from Autosomal Dominant Polycystic Kidney Disease Mutation Database: PKDB Database, available from http://pkd.mayo.edu.

Interesting features of the condition

ADPKD is also associated with a variety of other non-renal manifestations. These may be cystic or non-cystic. Cysts occur in the seminal vesicles in 40% of males although their clinical significance is uncertain. Arachnoid cysts are present in a small proportion of affected individuals and may predispose to subdural haematoma although this association requires further validation (Wijdicks et al. 2000).

The increased frequency of hypertension, cardiac valve abnormalities, ICA and other vascular anomalies in ADPKD suggests that they may represent primary manifestations of this disease. In support of this *Pkd1* has been shown to be widely expressed in the vasculature (Figure 10.3). Whether the same pathogenic mechanisms seen in renal cyst formation underlie the vascular abnormalties, e.g. abnormal cilial function, remains to be determined.

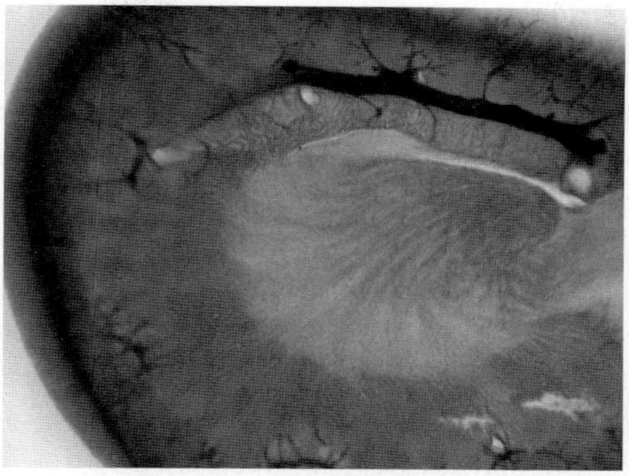

Figure 10.3 Renal and vascular expression of murine Pkd1 (for experimental methods see Boulter et al. 2001).

Rare cases of ADPKD have also been associated with a variety of hepatic ductal plate malformations including multiple biliary hamatoma and congential hepatic fibrosis (O'Brien et al. 2012). Other associations include connective tissue abnormalities and colonic diverticular disease. The latter association has only been reported in the ESRD population.

Genetics

The genetic basis of ADPKD has been well defined. Using linkage analysis in large multiplex families with autosomal dominant inheritance and classical features of the disease, several groups in the 1980s demonstrated genetic heterogeneity with a least two loci associated with disease (Kimberling et al. 1988; Reeders et al. 1985). Subsequent molecular analysis confirmed two loci, *PKD1* on chromosome 16p13.3 and *PKD2* on chromosome 4p21. An elusive third locus, *PKD3,* has not been identified. *PKD1* and *PKD2* were eventually cloned in 1994 and 1996, respectively (Mochizuki et al. 1996; Ward et al. 1994). *PKD1* spans ~50 kb of genomic DNA and contains 46 exons encoding a large ~14 kb mRNA. This predicts a large multi-domain transmembrane protein product, polycystin-1, of molecular mass >460 kDa. Exons 1–33 of *PKD1* are duplicated as six pseudogenes also on chromosome 16 with sequence identity of 98–99%. This has caused considerable difficulty with mutation analysis that has now been overcome using locus specific long-range PCR methods (Rossetti et al. 2007). *PKD2* is a 15-exon gene spanning 68 kb of genomic DNA. It produces a mRNA with a 2904 bp open reading frame that encodes a novel member of the transient receptor potential (TRP) channel family, TRPP2.

Mutations in these two genes have been identified in the majority of ADPKD cases analysed (Rossetti et al. 2007). Eight-five per cent of clinically well-characterised cases are due to mutations in *PKD1* with the remainder in *PKD2* (Rossetti et al. 2007). Using different methodologies, including Next Generation sequencing, mutation detection rates of between 63 and 89% have been achieved (Rossetti et al. 2007, 2012). The majority of mutations are private (i.e. unique to a single family) but ~30% are recurrent. The ADPKD mutation database (http://pkdb.mayo.edu/) is the most complete repository for all sequence variation in *PKD1* and *PKD2* containing information on nearly 1800 unique ADPKD pedigrees (Figure 10.2). *PKD1* is highly polymorphic with over 40% of variants having a likely neutral effect. Of the remainder, ~50% can be ascribed pathogenic status of which the majority are loss of function, e.g. nonsense or frame-shifting. *PKD2* contains less variation with ~25% being neutral whilst ~70% are pathogenic, again mostly loss of function.

Current evidence supports most ADPKD associated mutations being fully penetrant.

However, several families presenting with very mild disease or apparent non-penetrance have recently been reported where convincing evidence for incompletely penetrant or hypomorphic alleles has been provided (Rossetti et al. 2009). Co-inheritance of an additional hypomorphic allele or pathogenic mutation then results in disease. Somatic mosaicism may also be responsible for mild disease (Harris & Rossetti, 2010).

ADPKD is an example of a 'two-hit' disease (Harris, 2010). Strong evidence supports cyst formation resulting from the acquisition of a second somatic mutation in the normal allele in addition to the inherited germline mutation, i.e. bi-allelic inactivation. Somatic mutations in *PKD1* and *PKD2* have been identified in cyst-lining epithelial cells from both renal and hepatic cysts. This suggests that cyst initiation and hence disease severity may be dependent on the rate of stochastic somatic mutation of *PKD1* and *PKD2*, a process that may be hard to modify therapeutically. Other phenotypes associated with ADPKD, however, may not require a second somatic mutation and therefore haplo-insufficiency may play a role. For example, there is no evidence to support somatic mutation in the development of ICA.

Considerable intrafamilial variability is seen in ADPKD, suggesting the effect of environmental and genetic modifiers. Genetic background has been estimated to account for 32–42% of the variance in estimated GFR (eGFR) before ESRD and 43–78% of the variance in age at ESRD onset. Genetic variation at the *DKK3* locus has been shown to modify disease resulting from *PKD1* mutations and further studies are required to replicate this finding and identify other modifier loci (Liu et al. 2010; Paterson et al. 2005).

Underlying ciliary defect

ADPKD is characterised by PKD and hepatic ductal plate malformations. The latter include biliary hamartomas and PLD. These phenotypes are common to other ciliopathies. However, the condition is only rarely associated with other classical ciliopathy features. Rarely reported associations include congenital hepatic fibrosis and laterality defects (Bataille et al. 2011; O'Brien et al. 2012). Therefore ADPKD is clinically easy to recognise. Unlike most ciliopathies that are inherited as autosomal recessive or oligogenic traits, ADPKD is an autosomal dominant disease that typically presents in adult life. This later presentation may in part be due to the requirement for a stochastic 'second hit'.

Physiological effect of the ciliary defect

Polycystin-1 and TRPP2 form a multimeric ion channel complex in the primary cilium although both proteins are also predicted to have independent functions in non-cilial locations (for a review see Torres et al. 2007a). Interaction between the two proteins was initially shown to be essential for their membrane localisation and function as a calcium-permeable non-selective cation channel (Hanaoka et al. 2000). Further studies revealed that both proteins co-localise to the primary cilium although the largest pool of TRPP2 resides in the endoplasmic reticulum (Yoder et al. 2002). In the primary cilium, the polycystin complex is activated by ciliary bending, which in the context of the renal tubule may be initiated by urinary flow (Nauli et al. 2003). This results in an increase in intracellular calcium and the activation of multiple downstream signalling pathways (Chapin & Caplan, 2010). The localised increase in cilial calcium levels is amplified as TRPP2 is also a calcium-activated cation channel that mediates calcium release from intracellular stores, principally from the endoplasmic reticulum through interaction with two major intracellular calcium channels, the ryanodine receptor and the inositol 1,4,5-triphosphate receptor (Anyatonwu et al. 2007; Sammels et al. 2010). Conversely, a reduction in flow or activation of TRPP2 in model systems has been shown to result in the nuclear accumulation of several C-terminal polycystin-1 cleavage fragments. This translocation occurs with components of the Wnt pathway and transcriptional activators STAT6/p100 (reviewed in Chapin & Caplan, 2010). Polycstin-1 also functions as a G-protein coupled receptor to regulate AP-1 and NFAT signalling. Other pathways modulated by the polycystins include mTOR, p21 and canonical and non-canonical Wnt. These explain many of the cellular abnormalities seen in ADPKD cystic epithelia and animal models that include altered cell proliferation, differentiation, apoptosis and planar cell polarity.

Which of these abnormalities is the critical initiating step in cyst formation is not clear. However, the polycystin complex plays a key developmental role. Loss of polycystin function during nephrogenesis when tubular cell proliferation, differentiation and orientated cell division (required for tubule elongation) occur leads to a severe cystic phenotype (Fischer et al. 2006). When these processes are complete, loss of polycystin function does not lead to cyst formation unless cells enter a proliferative phase after injury that would normally lead to tissue repair (Happe et al. 2009).

Once cyst formation has occurred, fluid secretion is the critical driver to cyst expansion. cAMP is a key second messenger and in cystic epithelial cells drives both cell proliferation and fluid secretion, an effect modulated by intracellular calcium levels (Mangoo-Karim et al. 1989). Therefore attempts to reduce renal cAMP via blockade of the vasopressin and somatostain receptor signalling systems to modulate disease severity has been a priority, resulting in clinical trials to test this approach in PKD and PLD (Gattone et al. 2003; Hogan et al. 2012; Torres et al. 2012a).

Clinical management

The clinical management of ADPKD aims to reduce the morbidity and mortality associated with the renal and extrarenal manifestations of the disease. In addition it provides on-going support and counseling for individuals and their families living with the long-term medical and non-medical consequences of a genetic disease. In particular, care is aimed at preventing the long-term cardiovascular and metabolic consequences that occur with CKD and ESRD.

Therapies

Currently, there are no licensed therapies that specifically target the renal cystic disease. Many potential therapies are being evaluated in preclinical and clinical studies. Until recently there were no therapies that showed promise in modifying the natural history of the disease, in particular the progressive decline in renal function. This was in part due to the challenges of designing clinical trials in the absence of any validated end points that could be assessed in a clinically meaningful timescale. With the development of total renal volume as a surrogate for disease progression many clinical trials are planned or are in progress. The website Clinicaltrials.gov currently lists 46 studies using the search term ADPKD. Many of these use change in renal volume as a primary study endpoint and target the dysregulated signalling pathways such as mTOR or response to cAMP that have been identified in models of ADPKD. The results of the TEMPO 3:4 trial demonstrate that the vasopressin receptor V2 antagonist Tolvaptan which reduces renal cAMP, can modify the natural history of the disease resulting in a slowing down of the increase in renal volume and decline in GFR seen over a 3-year period (Torres et al. 2012a). Using similar end-points, mTOR inhibitors have not been shown to be of benefit (Serra et al. 2010). Clinical trial design, patient or renal characteristics in addition to the trial drug, however, may account for some of these differences and further trials, perhaps using combinations of different therapies at earlier stages of disease are likely to be required (Torres, 2010).

How drugs with a beneficial effect will be introduced in to clinical practice and who is likely to benefit most from their introduction remains to be determined. Indeed it is not clear whether long-term use, for example over decades, will also show the same benefit.

The long period of follow-up of ADPKD patients, when renal function is well preserved, offers an exciting therapeutic 'window' during which the course of the disease may be modified with the aim of reducing the number and age at which patients develop ESRD. It is therefore likely that many more trials will be conducted in the near future at all stages of disease.

Management of complications

Patients with ADPKD frequently present with symptoms related to acute or chronic renal pain, macroscopic haematuria and hypertension. Their management will be directed to these complications as well as the other features of the disease. However, uncertainty still remains over routine clinical care including how to monitor disease progression, choice of antihypertensive agent and level of blood pressure control, optimum age for diagnostic screening and the role of genetic testing. Once the diagnosis of ADPKD has been made, all patients should undergo a comprehensive evaluation to establish the extent of disease. This should adhere to local and national approved guidelines for CKD and take in to account disease specific complications; for example, see http://www.renal.org/Clinical/GuidelinesSection/Detection-Monitoring-and-Care-of-Patients-with-CKD.aspx.

This should include:

- Renal imaging to document the extent of the renal disease and provide a renal volume to assess prognosis. It can also be used to document the extent of any PLD
- Blood pressure monitoring to detect onset of hypertension (24-hour ambulatory blood pressure monitoring can also be offered if clinic readings are felt to be unreliable or in young individuals)
- Full urinalysis to document microalbuminuria or proteinuria and exclude infection
- Full biochemical and metabolic investigation to determine renal function and cardiovascular risk factors such as hyperlipidaemia
- Cardiac imaging if the cardiac examination is abnormal or if there is a family history of thoracic aortic disease
- Referral to a neurosurgeon if there is a personal or family history of ICA or early onset stroke.

Hypertension

Hypertension occurs early in ADPKD. Several studies are trying to address whether the level of control and choice of drug may modify disease progression (Chapman et al. 2010). Considerable evidence supports activation of the renin–angiotensin–aldosterone system (RAAS) in ADPKD and drugs that target this system are commonly used in ADPKD for cardiovascular risk prevention and reno-protection. There is also some historic data that suggests diuretics alone may be less effective in preserving renal function when compared to the use of RAAS blockade (Ecder et al. 2001). The MDRD Study has suggested that tight control of blood pressure may preserve renal function in ADPKD (Sarnak et al. 2005). It has also been shown that screening and treatment of hypertension in children with ADPKD may preserve renal function. This should be carried out by an experienced paediatrician (Cadnapaphornchai et al. 2008). Evidence-based guidelines for the screening and follow-up of at-risk children are still required. Currently, unless an at-risk child presents with symptoms requiring investigation, screening is not recommended until mid teens or later.

Loin pain

Loin pain is a common presenting symptom in ADPKD. It can usually be managed with simple analgesics once an underlying cause, such as nephrolithiais or cyst infection, has been excluded. Nephrotoxic drugs such as non-steroidal anti-inflammatory agents should be avoided. More protracted and severe pain can be managed using other agents under the guidance of a specialist pain clinic. Rarely cyst aspiration or surgical decompression is indicated. The latter has better long-term results. Once ESRD has developed nephrectomy may be offered. Other techniques such as selective embolisation to reduce renal mass are being evaluated.

If pain is due to cyst haemorrhage, symptomatic relief is usually sufficient. It is typically self-limiting. If prolonged or associated with haemodynamic instability, embolisation or surgical intervention may be required. Cyst infection is usually indicated by pain with fever. Imaging may identify a complex cyst although this is not specific for infection. Urine and blood cultures should be obtained. Cyst aspiration or drainage may be carried out if there is treatment failure after several weeks. Cyst infection can be difficult to treat and require several months of antimicrobial therapy if there is previous treatment failure. Pain due to nephrolithiasis is managed in the same manner as for non-ADPKD patients. Stones in ADPKD are usually uric acid or calcium oxalate.

Pregnancy

During pregnancy women with ADPKD should be offered enhanced surveillance for hypertension and other disease complications. Additional antenatal scans should also be offered if there are concerns about foetal growth and amniotic fluid volume. Transient increased foetal renal echogenicity and foetal renal cysts may also be incidentally discovered. This may warrant more detailed pre- and post-natal scanning.

End-stage renal disease

Once ESRD has developed patients with ADPKD can be offered all forms of renal replacement therapy including haemodialysis and peritoneal dialysis. Massive nephromegaly or hepatomegaly may preclude the latter. Renal transplantation is the treatment of choice, ideally using a living donor. Transplant outcomes in ADPKD are similar or better when compared to other non-diabetic groups. Native nephrectomy may also be required in some cases especially is there is severe pain or frequent haematuria, cyst infection or stone disease.

Other therapies

Other therapies or interventions for which there is no current evidence of clinical benefit include increased fluid intake and abstinence from caffeinated drinks. Both potentially modulate cAMP levels and may therefore have beneficial effects. High fluid intake may produce a similar physiological benefit to vasopressin receptor V2 blockade (Wang et al. 2011).

Genetic counselling

All individuals with a diagnosis of ADPKD or at risk of inheriting it should be offered genetic counselling. In particular at risk individuals should be made aware of the issues surrounding a new diagnosis especially if they are young and asymptomatic. Children who express a wish to find out more information may be seen with their parents or guardians although screening for ADPKD is not currently recommended for at-risk asymptomatic individuals until their mid teens or later or until an effective treatment becomes available. Blood pressure monitoring may be instituted alone. Issues that may be discussed include pre-symptomatic screening, health and insurance implications, employment, reproductive issues and living related transplantation. Obtaining a family history during genetic counselling is also a simple and effective way to identify other family members who might be at risk for ADPKD and wish to consider screening as well as having some predictive value for the underlying genotype (Barua et al. 2009; Robinson et al. 2012). Genetic testing may also be offered. With a clinically useful mutation detection rate, many more families will be able to have access to testing. Current indications include pre-symptomatic testing of at risk relatives and potential living related organ donors, diagnostic testing in the presence of atypical features, prenatal and pre-implantation testing.

Additional resources

GeneReviews	http://www.ncbi.nlm.nih.gov/sites/GeneTests
PKD Foundation	www.pkdcure.org/
PKD Charity	http://pkdcharity.org.uk/
Clinical Trials.gov	www.clinicaltrials.gov/
HALT	http://haltpkd.org/index.html
GeneTests	http://www.ncbi.nlm.nih.gov/sites/GeneTests
UKGTN	www.ukgtn.nhs.uk/

References

Anyatonwu, G. I., Estrada, M., Tian, X., Somlo, S. & Ehrlich, B. E. 2007. Regulation of ryanodine receptor-dependent calcium signaling by polycystin-2. *Proc Natl Acad Sci U.S.A.* **104**, 6454–6459.

Bae, K. T., Zhu, F., Chapman, A. B., Torres, V. E., Grantham, J. J., Guay-Woodford, L. M., et al. 2006. Magnetic resonance imaging evaluation of hepatic cysts in early autosomal-dominant polycystic kidney disease: The Consortium for Radiologic Imaging Studies of Polycystic Kidney Disease cohort. *Clin J Am Soc Nephrol* **1**, 64–69.

Bajwa, Z. H., Gupta, S., Warfield, C. A. & Steinman, T. I. 2001. Pain management in polycystic kidney disease. *Kidney Int* **60**, 1631–1644.

Bajwa, Z. H., Sial, K. A., Malik, A. B. & Steinman, T. I. 2004. Pain patterns in patients with polycystic kidney disease. *Kidney Int* **66**, 1561–1569.

Barua, M., Cil, O., Paterson, A. D., Wang, K., He, N., Dicks, E., et al. 2009. Family history of renal disease severity predicts the mutated gene in ADPKD. *J Am Soc Nephrol* **20**, 1833–1838.

Bataille, S., Demoulin, N., Devuyst, O., Audrezet, M. P., Dahan, K., Godin, M., et al. 2011. Association of PKD2 (polycystin 2) mutations with left–right laterality defects. *Am J Kidney Dis* **58**, 456–460.

Bonsib, S. M. 2009. Renal cystic diseases and renal neoplasms: A mini-review. *Clin J Am Soc Nephrol* **4**, 1998–2007.

Boulter, C., Mulroy, S., Webb, S., Fleming, S., Brindle, K. & Sandford, R. 2001. Cardiovascular, skeletal, and renal defects in mice with a targeted disruption of the Pkd1 gene. *Proc Natl Acad Sci U.S.A.* **98**, 12174–12179.

Cadnapaphornchai, M. A., Mcfann, K., Strain, J. D., Masoumi, A. & Schrier, R. W. 2008. Increased left ventricular mass in children with autosomal dominant polycystic kidney disease and borderline hypertension. *Kidney Int* **74**, 1192–1196.

Chapin, H. C. & Caplan, M. J. 2010. The cell biology of polycystic kidney disease. *J Cell Biol*, **191**, 701–710.

Chapman, A. B., Bost, J. E., Torres, V. E., Guay-Woodford, L., Bae, K. T., Landsittel, D., et al. 2012. Kidney volume and functional outcomes in autosomal dominant polycystic kidney disease. *Clin J Am Soc Nephrol* **7**, 479–486.

Chapman, A. B., Guay-Woodford, L. M., Grantham, J. J., Torres, V. E., Bae, K. T., Baumgarten, D. A., et al. 2003. Renal structure in early autosomal-dominant polycystic kidney disease (ADPKD): The Consortium for Radiologic Imaging Studies of Polycystic Kidney Disease (CRISP) cohort. *Kidney Int* **64**, 1035–1045.

Chapman, A. B., Torres, V. E., Perrone, R. D., Steinman, T. I., Bae, K. T., Miller, J. P., et al. 2010. The HALT polycystic kidney disease trials: Design and implementation. *Clin J Am Soc Nephrol* **5**, 102–109.

Chapman, A. B. & Wei, W. 2011. Imaging approaches to patients with polycystic kidney disease. *Semin Nephrol* **31**, 237–244.

Dicks, E., Ravani, P., Langman, D., Davidson, W. S., Pei, Y. & Parfrey, P. S. 2006. Incident renal events and risk factors in autosomal dominant polycystic kidney disease: A population and family-based cohort followed for 22 years. *Clin J Am Soc Nephrol* **1**, 710–717.

Ecder, T., Edelstein, C. L., Fick-Brosnahan, G. M., Johnson, A. M., Chapman, A. B., Gabow, P. A., et al. 2001. Diuretics versus angiotensin-converting enzyme inhibitors in autosomal dominant polycystic kidney disease. *Am J Nephrol* **21**, 98–103.

Fischer, E., Legue, E., Doyen, A., Nato, F., Nicolas, J. F., Torres, V., et al. 2006. Defective planar cell polarity in polycystic kidney disease. *Nat Genet* **38**, 21–23.

Gattone, V. H. 2nd, Wang, X., Harris, P. C. & Torres, V. E. 2003. Inhibition of renal cystic disease development and progression by a vasopressin V2 receptor antagonist. *Nat Med* **9**, 1323–1326.

Grantham, J. J., Chapman, A. B. & Torres, V. E. 2006a. Volume progression in autosomal dominant polycystic kidney disease: The major factor determining clinical outcomes. *Clin J Am Soc Nephrol* **1**, 148–157.

Grantham, J. J., Cook, L. T., Torres, V. E., Bost, J. E., Chapman, A. B., Harris, P. C., et al. 2008. Determinants of renal volume in autosomal-dominant polycystic kidney disease. *Kidney Int* **73**, 108–116.

Grantham, J. J., Torres, V. E., Chapman, A. B., Guay-Woodford, L. M., Bae, K. T., King Jr, B. F. et al. 2006b. Volume progression in polycystic kidney disease. *N Engl J Med* **354**, 2122–2130.

Hanaoka, K., Qian, F., Boletta, A., Bhunia, A. K., Piontek, K., Tsiokas, L., et al. 2000. Co-assembly of polycystin-1 and -2 produces unique cation-permeable currents. *Nature* **408**, 990–994.

Happe, H., Leonhard, W. N., Van Der Wal, A., Van De Water, B., Lantinga-Van Leeuwen, I. S., Breuning, M. H., et al. 2009. Toxic tubular injury in kidneys from Pkd1-deletion mice accelerates cystogenesis accompanied by dysregulated planar cell polarity and canonical Wnt signaling pathways. *Hum Mol Genet* **18**, 2532–2542.

Harris, P. C. 2010. What is the role of somatic mutation in autosomal dominant polycystic kidney disease? *J Am Soc Nephrol* **21**, 1073–1076.

Harris, P. C., Bae, K. T., Rossetti, S., Torres, V. E., Grantham, J. J., Chapman, A. B., et al. 2006. Cyst number but not the rate of cystic growth is associated with the mutated gene in autosomal dominant polycystic kidney disease. *J Am Soc Nephrol* **17**, 3013–3019.

Harris, P. C. & Rossetti, S. 2010. Determinants of renal disease variability in ADPKD. *Adv Chronic Kidney Dis* **17**, 131–139.

Harris, P. C. & Torres, V. E. 2011. *Polycystic Kidney Disease, Autosomal Dominant* [Online]. Available: http://www.ncbi.nlm.nih.gov/books/NBK1246/.

Hateboer, N. V., Dijk, M. A., Bogdanova, N., Coto, E., Saggar-Malik, A. K., San Millan, J. L., Torra, R., et al. 1999. Comparison of phenotypes of polycystic kidney disease types 1 and 2. European PKD1-PKD2 Study Group. *Lancet* **353**, 103–107.

Haynes, R., Kheradmand, F. & Winearls, C. G. 2011. Survival after starting renal replacement treatment in patients with autosomal dominant polycystic kidney disease: A single-centre 40-year study. *Nephron Clin Pract* **120**, c42–47.

Helal, I., Reed, B., Mcfann, K., Yan, X. D., Fick-Brosnahan, G. M., Cadnapaphornchai, M., et al. 2011. Glomerular hyperfiltration and renal progression in children with autosomal dominant polycystic kidney disease. *Clin J Am Soc Nephrol* **6**, 2439–2443.

Helal, I., Reed, B. & Schrier, R. W. 2012. Emergent early markers of renal progression in autosomal-dominant polycystic kidney disease patients: Implications for prevention and treatment. *Am J Nephrol* **36**, 162–167.

Higashihara, E., Nutahara, K., Kojima, M., Tamakoshi, A., Yoshiyuki, O., Sakai, H., et al. 1998. Prevalence and renal prognosis of diagnosed autosomal dominant polycystic kidney disease in Japan. *Nephron*, **80**, 421–427.

Hogan, M. C., Masyuk, T. V., Page, L., Holmes, D. R. 3rd, Li, X., Bergstralh, E. J., et al. 2012. Somatostatin analog therapy for severe polycystic liver disease: Results after 2 years. *Nephrol Dial Trans* **27**, 3532–3539.

Iglesias, C. G., Torres, V. E., Offord, K. P., Holley, K. E., Beard, C. M. & Kurland, L. T. 1983. Epidemiology of adult polycystic kidney disease, Olmsted County, Minnesota: 1935–1980. *Am J Kidney Dis* **2**, 630–639.

Irazabal, M. V., Huston, J. 3rd, **Kubly, V., Rossetti, S., Sundsbak, J. L., Hogan, M. C., et al.** 2011. Extended follow-up of unruptured intracranial aneurysms detected by presymptomatic screening in patients with autosomal dominant polycystic kidney disease. *Clin J Am Soc Nephrol* **6**, 1274–1285.

Kimberling, W. J., Fain, P. R., Kenyon, J. B., Goldgar, D., Sujansky, E. & Gabow, P. A. 1988. Linkage heterogeneity of autosomal dominant polycystic kidney disease. *N Engl J Med* **319**, 913–918.

Liu, M., Shi, S., Senthilnathan, S., Yu, J., Wu, E., Bergmann, C., et al. 2010. Genetic variation of DKK3 may modify renal disease severity in ADPKD. *J Am Soc Nephrol* **21**, 1510–1520.

Mangoo-Karim, R., Uchic, M. E., Grant, M., Shumate, W. A., Calvet, J. P., Park, C. H., et al. 1989. Renal epithelial fluid secretion and cyst growth: The role of cyclic AMP. *FASEB J* **3**, 2629–2632.

Meijer, E., Rook, M., Tent, H., Navis, G., Van Der Jagt, E. J., De Jong, P. E. & Gansevoort, R. T. 2010. Early renal abnormalities in autosomal dominant polycystic kidney disease. *Clin J Am Soc Nephrol* **5**, 1091–1098.

Mekahli, D., Woolf, A. S. & Bockenhauer, D. 2010. Similar renal outcomes in children with ADPKD diagnosed by screening or presenting with symptoms. *Pediatr Nephrol* **25**, 2275–2282.

Mochizuki, T., Wu, G., Hayashi, T., Xenophontos, S. L., Veldhuisen, B., Saris, J. J., Reynolds, D. M., et al. 1996. PKD2, a gene for polycystic kidney disease that encodes an integral membrane protein. *Science* **272**, 1339–1342.

Nauli, S. M., Alenghat, F. J., Luo, Y., Williams, E., Vassilev, P., Li, X., et al. 2003. Polycystins 1 and 2 mediate mechanosensation in the primary cilium of kidney cells. *Nat Genet* **33**, 129–137.

O'Brien, K., Font-Montgomery, E., Lukose, L., Bryant, J., Piwnica-Worms, K., Edwards, H., et al. 2012. Congenital hepatic fibrosis and portal hypertension in autosomal dominant polycystic kidney disease. *J Pediatr Gastroenterol Nutr* **54**, 83–89.

O'Neill, W. C., Robbin, M. L., Bae, K. T., Grantham, J. J., Chapman, A. B., Guay-Woodford, L. M., et al. 2005. Sonographic assessment of the severity and progression of autosomal dominant polycystic kidney disease: The Consortium of Renal Imaging Studies in Polycystic Kidney Disease (CRISP). *Am J Kidney Dis* **46**, 1058–1064.

Orskov, B., Romming Sorensen, V., Feldt-Rasmussen, B. & Strandgaard, S. 2010. Improved prognosis in patients with autosomal dominant polycystic kidney disease in Denmark. *Clin J Am Soc Nephrol* **5**, 2034–2039.

Paterson, A. D., Magistroni, R., He, N., Wang, K., Johnson, A., Fain, P. R., et al. 2005. Progressive loss of renal function is an age-dependent heritable trait in type 1 autosomal dominant polycystic kidney disease. *J Am Soc Nephrol* **16**, 755–762.

Pei, Y., Obaji, J., Dupuis, A., Paterson, A. D., Magistroni, R., Dicks, E., et al. 2009. Unified criteria for ultrasonographic diagnosis of ADPKD. *J Am Soc Nephrol* **20**, 205–212.

Pirson, Y., Chauveau, D. & Torres, V. 2002. Management of cerebral aneurysms in autosomal dominant polycystic kidney disease. *J Am Soc Nephrol* **13**, 269–276.

Reeders, S. T., Breuning, M. H., Davies, K. E., Nicholls, R. D., Jarman, A. P., Higgs, D. R., et al. 1985. A highly polymorphic DNA marker linked to adult polycystic kidney disease on chromosome 16. *Nature* **317**, 542–544.

Rizk, D., Jurkovitz, C., Veledar, E., Bagby, S., Baumgarten, D. A., Rahbari-Oskoui, F., et al. 2009. Quality of life in autosomal dominant polycystic kidney disease patients not yet on dialysis. *Clin J Am Soc Nephrol* **4**, 560–566.

Robinson, C., Hiemstra, T. F., Spencer, D., Waller, S., Daboo, L., Karet Frankl, F. E., et al. 2012. Clinical utility of PKD2 mutation testing in a polycystic kidney disease cohort attending a specialist nephrology out-patient clinic. *BMC Nephrol* **13**, 79.

Rossetti, S., Consugar, M. B., Chapman, A. B., Torres, V. E., Guay-Woodford, L. M., Grantham, J. J., et al. 2007. Comprehensive molecular diagnostics in autosomal dominant polycystic kidney disease. *J Am Soc Nephrol* **18**, 2143–2160.

Rossetti, S., Hopp, K., Sikkink, R. A., Sundsbak, J. L., Lee, Y. K., Kubly, V., et al. 2012. Identification of gene mutations in autosomal dominant polycystic kidney disease through targeted resequencing. *J Am Soc Nephrol* **23**, 915–933.

Rossetti, S., Kubly, V. J., Consugar, M. B., Hopp, K., Roy, S., Horsley, S. W., et al. 2009. Incompletely penetrant PKD1 alleles suggest a role for gene dosage in cyst initiation in polycystic kidney disease. *Kidney Int* **75**, 848–855.

Sammels, E., Devogelaere, B., Mekahli, D., Bultynck, G., Missiaen, L., Parys, J. B., et al. 2010. Polycystin-2 activation by inositol 1,4,5-trisphosphate-induced Ca^{2+} release requires its direct association with the inositol 1,4,5-trisphosphate receptor in a signaling microdomain. *J Biol Chem* **285**, 18794–18805.

Sarnak, M. J., Greene, T., Wang, X., Beck, G., Kusek, J. W., Collins, A. J., et al. 2005. The effect of a lower target blood pressure on the progression of kidney disease: Long-term follow-up of the modification of diet in renal disease study. *Ann Intern Med* **142**, 342–351.

Schievink, W. I., Torres, V. E., Piepgras, D. G. & Wiebers, D. O. 1992. Saccular intracranial aneurysms in autosomal dominant polycystic kidney disease. *J Am Soc Nephrol* **3**, 88–95.

Serra, A. L., Poster, D., Kistler, A. D., Krauer, F., Raina, S., Young, J., et al. 2010. Sirolimus and kidney growth in autosomal dominant polycystic kidney disease. *N Engl J Med* **363**, 820–829.

Torres, V. E. 2010. Treatment strategies and clinical trial design in ADPKD. *Adv Chronic Kidney Dis* **17**, 190–204.

Torres, V. E., Chapman, A. B., Devuyst, O., Gansevoort, R. T., Grantham, J. J., Higashihara, E., et al. 2012a. Tolvaptan in patients with autosomal dominant polycystic kidney disease. *N Engl J Med* **367**, 2407–2418.

Torres, V. E., Chapman, A. B., Perrone, R. D., Bae, K. T., Abebe, K. Z., Bost, J. E., et al. 2012b. Analysis of baseline parameters in the HALT polycystic kidney disease trials. *Kidney Int* **81**, 577–585.

Torres, V. E., Harris, P. C. & Pirson, Y. 2007a. Autosomal dominant polycystic kidney disease. *Lancet* **369**, 1287–1301.

Torres, V. E., King, B. F., Chapman, A. B., Brummer, M. E., Bae, K. T., Glockner, J. F., et al. 2007b. Magnetic resonance measurements of renal blood flow and disease progression in autosomal dominant polycystic kidney disease. *Clin J Am Soc Nephrol* **2**, 112–120.

Torres, V. E. & Watson, M. L. 1998. Polycystic kidney disease: Antiquity to the 20th century. *Nephrol Dial Trans* **13**, 2690–2696.

Vujic, M., Heyer, C. M., Ars, E., Hopp, K., Markoff, A., Orndal, C., et al. 2010. Incompletely penetrant PKD1 alleles mimic the renal manifestations of ARPKD. *J Am Soc Nephrol* **21**, 1097–1102.

Wang, C. J., Creed, C., Winklhofer, F. T. & Grantham, J. J. 2011. Water prescription in autosomal dominant polycystic kidney disease: A pilot study. *Clin J Am Soc Nephrol* **6**, 192–197.

Ward, C. J., Peral, B., Hughes, J., Thomas, S., Gamble, V., Maccarthy, A. B., et al. 1994. The polycystic kidney-disease-1 gene encodes a 14-kb transcript and lies within a duplicated region on chromosome-16. *Cell* **77**, 881–894.

Wijdicks, E. F., Torres, V. E. & Schievink, W. I. 2000. Chronic subdural hematoma in autosomal dominant polycystic kidney disease. *Am J Kidney Dis* **35**, 40–43.

Yoder, B. K., Hou, X. & Guay-Woodford, L. M. 2002. The polycystic kidney disease proteins, polycystin-1, polycystin-2, polaris, and cystin, are co-localized in renal cilia. *J Am Soc Nephrol* **13**, 2508–2516.

Chapter 11

Autosomal recessive polycystic kidney disease

Carsten Bergmann

Classification and differential diagnosis

In their seminal studies, Osathanondh and Potter systematically classified renal cystic diseases into four distinct types (Osathanondh & Potter, 1964). Potter syndrome type I is referred to as autosomal recessive polycystic kidney disease (ARPKD), type II as renal cystic dysplasia, type III as autosomal dominant polycystic kidney disease, and type IV occurs when a long-standing obstruction in either the kidney or ureter leads to cystic kidneys or hydronephrosis. Types II–IV can be part of many syndromes. While this classification still has an impact for pathoanatomical descriptions, it is hardly to be reconciled with clinical and genetic entities and is more and more replaced by the genetic nomenclature.

Accurate diagnosis is essential both in the management of patients with cystic kidneys and in counselling their families. When an effort is made to classify the wide array of different entities with renal cysts, it might be helpful to first distinguish between acquired and inherited forms. Knowledge about the family history and the clinical picture, together with the location and morphology of the cysts, and any possible extra-renal manifestations should help in making a diagnosis. Sometimes, cytogenetic studies and array-comparative genomic hybridisation CGH) may be useful to exclude rearrangements or aberrations such as large deletions or duplications (see diagnostic algorithm, Figure 11.1).

Inherited cystic kidney disorders mainly include ARPKD and autosomal dominant polycystic kidney disease, glomerulocystic kidney disease (GCKD), and entities comprising the medullary cystic kidney disease–nephronophthisis complex. Notably, cystic kidneys are an important feature of numerous genetic syndromes, such as the mainly recessively inherited ciliopathies Jeune, Bardet–Biedl, Joubert, and Meckel–Gruber syndromes or the dominant disorders tuberous sclerosis (TSC), von Hippel–Lindau (VHL) disease and branchio-oto-renal (BOR) syndrome.

The eponym

As the name implies, ARPKD is transmitted in an autosomal recessive fashion, i.e. virtually all individuals who have inherited two unfavourable, mutated *PKHD1* germline alleles from their parents will develop the disease. The parents usually each carry a heterozygous

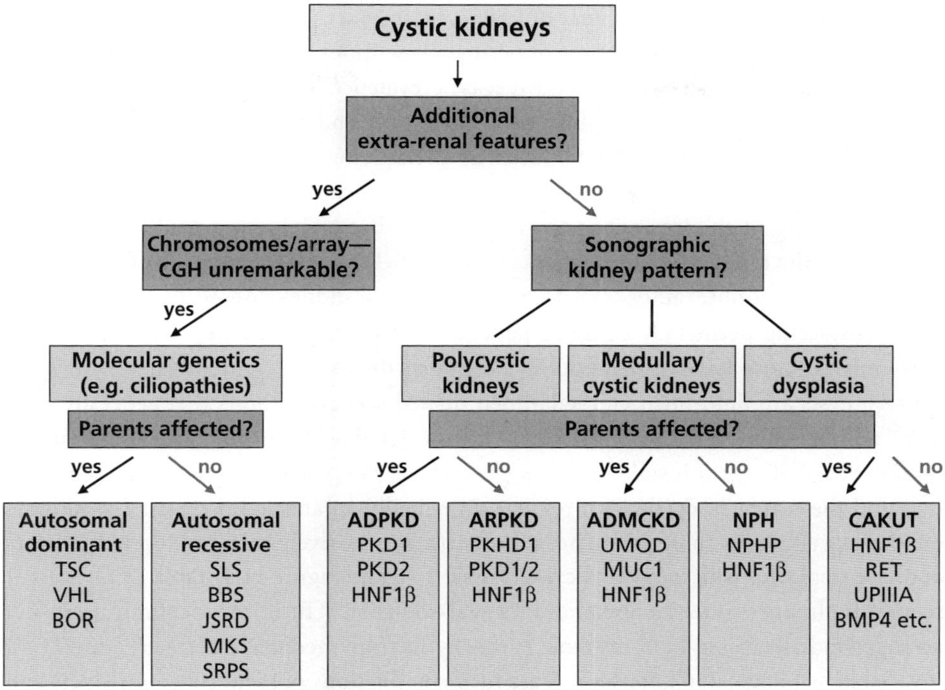

Figure 11.1 Diagnostic algorithm for cystic kidney disease.

Please note that such an algorithm can always only be a simplified approach and has to be modified individually. Conclusions also depend on parental age and family history. Dominant disorders may appear sporadically or recessively inherited in case of *de novo* mutations. TSC, tuberous sclerosis; VHL, von Hippel–Lindau syndrome; BOR, branchio-oto-renal syndrome; SLS, Senior–Løken syndrome; BBS, Bardet–Biedl syndrome; JSRD, Joubert syndrome-related disorders; MKS, Meckel–Gruber syndrome; SRPS, short rib–polydactyly syndromes (Jeune and others); ADPKD, autosomal dominant polycystic kidney disease; ARPKD, autosomal recessive polycystic kidney disease; ADMCKD, autosomal dominant medullary cystic kidney disease; NPH, nephronophthisis; CAKUT, congenital anomalies of the kidney and urinary tract.

PKHD1 mutation and are invariably healthy without developing cysts. The same applies to all other individuals bearing only one defective allele and an intact *PKHD1* gene copy *in trans*, i.e. there is no heterozygote manifestation in ARPKD. The gene's name *PKHD1* is an abbreviation that stands for 'polycystic kidney and hepatic disease 1' and is also sometimes used as a disease name, pointing to the fact that liver changes in terms of ductal plate malformation (DPM) with hyperplastic biliary ducts and congenital hepatic fibrosis (CHF) are obligatory in ARPKD.

Epidemiology and morphology

ARPKD is much rarer than its dominant counterpart autosomal dominant polycystic kidney disease with a suspected incidence among Caucasians of about 1 in 20,000 live births corresponding to a carrier frequency of approximately 1:70 in non-isolated populations. Isolated populations may have considerably higher prevalences. For Finland, Kaariainen and colleagues reported an incidence of 1:8000, for instance (Kaariainen et al. 1988). Some severely affected babies may die pre- or perinatally without a definitive diagnosis. In line, it is difficult to give exact figures on the disease prevalence which is further challenged by the fact that the patient cohorts published vary greatly and range from severely affected, perinatally demised patients mainly seen by gynaecologists and pathologists, to mild and moderately affected patients followed by paediatricians and their adult colleagues. Notably, among all children with polycystic kidney disease in departments of paediatric nephrology, the total number of patients with ARPKD equals the number of individuals affected with early-onset autosomal dominant polycystic kidney disease. In contrast, when only considering perinatally demised patients with polycystic kidney disease, ARPKD outnumbers autosomal dominant polycystic kidney disease.

Renal cysts are fluid-filled epithelia-lined dilated saccular lesions that generally arise from tubular segments. Usually, ARPKD can be reliably diagnosed pathoanatomically (Zerres et al. 2003), but histological changes may vary depending on the age of presentation and the extent of cystic involvement. Principally, in affected neonates the kidneys retain their reniform contour and are symmetrically, massively enlarged (up to ten times of the normal size) with many tiny cysts (Figure 11.2 and Figure 11.3, Table 11.1). Macroscopically, the cut surface demonstrates cortical extension of fusiform or cylindrical spaces arranged radially throughout the renal parenchyma from medulla to cortex (Figure 11.4). Invariable histological manifestations are fusiform dilations of renal collecting ducts and distal tubuli lined by columnar or cuboidal epithelium that usually remain in contact with the urinary system (unlike autosomal dominant polycystic kidney disease), whereas glomerular cysts (as in autosomal dominant polycystic kidney disease) or dysplastic elements (e.g. cartilage; as often in Meckel–Gruber syndrome or some other syndromic ciliopathies) are usually not evident in ARPKD kidneys (Figure 11.4). During early foetal development, a transient phase of proximal tubular cyst formation has been identified that is largely absent by birth, however (Nakanishi et al. 2000). With advancing clinical course the kidney structure might increasingly resemble the pattern observed in autosomal dominant polycystic kidney disease with renal cysts that vary considerably in size and appearance, often also accompanied by some degree of interstitial fibrosis (Avni et al. 2002).

Liver changes are invariably present from early embryonic development on. Defective remodelling of the ductal plate leads to dysgenesis of the hepatic portal triad with hyperplastic biliary ducts and congenital hepatic fibrosis (CHF) (Figure 11.4b) (Desmet, 1998). At later stages, fibrous septa may link different portal tracts by intersecting the hepatic parenchyma often leading to portal hypertension; however, the remaining liver parenchyma usually develops normally. In accordance, the cholestasis parameters such as gamma glutamyltransferase (γ-GT) are sometimes elevated, whereas other liver enzymes are characteristically within normal ranges. Biliary anomalies may develop at any stage of the physiological involution–remodelling process, and the timing or stage of development determines the resulting clinical and histological phenotype. Typically, cysts that arise from small interlobular bile ducts are detached from the biliary tree, while those that stem from malformation of medium- and large-sized bile ducts usually maintain connected.

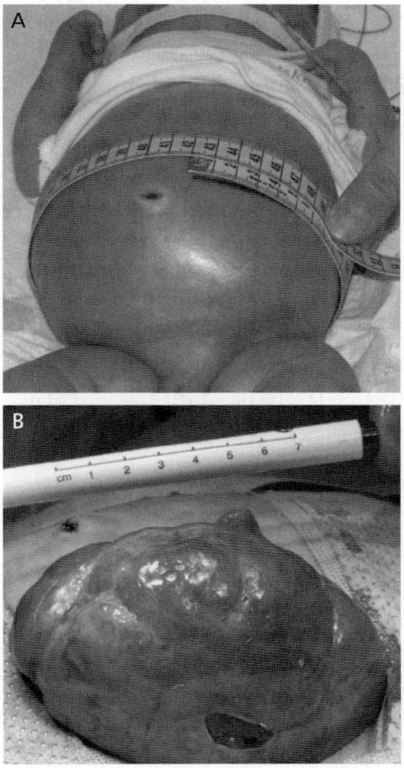

Figure 11.2 Baby with autosomal recessive polycystic kidney disease (ARPKD). (A) Distended abdomen due to voluminous kidneys that lead to respiratory problems. (B) Nephrectomised kidney of this girl.
Reproduced from *European Journal of Pediatrics*, Volume 17, 2011, pp. 1285–1300, Carsten Bergmann, Educational paper: Ciliopathies, Figure 3, Copyright © The Author(s) 2011. This article is published with open access at Springerlink.com. With kind permission from Springer Science and Business Media.

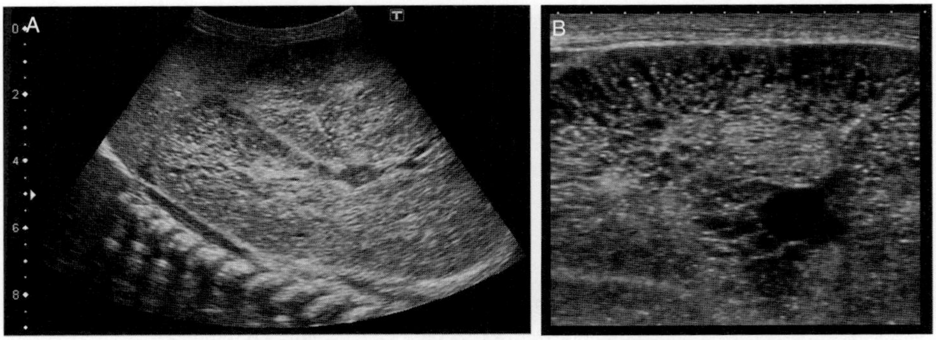

Figure 11.3 Renal ultrasound of a baby with ARPKD. Symmetrically enlarged echogenic kidneys (A) with fusiform dilations of collecting ducts and distal tubules arranged radially throughout the renal parenchyma from medulla to cortex (B).
Reproduced from *European Journal of Pediatrics*, Volume 17, 2011, pp. 1285–1300, Carsten Bergmann, Educational paper: Ciliopathies, Figure 3, Copyright © The Author(s) 2011. This article is published with open access at Springerlink.com. With kind permission from Springer Science and Business Media.

Table 11.1 Characteristics of autosomal recessive (ARPKD) and dominant (ADPKD) polycystic kidney disease

	ARPKD	**ADPKD**
Synonyms	Infantile polycystic kidney disease Potter type I	Adult polycystic kidney disease Potter type III
Incidence	~ 1:20,000	1:500–1000 (~ 2% early manifesting)
Pathology of kidneys		
Macroscopy	Massively, symmetrically enlarged kidneys (reniform)	Generally enlarged (also reniform), but usually to a lesser extent
Location of cysts	Dilated collecting ducts and distal tubuli	Cysts in all parts of the nephron (including glomerulus)
Ultrasound and diameter of cysts	At onset typical pepper–salt pattern in ultrasound, increased echogenicity of renal parenchyma throughout cortex and medulla due to tiny, sometimes invisible cysts (usually < 2 mm); with advancing age up to several cm similar to ADPKD pattern	Cysts of different size in cortex and medulla (usually several larger cysts in adults); at onset often small; however, sometimes already several cm early in childhood
Pathology of liver	Mandatory: Ductal plate malformation/congenital hepatic fibrosis with hyperplastic biliary ducts and portal fibrosis (may impress as Caroli disease)	'Liver cysts' common in adults, but rare in children. Occasionally, ductal plate malformation/ congenital hepatic fibrosis
Associated anomalies	Rarely pancreatic cysts and/or fibrosis; single case reports with intracranial aneurysms	Pancreatic cysts and/or cysts in other epithelial organs; intracranial aneurysms in ~ 8%, familially clustered
Main clinical manifestations	Peri-/neonatal period: Respiratory distress (in 30–50%). With prolonged survival renal insufficiency, portal hypertension, and other variable co-morbidities	General onset third to fifth decade with arterial hypertension, proteinuria, haematuria, and/or renal insufficiency, ~ 2% early manifestation in childhood (rarely with perinatal respiratory distress)
Risk for siblings	25%	50% (except for rare cases of spontaneous mutation with virtually no risk)
Risk for own children	< 1% (unless unaffected parent is related to his/her affected partner, or ARPKD is known in the unaffected partner's family)	50% (also for patients with a spontaneous mutation)
Manifestation in affected family members	Often similar clinical course in siblings (in ~ 20% extensive intrafamilial variability)	Variable; however, often similar within the same family; in case of early manifestation ~ 50% recurrence risk

Table 11.1 (continued) Characteristics of autosomal recessive (ARPKD) and dominant (ADPKD) polycystic kidney disease

	ARPKD	ADPKD
Parental kidneys	No alterations	Except for rare cases of spontaneous mutation, usually one parent is affected and shows renal cysts (be careful when parents are too young for definite clinical diagnosis/< 30–40 years)
Prognosis	In perinatal cases with respiratory distress usually poor, for those surviving the neonatal period much better with renal death in ~ 15% in childhood, often severe complications (e.g. oesophageal varices) due to portal hypertension, if possible transplantation (often combined kidney–liver TX)	In early manifesting cases often better than in ARPKD. In 'adult' cases, chronic renal failure in ~ 50% by age of 60 years; median age of ESRD onset (54 vs. 74 years in PKD1 vs. PKD2)

While there are overlaps between each subgroup, bile duct hamartomas, ARPKD and other ciliopathies (e.g. Bardet–Biedl, Joubert, and Meckel–Gruber syndrome) are mainly manifestations of ductal plate malformation of the small interlobular bile ducts, whereas medium-sized intrahepatic ducts are generally afflicted in autosomal dominant polycystic kidney disease and polycystic liver disease. Caroli's disease is usually the result of ductal plate malformation of large intrahepatic bile ducts, while choledochal cysts are thought to represent large extrahepatic ductal plate malformation (Desmet, 1998).

Clinical features and management

Kidney- and liver-related medical problems are central to the care of children with ARPKD. While there are still no disease-specific treatment options, it is crucial, however, as with other forms of chronic kidney disease, to prevent and effectively manage complications, such as arterial hypertension and urinary tract infections, to slow the progression to end-stage renal disease (ESRD). Renal insufficiency and ESRD in ARPKD is treated with standard medical management of chronic renal failure and renal replacement therapy as outlined elsewhere.

The clinical spectrum in ARPKD

While autosomal dominant polycystic kidney disease is usually a disease of adults with less than 5% of patients displaying an early manifesting clinical course, ARPKD is typically an infantile disease. However, the clinical spectrum is much more variable than generally presumed (Adeva et al. 2006). Nevertheless, despite dramatic advances in neonatal and intensive care over the past decades, the short-term and long-term morbidity and mortality of ARPKD remain substantial. Ages at diagnosis and initial clinical features are listed

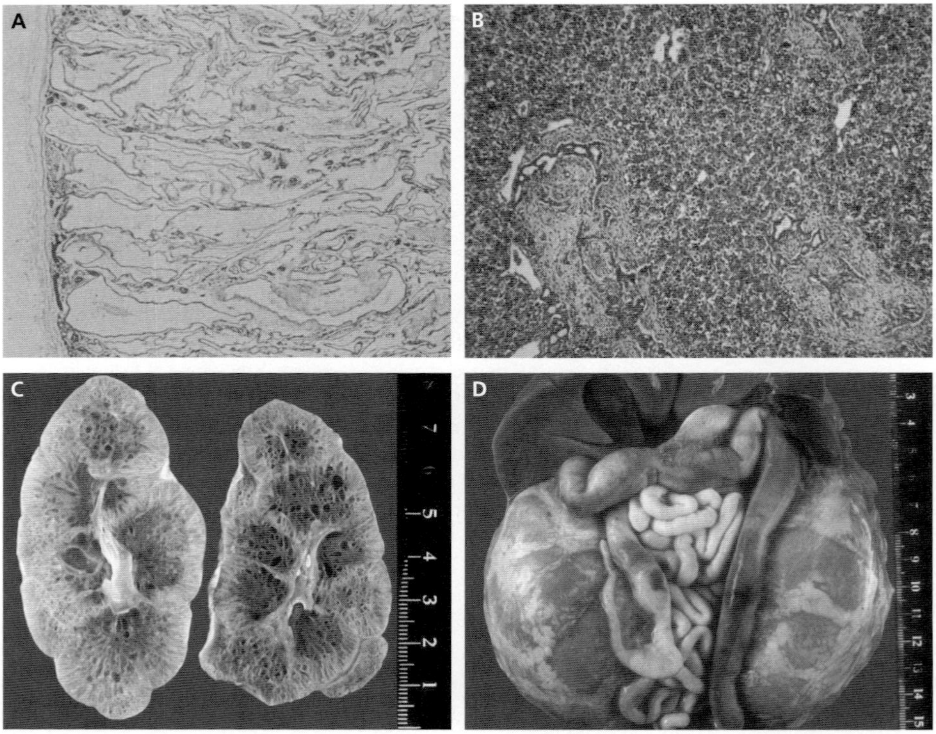

Figure 11.4 Microscopic and macroscopic appearance of ARPKD. (A) Microscopically, fusiform dilations of renal collecting ducts and distal tubuli lined by columnar or cuboidal epithelium. These dilated collecting ducts run perpendicular to the renal capsule. (B) Obligatory hepatobiliary changes in ARPKD subsumed as ductal plate malformation (DPM) and characterised by dysgenesis of the hepatic portal triad with hyperplastic biliary ducts and congenital hepatic fibrosis (CHF). (C) Cross section of ARPKD kidneys reveals the cortical extension of fusiform or cylindrical spaces arranged radially throughout the renal parenchyma from medulla to cortex. (D) Abdominal situs of an ARPKD patient with symmetrically enlarged kidneys that maintain their reniform configuration.

in Table 11.2, which summarises results of various clinical studies on ARPKD. Notably, these studies differ widely by their selection criteria of patients and their mode of analysis of data. Patients from most of these surveys were recruited from paediatric departments, some from specialised single-centres. As a consequence, individuals with an early lethal form of ARPKD were underrepresented. Most of the individuals in the study by Roy and colleagues (Roy et al. 1997) had previously been reported by Kaplan and co-workers who exclusively included patients with pathoanatomically proven CHF (Kaplan et al. 1989). Kaariainen et al. analysed mainly data obtained from Finnish death registers (Kaariainen et al. 1988). Inclusion criteria of the study by Cole et al. were diagnosis within the first year of life and survival of the neonatal period (Cole et al. 1987).

Table 11.2 Summary of findings obtained in clinical studies of ARPKD patients

	Gunay-Aygun et al. (2010, 2012)	**Bergmann et al. (2005)**	**Guay-Woodford and Desmond (2003)**	**Capisonda et al. (2003)**	**Roy et al. (1997)**	**Zerres et al. (1996)**	**Gagnadoux et al. (1989)**	**Kaplan et al. (1989)**	**Kääriäinen et al. (1988)**	**Cole et al. (1987)**
Patients (*n*)	73 (63 unrelated patients)	186 (164)	166	31	52	115	33	55	73 (18 neonatal survivors)	17
Age at diagnosis	42% perinatal 58% non-perinatal (median, 2.9 y)	23% prenatal 31% < 1 m 16% 1–12 m 30% > 1 y	46% prenatal 27% < 1 m 11% 1–12 m 16% > 1 y	32% prenatal 23% < 1 m 19% 1–12 m 26% > 1 y	85% < 1 y 15% > 1 y	10% prenatal 41% < 1 m 23% 1–12 m 26% > 1 y	33% < 1 m 55% 1–18 m 12% 6–11 y	42% < 1 m 42% 1–12 m 16% < 1 y	72% < 1 m 6% < 1 y 22% > 1 y	100% 1–12 m (inclusion criteria)
Renal function	25% of perinatally symptomatic patients ESRD by 11 y 25% of non-perinatally symptomatic patients ESRD by 32 y	86% GFR < 3rd centile for age median age CRF 4.0 y 29% ESRD (by 10y) 58% ESRD (by 20y)	42% GFR < 3rd centile for age 13% ESRD	51% GFR < 80 ml/min/1.73 m^2 16% ESRD	33% ESRD (by 15y)	72% GFR < 3rd centile for age 10% ESRD	42% GFR < 80 ml/min/1,73m^2 21% ESRD	58% SC > 100 μmol/ml	82% GFR < 90 ml/min/1,73m^2	35% GFR < 40 ml/min/1,73m^2 29% ESRD
Kidney length	Perinal patients: 6.3 ± 3.3 SD Non-perinal patients: 4.5 ± 3.7 SD	92% > 2 SD	NA	NA	NA	68% > 2 SD	100% > 2 SD	NA	NA	NA
Hypertension (% on drug treatment)	NA	76% (80% M/72% F) medication started at median age of 3 y (53% during first 6 m)	65%	55%	60% (by 15y)	70%	76%	65%	61%	100% (drug treatment or BP > 95th centile)

Table 11.2 (continued) Summary of findings obtained in clinical studies of ARPKD patients

	Gunay-Aygun et al. (2010, 2012)	Bergmann et al. (2005)	Guay-Woodford and Desmond (2003)	Capisonda et al. (2003)	Roy et al. (1997)	Zerres et al. (1996)	Gagnadoux et al. (1989)	Kaplan et al. (1989)	Kääriäinen et al. (1988)	Cole et al. (1987)
Growth retardation	NA	16% < 2 SD (23% M/10% F)	24% < 2 SD	NA	NA	25% < 2 SD	18% < 4 SD	NA	6% < 2,5 SD	NA
Anaemia	NA	14% (9% M/19% F)	NA	NA	NA	NA	NA	NA	NA	NA
Evidence of portal HTN #	64% splenomegaly 30% oesophageal varices	44% (41% M/47% F) * 38% splenomegaly 15% oesophageal varices 2% ascites	15%	37%	23% (8/35)	46%	39%	47%	50% (hepatomegaly)	35%
Survival rate	NA	1y 85% 5y 84% 10y 82%	1y 79% 5y 75%	1y 87% 9y 80%	NA	1y 89% 3y 88%	1y 91%	1y 79% 10y 51% 15y 46%	1y 19%	1y 88%
Death rate in the first year of life	NA	15%	8%	13%	26%	9%	9%	24%	22%	12%

ESRD, end-stage renal disease; SD, standard deviation; NA, not available; BP, blood pressure.

Based on sonographic evidence of hepatomegaly, splenomegaly, and directional reversal of portal vein flow, or clinical, radiological or endoscopical evidence of oesophageal varices or ascites.

* Manifestation of clinical signs of congenital hepatic fibrosis was positively correlated with age. In 87% increased echogenicity of the liver has been reported (89% M/84% F). Cystic changes of the liver probably representing Caroli's disease with dilated larger intrahepatic bile ducts have been noted in 27 individuals equaling 16% of the total cohort (17% M/15% F). Within this survey only two boys exhibited impaired hepatocellular function (1%) underscoring that liver function is usually retained in ARPKD. In six patients (4 M/2 F) liver transplantation (LTX) was performed (mean age 13.8 y). In four cases it was done in parallel with NTX (combined LNTX).

Overall, the majority of cases are identified late in pregnancy or at birth. Severely affected foetuses display a 'Potter' oligohydramnios phenotype with pulmonary hypoplasia, a characteristic facies, and contracted limbs with club feet. As many as 30–50% of affected neonates die from respiratory insufficiency shortly after birth. Respiratory distress in these severely affected children is mainly caused by two reasons: First, mechanistically related by the massively enlarged kidneys that push the diaphragm upwards and lead to thoracic compression. Another cause is the critical degree of pulmonary hypoplasia as a consequence of oligo-/anhydramnios due to *in utero* renal dysfunction. In contrast, end-stage renal failure itself is only very rarely a cause for neonatal demise. Hyponatraemia related to a urine dilution defect is often present in the newborn period, but usually resolves over time (Guay-Woodford & Desmond, 2003; Kaplan et al. 1989). Advances in mechanical ventilation and other supportive measures as well as further improvements in renal replacement therapies have increased the survival rates of ARPKD patients with many of them reaching adulthood nowadays. A few may even be clinically asymptomatic until advanced adulthood and have a normal life expectancy (Adeva et al. 2006). While these cases clearly support some alertness and a much broader clinical spectrum than expected in former times, overall, those only very mildly afflicted individuals are exceptions to the rule. Usually, a wide range of associated co-morbidities evolve in ARPKD, such as systemic hypertension, ESRD and clinical manifestations of CHF (Bergmann et al. 2005; Guay-Woodford & Desmond, 2003; Roy et al. 1997). Therefore, ARPKD is still a disease with a severely diminished life expectancy and an important cause of renal- and liver-related morbidity and mortality in children. In a study of almost 200 ARPKD patients with known *PKHD1* mutational status, the survival rate of those patients who survived the critical newborn period was 94% at 5 years and 92% at 10 years of age (Bergmann et al. 2005), while Dell and Avner reported a 10-year survival rate of 82% for patients who survived the first year of life (Dell & Avner, 1993).

Cysts

Ultrasound scans show that children with ARPKD typically have characteristic bilateral large hyper-echoic kidney masses with poor corticomedullary differentiation. Cysts are usually fusiform and tiny often giving the so-called 'pepper–salt' pattern at early stages. Macrocysts are uncommon in small infants, although they may be observed with advanced clinical course when ultrasonographic patterns of ARPKD and autosomal dominant polycystic kidney disease often adjust and become difficult to distinguish (Avni et al. 2002; Nicolau et al. 2000). Data for kidney length measured by ultrasound related to age revealed that 92% had a kidney length above or on the 97th centile for age (Bergmann et al. 2005). In this study, in no cases was the kidney size decreased, ranging from 0 to +17 SDs.

Chronic renal failure

In our survey, which included mainly patients from tertiary hospitals with departments of paediatric nephrology, chronic renal failure was first detected at a mean age of 4 years (Bergmann et al. 2005). Infants with ARPKD may have a transient improvement in their glomerular filtration rate (GFR) due to renal maturation in the first 6 months of life (Cole et al. 1987). However, subsequently, a progressive but highly variable decrease in renal function occurs. The management of children with declining renal function should follow the standard guidelines established for chronic renal insufficiency in other paediatric patients (Warady et al. 1999). In our study (Bergmann et al. 2005), ESRD occurred in 29% of patients at 10 years and 58% at 20 years, which is much lower than the figures reported by previous studies that proposed rates of approximately 50% of ARPKD patients progressing to ESRD within the first decade of life (Cole et al. 1987; Roy et al. 1997). Renal transplantation is the treatment of choice for individuals with ESRD. In cases of massively enlarged kidneys, native nephrectomies may be warranted to allow allograft placement. In a few cases reported in the literature uni- or bilateral nephrectomy led to respiratory improvement, better enteral nutrition and more effective peritoneal dialysis. Early renal replacement therapy by renal transplantation after bilateral nephrectomy was reported in two infants at the age of 9 and 15 months of life (Prelog et al. 2006; Spechtenhauser et al. 1999); however, is usually not an option in the neonatal period. Recently, we described a newborn with ARPKD and huge kidneys in whom unilateral nephrectomy was done as rescue therapy and haemodialysis was performed for several days before the girl died from complications of pulmonary hypertension (Arbeiter et al. 2008). Overall, uni- or bilateral nephrectomy should be considered in ARPKD patients with massively enlarged kidneys to be performed early with consecutive renal replacement therapy.

Hepatobiliary complications

All individuals with ARPKD invariably show evidence of congenital hepatic fibrosis. In keeping with a generally prolonged survival in ARPKD, the hepatobiliary complications may come to dominate the clinical picture in some patients. A serious, potentially lethal complication in ARPKD is ascending suppurative cholangitis that may cause fulminant hepatic failure. It always requires diligent evaluation with aggressive antimicrobial treatment. It is noteworthy that ARPKD patients may not display the typical clinical findings of cholangitis; thus, every patient with unexplained recurrent sepsis, particularly with Gram-negative organisms, should be critically evaluated for this diagnosis (Kashtan et al. 1999). While hepatocellular function is usually preserved, sequelae of portal hypertension may lead to haematemesis or melena due to bleeding oesophageal varices and/or hypersplenism with consequent pancytopenia. Primary management of variceal bleeding may include endoscopic approaches, such as sclerotherapy or variceal banding. In other patients, portosystemic shunting or a combined liver and kidney transplantation (CLKT) might be considered a viable therapeutic option. Currently, there is no consensus as to the indication of CLKT and data on long-term outcome are scarce. We recently analysed in detail a cohort of eight ARPKD patients with known *PKHD1* mutational status undergoing CLKT in a single specialised centre (Brinkert, 2013). Patient survival after CLKT was 100% and liver and kidney graft survival was 72% and 88%, respectively. Liver and kidney function were stable in all patients with median eGFR of 95 ml/min/1.73 m^2 (range 68–133 ml/min/1.73 m^2). Further data demonstrated significantly better growth in these patients after CLKT. In accord with our results, Chapal and colleagues conclude that pre-emptive liver transplantation might be a therapeutic option in ARPKD patients with severe portal hypertension and/or Caroli's disease evaluated for renal transplantation (Chapal et al. 2012). Although the decision for a combined transplant against the background of normal liver synthesis remains difficult, our data and that of others document a potentially favourable outcome of CLKT if performed in specialised institutions.

Renal–hepatobiliary morbidity

Another aspect discussed in clinical studies on ARPKD is the renal–hepatobiliary morbidity pattern (Bergmann et al. 2005; Guay-Woodford & Desmond, 2003; Gunay-Aygun et al. 2010, 2012). Although most patients show a comparable degree of severity with regard to liver and kidney affection (i.e. in a patient with renal insufficiency it is highly likely that also the liver is considerably affected and *vice versa*), there is no direct correlation or interdependency between those two organs. A recent study by Gunay-Aygun and colleagues found spleen volume to have an inverse correlation with platelet count and prothrombin time (Gunay-Aygun et al. 2012). Platelet count was the best predictor of spleen volume and the severity of portal hypertension, but did not correlate with renal function. Single ARPKD patients may even present with an organ-specific phenotype, i.e. either an (almost) exclusive renal phenotype or a predominant or mere liver phenotype. In accordance, it could be demonstrated that *PKHD1* mutations can cause isolated congenital hepatic fibrosis or Caroli's disease (Bergmann et al. 2005; Rossetti et al. 2003). It is noteworthy that two transgenic mouse models for *Pkhd1* display an isolated liver phenotype without any renal involvement (Moser et al. 2005).

Hypertension

Arterial hypertension usually develops in the first few months of life and affects up to 80% of children with ARPKD (Table 11.2). Hypertension can be difficult to control in these children and may require multi-drug treatment. To prevent sequelae of hypertension (e.g. cardiac hypertrophy, congestive heart failure) and deterioration of renal function, careful blood pressure monitoring is essential and systemic hypertension needs to be treated early and aggressively. Angiotensin converting enzyme (ACE) inhibitors are regarded as treatment of choice. Further appropriate drugs generally effective are AT II receptor inhibitors, calcium channel blockers, beta blockers (particularly in those patients with signs of CHF and portal hypertension), and diuretics (especially loop agents) (Guay-Woodford & Desmond, 2003; Jafar et al. 2005). The precise pathogenesis of hypertension in PKD still remains to be elucidated. At least in part it appears to be mediated by activation of the intrarenal renin–angiotensin–aldosterone system (RAAS), reduced renal blood flow, and increased sodium retention(Chapman & Schrier, 1991). However, there is controversial data concerning whether or not RAAS is activated in the first place, or at least inappropriately with respect to the prevailing blood pressure and sodium state (Ritz, 2006). Several studies postulated that the demonstrated link between renal structural severity and hypertension is likely due to upregulation of the RAAS machinery (Chapman et al. 1990). As also true for most other pathomechanisms in PKD, data obtained on the pathogenesis of hypertension mainly originate from studies of autosomal dominant polycystic kidney disease and only rarely from its recessive counterpart. Larger kidneys with a greater number of cysts may predispose for arterial hypertension through excess renal angiotensin II production. This hypothesis is corroborated by the finding of high renin concentrations in cyst fluid and the ability of cystic epithelia to synthesise renin (Torres et al. 1992). In turn, angiotensin II acts as a growth factor for renal tubular cells and boosts the mitogenic actions of epidermal growth factor (EGF), which may further sustain faster renal growth in hypertensive children with enlarged kidneys (Chatterjee et al. 1997). In clear contradiction to inappropriate RAAS stimulation in autosomal dominant polycystic kidney disease is a study by Doulton and co-workers (Doulton et al. 2006) in which there was no significant difference between hypertensive patients and individuals with essential hypertension excluding absolute or relative over-activity of the RAAS in autosomal dominant polycystic kidney disease. These authors also re-emphasised the fact that an impressive lowering of the blood pressure and amplified blood pressure response to angiotensin converting enzyme inhibition can be achieved by restricted sodium intake. Conclusively, the pathophysiology of hypertension in ARPKD is not clearly understood. Although peripheral vein renin values are usually not elevated in hypertensive ARPKD patients, the pathogenesis of hypertension appears to be mediated, at least in part, by dysregulation of renal sodium transport and activation of the RAAS that lead to increased intravascular volume (Kaplan et al. 1989).

Intracerebral aneurysm

In contrast to autosomal dominant polycystic kidney disease in which cardiovascular comorbidities and in particular intracerebral aneurysms (ICAs) play a significant role, patients with ARPKD are thought to be less likely affected. However, when assuming similar underlying pathomechanisms for recessive and dominant PKD and given the fact that most cardiovascular events in autosomal dominant polycystic kidney disease occur in elderly patients, it is important to discuss that lower prevalence figures in ARPKD may also be (at least in part) explained by the younger age of most ARPKD patients. Thus, a certain kind of alertness for adult patients with ARPKD and a closer look at the available data might be warranted. Unfortunately, general recommendations neither exist for adults nor adolescents with autosomal dominant polycystic kidney disease and the question of screening for intracerebral aneurysms in patients known to be affected with autosomal dominant polycystic kidney disease is still a matter of ongoing debate and challenging in every individual case (Pirson et al. 2002). An overall prevalence of about 8% of asymptomatic intracerebral aneurysms has been estimated by a couple of large prospective series (Chapman et al. 1992). This number equals a rate four to five times above that found in the general population (Rinkel et al. 1998). Prevalence increases with age and was 23.3% among 60- to 69-year-old autosomal dominant polycystic kidney disease patients (Xu et al. 2011). ICAs are also known to be familially clustered. Xu and colleagues described a relative risk of 1.97 in autosomal dominant polycystic kidney disease patients with a positive family history of intracerebral aneurysms. Pirson and colleagues concluded that the prevalence of asymptomatic intracerebral aneurysms is about 6% in autosomal dominant polycystic kidney disease patients in the absence of a positive family history of intracerebral aneurysms or sub-arachnoid haemorrhage and approximately 16% in those patients with a family history (Pirson et al. 2002). These authors suggested a balanced and reasonable view: patients without a family history of intracerebral aneurysms should not be screened unless they firmly request it. In contrast, in autosomal dominant polycystic kidney disease patients with a positive family history of ICAs the pros and cons of screening should be explained in detail. In accord with this, the data of a recent study by Irazabal and colleagues supports very selective screening for unruptured ICAs in autosomal dominant polycystic kidney disease, whereas widespread screening is not indicated. It is important to note that growth and rupture risk among autosomal dominant polycystic kidney disease patients are not higher than in the general population (Irazabal et al. 2011). To return to the question of screening in children and adolescents with polycystic kidney disease, it is good to know that aneurysm rupture is usually very uncommon among children and adolescents. Mariani and colleagues pointed out that there is only little chance of detecting an ICA before age 30 and, thus, do not recommend screening before the third decade (Deget et al. 1995; Mariani et al. 1999). Chapman and Guay-Woodford tend towards consideration of screening at the age of 20 years, especially when there is a positive family history (Chapman & Guay-Woodford, 2006).

Phenotypic variability among affected siblings

It is well known that affected sibships can be used to set up genotype–phenotype correlations. While most sibships display comparable clinical courses, about 20% of ARPKD multiplex pedigrees exhibit gross intrafamilial phenotypic variability with peri-/neonatal demise in one and survival into childhood or even adulthood in another affected sib (Deget et al. 1995). An even higher proportion of 20 out of 48 sibships (42%) was present in a study cohort among families with at least one neonatal survivor per family representative for the spectrum of patients usually followed by departments of paediatric nephrology (Bergmann et al. 2005). Adjusted for differing family sizes the risk for perinatal demise of a further affected child in this study was 37% (22 perinatally deceased children from a total of 59 patients excluding the moderately affected index cases). Irrespective of the exact risk figures, some shortcomings of the categorisation into severe and moderate phenotypes, and the fact that the numbers obtained in our mentioned survey were clearly biased by the study design, some alertness is definitely warranted in predicting the clinical outcome of a further affected child in a family with an increased risk for ARPKD. Overall, the rate of clinically discordant siblings is alarming and is information that should be shared with afflicted families. While the process of genetic counselling itself is meant to be non-directive, it is clearly reasonable to recommend to all of those families to deliver in an experienced, well-equipped clinic with paediatric intensive care units and interdisciplinary teams of obstetricians and paediatricians.

What becomes clear from this data with striking phenotypic variability even among affected siblings of the same family is the fact that phenotypes cannot be simply explained on the basis of the underlying *PKHD1* genotype. In addition, modifying alleles, environmental factors and other mechanisms such as epigenetics potentially influence the clinical course. However, conclusive data of underlying mechanisms is still largely lacking and a matter of ongoing research. Second-site modifiers are expected to exert an aggravating effect in a mainly epistatic way. In this scenario, altered dosage of disease proteins may disturb cell homeostasis and network integrity contributing to early and more severe disease expression (Garcia-Gonzalez et al. 2007; Hopp et al. 2012). Recently, we and others could show that in some PKD families with variable expressivity only the severely affected patients harboured further mutations in addition to their expected familial germline defect that were assumed to aggravate the phenotype (Bergmann et al. 2011; Rossetti et al. 2009).

Diagnosis and genetics

Marquardt is thought to be the first who postulated genetic heterogeneity of polycystic kidney disease when stating: 'In surviving individuals, cystic kidneys are inherited dominantly. In non-viable individuals, cystic kidneys are recessive.' (Blyth & Ockenden, 1971). It took more than 35 years from that point of view when Blyth and Ockenden demonstrated in a systemic analysis that the age at presentation alone is no reliable criterion for defining genetic heterogeneity. Parental renal ultrasound is still the most important classification criterion in most cases to distinguish between ARPKD and autosomal dominant polycystic kidney disease. However, PKD becomes increasingly complex from a clinical and genetic point of view.

Given its autosomal recessive mode of inheritance the recurrence risk for subsequent pregnancies of parents of an affected child is 25%. Overall, males and females seem to be equally affected. As indicated by formal genetics, unaffected siblings harbour a two-thirds risk of being a carrier for ARPKD. However, most healthy siblings, other close relatives and patients themselves seeking genetic counselling for their own family planning usually can be reassured, given that the risk for own children with ARPKD will be comparably low when neither the partner is related with the index family nor a case of ARPKD is known in the partner's pedigree (for offspring of patients 1:140; for offspring of patients' healthy siblings 1:420; for offspring of patients' healthy uncles/aunts 1:560, when using a heterozygosity rate of 1:70, respectively).

The main gene mutated in ARPKD is *PKHD1* that is amongst the largest disease genes characterised to date in the human genome, extending over a genomic segment of about 470 kb and evidence for 86 exons (Onuchic et al. 2002; Ward et al. 2002). The longest *PKHD1* transcript contains 67 exons with an open reading frame (ORF) composed of 66 exons (ATG start codon in exon 2) that encodes a protein of 4074 amino acids. In accordance with the disease phenotype, the gene is highly expressed in foetal and adult kidney and at lower levels in the liver (Nagasawa et al. 2002; Onuchic et al. 2002). Weak expression is present in other tissues too, among them pancreas and arterial wall. The predicted full-length protein (fibrocystin/polyductin) represents a novel putative type I integral membrane protein with a signal peptide at the amino terminus of its extensive, highly glycosylated extracellular domain, a single transmembrane-spanning segment, and a short cytoplasmic C-terminal tail (192 amino acids) containing potential protein kinase

A phosphorylation sites. The ~3860 amiono acid extracellular portion contains several immunoglobulin-like, plexin, transcription factor (IPT) and IPT-like domains that can be found in cell surface receptors and in the Rel family of transcription factors. Between the IPT domains and the transmembrane segment, multiple parallel beta-helix 1 (PbH1) repeats are present, a motif that can be observed in polysaccharidases and may bind to carbohydrate moieties such as glycoproteins on the cell surface and/or in the basement membrane. Based on the structural features of the deduced protein and on the human ARPKD phenotype, fibrocystin might be involved in cellular adhesion, repulsion and proliferation. In addition, the domain and structural analyses suggest that the *PKHD1* potential products may be involved in intercellular signalling and function as a receptor, ligand and/or membrane-associated enzyme (Onuchic et al. 2002). While the definite role of fibrocystin currently remains uncertain, there is increasing evidence from mutational and functional data that it functions as a part of the polycystin complex (Bergmann et al. 2011; Kim et al. 2008; Wang et al. 2007; Wu et al. 2006). In common with most other cystoproteins, fibrocystin has been shown to be localised to primary cilia with concentration in the basal body area (Masyuk et al. 2003; Menezes et al. 2004; Wang et al. 2004; Ward et al. 2003; Zhang et al. 2004). The lack of a clear fibrocystin homologue in *Caenorhabditis elegans* as well as in other phyla suggests that this protein has been a relatively late evolutionary addition to the ancestral polycystin pathway. By immunoprobing human metanephroi and kidney epithelial lines during acquisition of epithelial polarity, fibrocystin becomes localised to the apical zone of nephron precursor cells and then to basal bodies at the origin of primary cilia in fully differentiated epithelia. These striking patterns of subcellular localisation and known interactions with polycystin-2, CAML and some other proteins place fibrocystin at key sites of microtubule organisation. In line with its proposed role as a ciliary-localised membrane protein, Follit and co-workers demonstrated an 18-residue motif in the cytoplasmic tail of fibrocystin to be sufficient for ciliary targeting (Follit et al. 2010).

Some data suggest the existence of different, partly secreted fibrocystin isoproteins and Notch-like post-translational processing (Garcia-Gonzalez et al. 2007; Hiesberger et al. 2006; Kaimori et al. 2007; Masyuk et al. 2003; Menezes et al. 2004; Wang et al. 2004; Ward et al. 2003; Zhang et al. 2004). Human *PKHD1* and its murine orthologue undergo a complex and extensive pattern of alternative splicing, generating transcripts variable in size. While alternative splicing was initially thought to play a crucial role in disease manifestation and clinical expressivity, a recent study does not support this (Bakeberg et al. 2011). This issue can be regarded as not finally solved and it is presently unknown if and when how many alternative *PKHD1* transcripts are actually translated into protein and do have biological function(s). Kaimori and colleagues detected the expected full-length fibrocystin product (>400 kDa) and a C-terminally tagged 80–90 kDa product in the plasma membrane when using a cell surface biotinylation assay (Kaimori et al. 2007). Intriguingly, Hiesberger and co-workers could demonstrate that regulated intramembrane proteolysis (RIP) is induced by primary cilia dependent Ca^{2+} signalling and generates a C-terminal fibrocystin fragment that can signal directly to the nucleus (Hiesberger et al. 2006). Finally, it will be important to establish which isoforms are essential for renal and hepatobiliary integrity to better understand the role of fibrocystin in the aetiology of ARPKD. The distribution of mutations over the entire *PKHD1* gene suggests that the longest ORF transcript is necessary for proper fibrocystin function in kidney and liver. Thus, it might be proposed that a critical amount of the full-length protein is required for normal function. Alternatively, it might be hypothesised that mutations disrupt a critical functional stoichiometric or temporal balance between the different protein products, which is normally maintained by elaborate, tightly regulated splicing patterns.

The large size of *PKHD1* has posed significant challenges to DNA-based diagnostic testing in times of Sanger sequencing, but has now much improved with the utilisation and implementation of new sequencing technologies (Next Generation Sequencing/NGS) in routine diagnostic settings. Other challenges are set by the extensive allelic heterogeneity with a high level of mis-sense mutations and private mutations in 'non-isolate' populations (Bergmann et al. 2003, 2004a,b, 2005; Losekoot et al. 2005; Rossetti et al. 2003). Mutation detection rates of about 80% for the entire clinical spectrum of ARPKD patients ranging from individuals with perinatal demise to moderately affected adults have been shown (Bergmann et al. 2004b, 2005; Losekoot et al. 2005). The power of *PKHD1* mutation analysis is further strengthened by the observation that in more than 95% of families screened at least one mutation could be identified. However, the molecular defect still remains to be determined in a considerable proportion of chromosomes. One major cause of missing mutations is the limited sensitivity of screening methods used in some of the older studies such as denaturing high-performance liquid chromatography or single-strand conformation polymorphism. Moreover, some silent exonic changes and a subset of adjacent intronic sequence variations may also have an effect on *PKHD1* splicing, e.g. by affecting splice enhancer or silencer sites (ESE/ISE or ESS/ISS) (Baralle & Baralle, 2005). However, functional and mRNA studies are usually needed to prove any possible pathogenic effect of such changes and these studies are considerably hampered by the size and limited expression pattern of *PKHD1* (e.g. not expressed in lymphocytes) (Bergmann et al. 2006). We and others could show that missing mutations alternatively reside in regulatory elements and genomic rearrangements occur in the *PKHD1* gene (Bergmann et al. 2005; Zvereff et al. 2010; own unpublished data).

In patients without a detectable *PKHD1* mutation, misdiagnosis of *PKHD1*-linked ARPKD has to be considered. Polycystic kidney disease has become much more complex in recent times with increasing evidence for a genetic network and mutations in multiple cilia-related disease genes that may mimic ARPKD. First, it is known that about 2% of autosomal dominant polycystic kidney disease patients with a mutation in *PKD1* or *PKD2* show an early and severe phenotype with considerable perinatal morbidity and mortality that can be clinically indistinguishable from ARPKD. Furthermore, mutations in *PKD1* and *PKD2* can also be inherited in a recessive way (Bergmann, 2012). Finally, the phenotype of ARPKD can also be mimicked by mutations in *HNF1β* and genes typically causing other ciliopathies (e.g. nephronophthisis, but also other usually more syndromic ciliopathies). Therefore, a certain kind of alertness is important to circumvent possible pitfalls in genetic diagnostics. Genetic heterogeneity and prediction of the pathogenicity of mis-sense variants remain challenging. When only *PKHD1* sequencing data is available, caution is required for clinical decisions especially when only novel or rare mis-sense changes were found. Due to these different aspects, we established a novel genetic diagnostic testing approach based on Next Generation Sequencing that allows simultaneous investigation of all genes known for cystic and polycystic kidney disease and other ciliopathies. This analysis provides a thorough and complete result of all genes of interest and avoids otherwise possible misdiagnoses, especially with regard to prenatal testing.

Given the recurrence risk of 25%, frequently devastating early manifestations of ARPKD and comparable clinical courses among affected siblings, many parents of ARPKD children seek early and reliable prenatal diagnosis to guide future family planning. Typically, ARPKD patients are identified by ultrasound only late in pregnancy or at birth. However, even with state-of-the-art technology, foetal sonography at the time when termination of pregnancy is usually performed frequently fails to detect enlargement and increased echogenicity of kidneys or oligohydramnios secondary to poor foetal urine output (Zerres et al. 1988). Therefore, an early and reliable prenatal diagnosis for ARPKD in 'at risk' families is only feasible by molecular genetic analysis. Indirect, haplotype-based linkage analysis has often been performed for ARPKD in the past in terms of prenatal diagnosis. However, due to the aforementioned reasons this is now regarded as risky without knowledge of the *PKHD1* mutational status and should only be performed in those families in which the diagnosis has been proven. Interested families should also be informed on the possibility of preimplantation genetic diagnosis at some single diagnostic centres and that this procedure always requires a lot of coordination and work-up in advance.

To set up genotype–phenotype correlations for *PKHD1* is hampered by multiple allelism and the high rate of different compound heterozygotes. Genotype–phenotype correlations can be drawn for the type of mutation rather than for the site of individual mutations (Bergmann et al. 2003). All patients carrying two truncating mutations display a severe phenotype with peri- or neonatal demise while patients surviving the neonatal period bear at least one missense mutation. Although the converse did not apply and some missense changes are obviously as devastating as truncating mutations, it is not surprising that missense changes are more frequently observed among patients with a moderate clinical course, whereas chain-terminating mutations are more commonly associated with a severe phenotype. Loss of function probably explains the usually uniform and early demise of patients carrying two truncating alleles. This 'frameshift rule' based on the assumption that a truncated ORF will always constitute a null mutation has also been postulated for other disorders (Muntoni et al. 2003). This uniformity is probably attributable to ablation of the message by nonsense-mediated decay. As regards fibrocystin, a critical amount of the full-length protein seems to be required for normal function that obviously cannot be compensated by alternative isoforms, which might be generated by re-initiation of translation at a downstream ATG codon as a possible mechanism for the evasion of nonsense-mediated decay. No significant clinical differences could be observed between patients with two missense mutations and those patients harbouring a truncating mutation *in trans*, thus, the milder mutation obviously defines the phenotype (Bergmann et al. 2005).

Physiological effects of the underlying ciliary defect

In accordance with the idea of a common network in PKD, the cystogenic process of different genetic entities has been shown to share common phenotypic abnormalities compatible with cellular dedifferentiation and re-expression of proteins usually found during developmental stages, increased proliferation and apoptosis rates, disorganisation of the extracellular matrix, and aberrant protein sorting and fluid-transport characteristics. Without doubt primary cilia and their associated cellular organisms as basal bodies and centrosomes play a major role in this network in which cystoproteins converge into the same signalling cascades. However, there is increasing evidence that PKD and other ciliopathies do not only depend on proper cilia function and structure, but also other mechanisms such as trafficking and quality control processes within the cell (Fedeles et al. 2011). Almost every third protein encoded by the human genome passes through the translocation and quality control machinery in the endoplasmic reticulum. This especially holds true for secretory proteins and transmembrane proteins, as all PKD genes are. In line with other data from mice (Garcia-Gonzalez et al. 2007) and patients (Bergmann et al. 2011; Kleffmann et al. 2012), Fedeles and co-workers recently presented convincing evidence for a dosage-sensitive network in which the severity of PKD is controlled by a network of different genes/proteins to which the genes/proteins for polycystic liver disease that play crucial roles in trafficking and quality control processes also belong to.

The importance of primary renal cilia as critical organelles for architectural homeostasis of the kidney has been partly explained by their function in sensing environmental cues such as tubular luminal flow which in turn triggers transient Ca^{2+} currents with a strong increase of intracellular Ca^{2+} levels. (Nauli et al. 2003; Pazour & Rosenbaum, 2002; Praetorius & Spring, 2001). Coordinated cross-talk between Ca^{2+} and cAMP is crucial to maintain a differentiated kidney epithelium with controlled fluid secretion and cell proliferation (Belibi et al. 2004; Yamaguchi et al. 2004). Cellular proliferation is directed by mitogen-activated protein kinase-extracellular signal-regulated kinase (MAPK/ERK)/ signalling among others. Its importance in the pathogenesis of PKD is emphasised by increased renal levels of phosporylated Raf-1 and ERK in orthologous animal models for PKD and the development of PKD in H-Ras transgenic mice. Alterations in sorting and maintenance of the epidermal growth factor receptor (EGFR) on the basolateral surface of renal epithelial cells has been demonstrated with apical EGFR expression as a known contributor to disease severity in PKD (Wilson, 2011). At least some of the PKD proteins may assist in planar cell polarity. Other pathways already shown or hypothesised to play a central role in PKD are Wnt, hedgehog, JAK-STAT, transforming growth factor-β (TGF-β), and Notch.

Treatment prospects

A number of different studies have extended our understanding of the pathophysiology of PKD and some promising trials have identified potential approaches to influence the disease process by targeting downstream cellular changes (Torres, 2010). Currently, there is careful optimism and hope for treatment options and rational personalised therapies of patients with PKD to curb or ameliorate the clinical course. The use of vasopressin V_2 receptor (V_2R) antagonists is currently regarded to have the greatest potential. The concept is based on the common characteristic of all forms of cystic kidney disease to be inable to concentrate urine properly. Most probably as an attempt to compensate this inability, V_2R mRNA expression has been shown to be upregulated in PKD kidneys. Via V_2 receptors vasopressin is the major adenyl cyclase agonist in the principal cells of renal collecting ducts generating cAMP. Given the central role that Ca^{2+} and cAMP play in the pathogenesis of PKD, therapeutic approaches aim in either increasing intracellular Ca^{2+} concentration or reducing renal cAMP or both. To inhibit the accumulation of renal cAMP as a known promoter of renal cystic enlargement, several studies have been performed with V_2R antagonists. An advantage and disadvantage is the almost exclusive expression of the V_2R in renal collecting duct principal cells and endothelial cells. V_2R antagonists are associated with a higher discontinuation rate due to disease-unrelated adverse effects, they do inhibited cyst growth and slow the increase in total kidney volume and the decline of kidney function (Torres et al. 2012).

Acknowledgements

CB is an employee of Bioscientia, a member of Sonic Healthcare. He received support from the Deutsche Forschungsgemeinschaft (DFG), German Kidney Foundation (Deutsche Nierenstiftung), and PKD Foundation.

References

Adeva, M., El-Youssef, M. & Rossetti, S. 2006. Clinical and molecular characterization defines a broadened spectrum of autosomal recessive polycystic kidney disease (ARPKD). *Medicine (Baltimore)* **85**, 1–21.

Arbeiter, A., Buscher, R., Bonzel, K. E., Wingen, A. M., Vester, U., Wohlschlager, J., et al. 2008. Nephrectomy in an autosomal recessive polycystic kidney disease (ARPKD) patient with rapid kidney enlargement and increased expression of EGFR. *Nephrol Dial Transplant* **23**, 3026–3029.

Avni, F. E., Guissard, G., Hall, M., Janssen, F., DeMaertelaer, V. & Rypens, F. 2002. Hereditary polycystic kidney diseases in children: Changing sonographic patterns through childhood. *Pediatr Radiol* **32**, 169–174.

Bakeberg, J. L., Tammachote, R., Woollard, J. R., Hogan, M. C., Tuan, H. F., Li, M., et al. 2011. Epitope-tagged Pkhd1 tracks the processing, secretion, and localization of fibrocystin. *J Am Soc Nephrol* **22**, 2266–2277.

Baralle, D. & Baralle, M. 2005. Splicing in action: Assessing disease causing sequence changes. *J Med Genet* **42**, 737–748.

Belibi, F. A., Reif, G., Wallace, D. P., Yamaguchi, T., Olsen, L., Li, H., et al. 2004. Cyclic AMP promotes growth and secretion in human polycystic kidney epithelial cells. *Kidney Int* **66**, 964–973.

Bergmann, C. 2012. Educational paper: Ciliopathies. *Eur J Pediatr* **171**, 1285–1300.

Bergmann, C., Frank, V., Kupper, F., Schmidt, C., Senderek, J. & Zerres, K. 2006. Functional analysis of PKHD1 splicing in autosomal recessive polycystic kidney disease. *J Hum Genet* **51**, 788–793.

Bergmann, C., Küpper, F., Schmitt, C. P., Vester, U., Neuhaus, T. J., Senderek, J., et al. 2005 Multi-exon deletions of the PKHD1 gene cause autosomal recessive polycystic kidney disease (ARPKD). *J Med Genet* **42**, e63.

Bergmann, C., Senderek, J., Schneider, F., Dornia, C., Kupper, F., Eggermann, T., et al. 2004a. PKHD1 mutations in autosomal recessive polycystic kidney disease (ARPKD). *Hum Mutat* **23**, 453–463.

Bergmann, C., Senderek, J., Sedlacek, B., Pegiazoglou, I., Puglia, P., Eggermann, T., et al. 2004b. PKHD1 mutations in families requesting prenatal diagnosis for autosomal recessive polycystic kidney disease (ARPKD). *Hum Mutat* **23**, 487–495.

Bergmann, C., Senderek, J., Sedlacek, B., Pegiazoglou, I., Puglia, P., Eggermann, T., et al. 2003. Spectrum of mutations in the gene for autosomal recessive polycystic kidney disease (ARPKD/PKHD1). *J Am Soc Nephrol* **14**, 76–89.

Bergmann, C., Senderek, J. & Windelen, E. 2005. Clinical consequences of PKHD1 mutations in 164 patients with autosomal recessive polycystic kidney disease (ARPKD). *Kidney Int* **67**, 829–848.

Bergmann, C., von Bothmer, J., Ortiz Bruchle, N., Venghaus, A., Frank, V., Fehrenbach, H., et al. 2011. Mutations in multiple PKD genes may explain early and severe polycystic kidney disease. *J Am Soc Nephrol* **22**, 2047–2056.

Blyth, H. & Ockenden, B. G. 1971. Polycystic disease of kidney and liver presenting in childhood. *J Med Genet* **8**, 257–284.

Brinkert, F., Lehnhardt, A., Montoya, C., Helmke, K., Schaefer, H., Fischer, L., et al. 2013. Combined liver-kidney transplantation for children with autosomal recessive polycystic kidney disease (ARPKD): indication and outcome Transpl Int (in press).

Capisonda, R., Phan, V., Traubuci, J., Daneman, A., Balfe, J. W. & Guay-Woodford, L. M. 2003. Autosomal recessive polycystic kidney disease: outcomes from a single-center experience. *Pediatr Nephrol* **18**, 119–126.

Chapal, M., Debout, A., Dufay, A., Salomon, R., Roussey, G., Burtey, S., et al. 2012. Kidney and liver transplantation in patients with autosomal recessive polycystic kidney disease: A multicentric study. *Nephrol Dial Transplant* **27**, 2083–2088.

Chapman, A. B. & Guay-Woodford, L. M. 2006. The family and ADPKD. *The Polycystic Kidney Research (PKR) Foundation* 16. Available from: http://www.pkdcure.org/document.doc?id=98.

Chapman, A. B., Johnson, A., Gabow, P. A. & Schrier, R. W. 1990. The renin–angiotensin–aldosterone system and autosomal dominant polycystic kidney disease. *New Engl J Med* **323**, 1091–1096.

Chapman, A. B., Rubinstein, D., Hughes, R., Stears, J. C., Earnest, M. P., Johnson, A. M., et al. 1992. Intracranial aneurysms in autosomal dominant polycystic kidney disease. *New Engl J Med* **327**, 916–920.

Chapman, A. B. & Schrier, R. W. 1991. Pathogenesis of hypertension in autosomal dominant polycystic kidney disease. *Semin Nephrol* **11**, 653–660.

Chatterjee, P. K., Weerackody, R. P., Mistry, S. K., Hawksworth, G. M. & McLay, J. S. 1997. Selective antagonism of the AT1 receptor inhibits angiotensin II stimulated DNA and protein synthesis in primary cultures of human proximal tubular cells. *Kidney Int* **52**, 699–705.

Cole, B. R., Conley, S. B. & Stapleton, F. B. 1987. Polycystic kidney disease in the first year of life. *J Pediatr* **111**, 693–699.

Deget, F., Rudnik-Schoneborn, S. & Zerres, K. 1995. Course of autosomal recessive polycystic kidney disease (ARPKD) in siblings: A clinical comparison of 20 sibships. *Clin Genet* **47**, 248–253.

Dell, K. M. R. & Avner, E. D. 1993. Polycystic kidney disease, autosomal recessive.

Desmet, V. J. 1998. Ludwig symposium on biliary disorders—part I Pathogenesis of ductal plate abnormalities. *Mayo Clin Proc* **73**, 80–89.

Doulton, T. W., Saggar-Malik, A. K., He, F. J., Carney, C., Markandu, N. D., Sagnella, G. A., et al. 2006. The effect of sodium and angiotensin-converting enzyme inhibition on the classic circulating renin–angiotensin system in autosomal-dominant polycystic kidney disease patients. *J Hypertens* **24**, 939–945.

Fedeles, S. V., Tian, X., Gallagher, A. R., Mitobe, M., Nishio, S., Lee, S. H., et al. 2011. A genetic interaction network of five genes for human polycystic kidney and liver diseases defines polycystin-1 as the central determinant of cyst formation. *Nat Genet* **43**, 639–647.

Follit, J. A., Li, L., Vucica, Y. & Pazour, G. J. 2010. The cytoplasmic tail of fibrocystin contains a ciliary targeting sequence. *J Cell Biol* **188**, 21–28.

Gagnadoux, M. F., Habib, R., Levy, M., Brunelle, F. & Broyer M. 1989. Cystic renal diseases in children. *Adv Nephrol Necker Hosp* **18**, 33–57.

Garcia-Gonzalez, M. A., Menezes, L. F., Piontek, K. B., Kaimori, J., Huso, D. L., Watnick, T., et al. 2007. Genetic interaction studies link autosomal dominant and recessive polycystic kidney disease in a common pathway. *Hum Mol Genet* **16**, 1940–1950.

Guay-Woodford, L. M. & Desmond, R. A. 2003. Autosomal recessive polycystic kidney disease: The clinical experience in North America. *Pediatrics* **111**, 1072–1080.

Gunay-Aygun, M., Font-Montgomery, E., Lukose, L., Gerstein, M. T., Piwnica-Worms, K., Choyke, P., et al. 2012. Characteristics of congenital hepatic fibrosis in a large cohort of patients with autosomal recessive polycystic kidney disease. *Gastroenterology* **144**, 112–121.

Gunay-Aygun, M., Font-Montgomery, E., Lukose, L., Tuchman, M., Graf, J., Bryant, J. C., et al. 2010. Correlation of kidney function, volume and imaging findings, and PKHD1 mutations in 73 patients with autosomal recessive polycystic kidney disease. *Clin J Am Soc Nephrol* **5**, 972–984.

Hiesberger, T., Gourley, E., Erickson, A., Koulen, P., Ward, C. J., Masyuk, T. V., et al. 2006. Proteolytic cleavage and nuclear translocation of fibrocystin is regulated by intracellular Ca^{2+} and activation of protein kinase C. *J Biol Chem* **281**, 34357–34364.

Hopp, K., Ward, C. J., Hommerding, C. J., Nasr, S. H., Tuan, H. F., Gainullin, V. G., et al. 2012. Functional polycystin-1 dosage governs autosomal dominant polycystic kidney disease severity. *J Clin Invest* **122**, 4257–4273.

Irazabal, M. V., Huston, J., 3rd, Kubly, V., Rossetti, S., Sundsbak, J. L., Hogan, M. C., et al. 2011. Extended follow-up of unruptured intracranial aneurysms detected by presymptomatic screening in patients with autosomal dominant polycystic kidney disease. *Clin J Am Soc Nephrol* **6**, 1274–1285.

Jafar, T. H., Stark, P. C., Schmid, C. H., Strandgaard, S., Kamper, A. L., Maschio, G., et al. 2005. The effect of angiotensin-converting-enzyme inhibitors on progression of advanced polycystic kidney disease. *Kidney Int* **67**, 265–271.

Kaariainen, H., Koskimies, O. & Norio, R. 1988. Dominant and recessive polycystic kidney disease in children: Evaluation of clinical features and laboratory data. *Pediatr Nephrol* **2**, 296–302.

Kaimori, J. Y., Nagasawa, Y., Menezes, L. F., Garcia-Gonzalez, M. A., Deng, J., Imai, E., et al. 2007. Polyductin undergoes notch-like processing and regulated release from primary cilia. *Hum Mol Genet* **16**, 942–956.

Kaplan, B. S., Fay, J., Shah, V., Dillon, M. J. & Barratt, T. M. 1989. Autosomal recessive polycystic kidney disease.*Pediatr Nephrol* **3**, 43–49.

Kashtan, C. E., Primack, W. A. & Kainer, G. 1999. Recurrent bacteremia with enteric pathogens in recessive polycystic kidney disease. *Pediatr Nephrol* **13**, 678–682.

Kim, I., Fu, Y., Hui, K., Moeckel, G., Mai, W., Li, C., et al. 2008. Fibrocystin/polyductin modulates renal tubular formation by regulating polycystin-2 expression and function. *J Am Soc Nephrol* **19**, 455–468.

Kim, I., Li, C., Liang, D., Chen, X. Z., Coffy, R. J., Ma, J., et al. 2008. Polycystin-2 expression is regulated by a PC2-binding domain in the intracellular portion of fibrocystin. *J Biol Chem* **283**, 31559–31566.

Kleffmann, J., Frank, V., Ferbert, A. & Bergmann, C. 2012. Dosage-sensitive network in polycystic kidney and liver disease: Multiple mutations cause severe hepatic and neurological complications. *J Hepatol* **57**, 476–477.

Losekoot, M., Haarloo, C., Ruivenkamp, C., White, S. J., Breuning, M. H. & Peters, D. J. 2005. Analysis of missense variants in the PKHD1-gene in patients with autosomal recessive polycystic kidney disease (ARPKD). *Hum Genet* **118**, 185–206.

Mariani, L., Bianchetti, M. G., Schroth, G. & Seiler, R. W. 1999. Cerebral aneurysms in patients with autosomal dominant polycystic kidney disease—to screen, to clip, to coil? *Nephrol Dial Transplant* **14**, 2319–2322.

Masyuk, T. V., Huang, B. Q., Ward, C. J., Masyuk, A. I., Yuan, D., Splinter, P. L., et al. 2003. Defects in cholangiocyte fibrocystin expression and ciliary structure in the PCK rat. *Gastroenterology* **125**, 1303–1310.

Menezes, L. F., Cai, Y., Nagasawa, Y., Silva, A. M., Watkins, M. L., Da Silva, A. M., et al. 2004. Polyductin, the PKHD1 gene product, comprises isoforms expressed in plasma membrane, primary cilium, and cytoplasm. *Kidney Int* **66**, 1345–1355.

Moser, M., Matthiesen, S. & Kirfel, J. 2005. A mouse model for cystic biliary dysgenesis in autosomal recessive polycystic kidney disease (ARPKD). *Hepatology* **41**, 1113–1121.

Muntoni, F., Torelli, S. & Ferlini, A. 2003. Dystrophin and mutations: One gene, several proteins, multiple phenotypes. *Lancet Neurol* **2**, 731–740.

Nagasawa, Y., Matthiesen, S., Onuchic, L. F., Hou, X., Bergmann, C., Esquivel, E., et al. 2002. Identification and characterization of Pkhd1, the mouse orthologue of the human ARPKD gene. *J Am Soc Nephrol* **13**, 2246–2258.

Nakanishi, K., Sweeney, W. E., Jr., Zerres, K., Guay-Woodford, L. M. & Avner, E. D. 2000. Proximal tubular cysts in fetal human autosomal recessive polycystic kidney disease. *J Am Soc Nephrol* **11**, 760–763.

Nauli, S. M., Alenghat, F. J., Luo, Y., Williams, E., Vassilev, P., Li, X., et al. 2003. Polycystins 1 and 2 mediate mechanosensation in the primary cilium of kidney cells. *Nat Genet* **33**, 129–137.

Nicolau, C., Torra, R., Badenas, C., Perez, L., Oliver, J. A., Darnell, A., et al. 2000. Sonographic pattern of recessive polycystic kidney disease in young adults. Differences from the dominant form. *Nephrol Dial Transplant* **15**, 1373–1378.

Onuchic, L. F., Furu, L. & Nagasawa, Y. 2002. PKHD1, the polycystic kidney and hepatic disease 1 gene, encodes a novel large protein containing multiple immunoglobulin-like plexin-transcription-factor domains and parallel beta-helix 1 repeats. *Am J Hum Genet* **70**, 1305–1317.

Osathanondh, V. & Potter, E. L. 1964. Pathogenesis of polycystic kidneys. survey of results of microdissection. *Arch Pathol* **77**: 510–512.

Pazour, G. J. & Rosenbaum, J. L. 2002. Intraflagellar transport and cilia-dependent diseases. *Trends Cell Biol* **12**, 551–555.

Pirson, Y., Chauveau, D. & Torres, V. 2002. Management of cerebral aneurysms in autosomal dominant polycystic kidney disease. *J Am Soc Nephrol* **13**, 269–276.

Praetorius, H. A. & Spring, K. R. 2001. Bending the MDCK cell primary cilium increases intracellular calcium. *J Membr Biol* **184**, 71–79.

Prelog, M., Bergmann, C., Ausserlechner, M. J., Fischer, H., Margreiter, R., Gassner, I., et al. 2006. Successful transplantation in a child with rapid progression of autosomal recessive polycystic kidney disease associated with a novel mutation. *Pediatr Transplant* **10**, 362–366.

Rinkel, G. J., Djibuti, M. Algra, A. & van Gijn, J. 1998. Prevalence and risk of rupture of intracranial aneurysms: A systematic review. *Stroke* **29**, 251–256.

Ritz, E. 2006. Hypertension in autosomal dominant polycystic kidney disease: Is renin acquitted as a culprit? *J Hypertens* **24**, 1023–1025.

Rossetti, S., Kubly, V. J., Consugar, M. B., Hopp, K., Roy, S., Horsley, S. W., et al. 2009. Incompletely penetrant PKD1 alleles suggest a role for gene dosage in cyst initiation in polycystic kidney disease. *Kidney Int* **75**, 848–855.

Rossetti, S., Torra, R., Coto, E., Consugar, M., Kubly, V., Malaga, S., et al. 2003. A complete mutation screen of PKHD1 in autosomal-recessive polycystic kidney disease (ARPKD) pedigrees. *Kidney Int* **64**, 391–403.

Roy, S., Dillon, M. J., Trompeter, R. S. & Barratt, T.M. 1997. Autosomal recessive polycystic kidney disease: Long-term outcome of neonatal survivors. *Pediatr Nephrol* **11**, 302–306.

Spechtenhauser, B., Hochleitner, B. W., Ellemunter, H., Simma, B., Hormann, C., Konigsrainer, A., et al. 1999. Bilateral nephrectomy, peritoneal dialysis and subsequent cadaveric renal transplantation for treatment of renal failure due to polycystic kidney disease requiring continuous ventilation. *Pediatr Transplant* **3**, 246–248.

Torres, V. E. 2010. Treatment strategies and clinical trial design in ADPKD. *Adv Chronic Kidney Dis* **17**, 190–204.

Torres, V. E., Chapman, A. B., Devuyst, O., Gansevoort, R. T., Grantham, J. J., Higashihara, E., et al. 2012. Tolvaptan in patients with autosomal dominant polycystic kidney disease. *New Engl J Med* **367**, 2407–2418.

Torres, V. E., Donovan, K. A., Scicli, G., Holley, K. E., Thibodeau, S. N., Carretero, O. A., et al. 1992. Synthesis of renin by tubulocystic epithelium in autosomal-dominant polycystic kidney disease. *Kidney Int* **42**, 364–373.

Wang, S., Luo, Y., Wilson, P. D., Witman, G. B. & Zhou, J. 2004. The autosomal recessive polycystic kidney disease protein is localized to primary cilia, with concentration in the basal body area. *J Am Soc Nephrol* **15**, 592–602.

Wang, S., Zhang, J., Nauli, S. M., Li, X., Starremans, P. G., Luo, Y., et al. 2007. Fibrocystin/polyductin, found in the same protein complex with polycystin-2, regulates calcium responses in kidney epithelia. *Mol Cell Biol* **27**, 3241–3252.

Warady, B. A., Alexander, S. R., Watkins, S., Kohaut, E. & Harmon, W. E. 1999. Optimal care of the pediatric end-stage renal disease patient on dialysis. *Am J Kidney Dis* **33**, 567–583.

Ward, C. J., Hogan, M. C., Rossetti, S., Walker, D., Sneddon, T., Wang, X., et al. 2002. The gene mutated in autosomal recessive polycystic kidney disease encodes a large, receptor-like protein. *Nat Genet* **30**, 259–269.

Ward, C. J., Yuan, D., Masyuk, T. V., Wang, X., Punyashthiti, R., Whelan, S., et al. 2003. Cellular and subcellular localization of the ARPKD protein; fibrocystin is expressed on primary cilia. *Hum Mol Genet* **12**, 2703–2710.

Wilson, P. D. 2011. Apico-basal polarity in polycystic kidney disease epithelia. *Biochim Biophys Acta* **1812**, 1239–1248.

Wu, Y., Dai, X. Q., Li, Q., Chen, C. X., Mai, W., Hussain, Z., et al. 2006. Kinesin-2 mediates physical and functional interactions between polycystin-2 and fibrocystin. *Hum Mol Genet* **15**, 3280–3292.

Xu, H. W., Yu, S. Q., Mei, C. L. & Li, M. H. 2011. Screening for intracranial aneurysm in 355 patients with autosomal-dominant polycystic kidney disease. *Stroke* **42**, 204–206.

Yamaguchi, T., Wallace, D. P., Magenheimer, B. S., Hempson, S. J., Grantham, J. J. & Calvet, J. P. 2004. Calcium restriction allows cAMP activation of the B-Raf/ERK pathway, switching cells to a cAMP-dependent growth-stimulated phenotype. *J Biol Chem* **279**, 40419–40430.

Zerres, K., Hansmann, M., Mallmann, R. & Gembruch, U. 1988. Autosomal recessive polycystic kidney disease. Problems of prenatal diagnosis. *Prenat Diagn* **8**, 215–229.

Zerres, K., Rudnik-Schöneborn, S., Deget, F., Holtkamp, U., Brodehl, J., Geisert, J., et al. 1996. Autosomal recessive polycystic kidney disease in 115 children: clinical presentation, course and influence of gender. *Acta Paediatr* **85**, 437–445.

Zerres, K., Rudnik-Schoneborn, S., Senderek, J., Eggermann, T. & Bergmann, C. 2003. Autosomal recessive polycystic kidney disease (ARPKD). *J Nephrol* **16**, 453–458.

Zhang, M. Z., Mai, W., Li, C., Cho, S. Y., Hao, C., Moeckel, G., et al. 2004. PKHD1 protein encoded by the gene for autosomal recessive polycystic kidney disease associates with basal bodies and primary cilia in renal epithelial cells. *Proc Natl Acad Sci U.S.A.* **101**, 2311–2316.

Zvereff, V., Yao, S., Ramsey, J., Mikhail, F. M., Vijzelaar, R. & Messiaen, L. 2010. Identification of PKHD1 multiexon deletions using multiplex ligation-dependent probe amplification and quantitative polymerase chain reaction. *Genet Test Mol Biomarkers* **14**, 505–510.

Chapter 12

Primary ciliary dyskinesia

Claire Hogg

History of the eponym

Kartagener became synonymous with, what we now know as, primary ciliary dyskinesia (PCD) when he described his classical triad of sinusitis, bronchiectasis and situs inversus (Kartagener, 1933), but he was unable to connect it to ciliary dysfunction. It was Bjorn Afzelius, who made his observations on the ultrastructure of sperm tails in infertile men, which led to the realisation that abnormalities of cilia were the underlying cause of the syndrome (Afzelius, 1976). The re-naming of Kartagener's syndrome to immotile cilia syndrome and, more recently, PCD was appropriate to include all patients, whether they had all features of the triad or not. Furthermore, the current nomenclature references the basic ciliary biology underlying the clinical disease and has led to an explosion of interest in the spectrum of ciliopathy diseases in the 21st century.

Epidemiology and ciliary biology

The full complexities of ciliary biology are beyond the scope of this chapter, but a basic knowledge of the ciliated epithelium is essential to understand the disease epidemiology. The apical surface of several of our tissues is lined with microscopic tubular structures called cilia including the entire respiratory tract from the nose, eustachian tubes and upper and lower airways. The classical structure of the axoneme of respiratory (motile) cilia is of nine outer doublets, each with inner and outer dynein arms connected by radial spoke to a central doublet pair, the so-called '9 + 2' arrangement (Figure 12.1). The outer ring of microtubules are interconnected by nexin links which help maintain axonemal integrity (Carlen et al. 2003; Plesec et al. 2008), and are now known to be part of the dynein regulatory complex (N-DRC) (Heuser et al. 2009). The axoneme extends from the cell surface, is anchored by the basal body and the entire structure is covered in the cell membrane which is contiguous with the plasmalemma. The motor unit of the cilium is the dynein complex, formed from heavy, intermediate and light chains, and which contain ATP-ases that allow the sliding motion of the microtubules relative to each other (Chilvers & O'Callaghan, 2000). This effects a cycle of beating movement that results in mucociliary clearance, the primary defence mechanism within the airway.

Ciliary beat is only part of the axonemal function. Complex intraflagellar transport mechanisms operate within the axoneme with retrograde recycling of proteins and

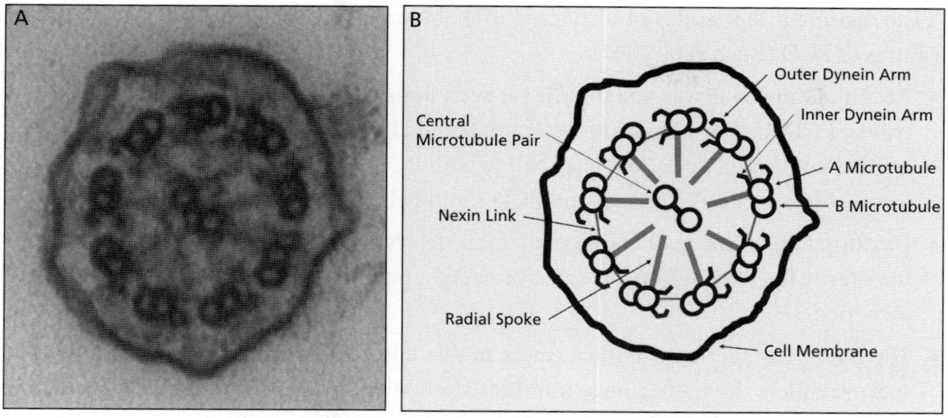

Figure 12.1 Cross section of normal motile respiratory cilia showing the typical '9 + 2' arrangement of microtubular doublet pairs. (a) The appearances at TEM. (b) A diagramatic version to highlight the individual structures of the ciliary axoneme in the normal cilia.

nutritive functions (Bisgrove & Yost, 2006). Little is known about the impact of the ultrastructural anomalies in PCD on these mechanisms, but some ciliopathies have clearly disordered intraflagellar transport systems such as retinitis pigmentosa (Armengot et al. 2012) and Bardet–Biedl syndrome.

In PCD, genetic defects of the production of ciliary proteins, in which over 250 genes are involved, results in an array of ultrastructural defects (Mizuno et al. 2012), the most common being absence of the dynein arms (Figure 12.2). The result is a disruption of the co-ordinated beat pattern and accumulation of mucus and potential pathogens resulting in the typical symptoms of PCD.

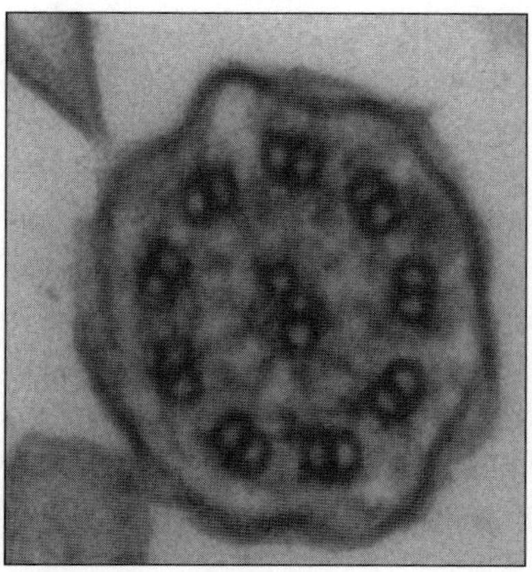

Figure 12.2 Transmission electron micrograph of respiratory cilia showing the absence of outer dynein arms in a patient with PCD. Note that the inner dynein arms are present on some doublet pairs whilst they appear to be absent on others in this micrograph.

The disrupted mechanics of ultrastructural abnormalities result in the typical clinical features of PCD (Bush et al. 2007):

- Accumulation of airway and middle ear secretions results in the typical clinical phenotype of PCD. Inability to clear foetal lung liquid at birth is often the earliest sign of ciliary immotility and up to 75% of all babies with PCD present with respiratory distress at term. This is a cardinal feature and PCD should always be excluded in such cases.
- Dysfunctional cilia in the Fallopian tube delays transport of the normal ovum to the uterus for fertilisation. The risk of ectopic pregnancy or sub-fertility is therefore increased.
- The tail of the spermatozoa is a single motile cilium and in many patients immotile sperm renders them effectively infertile. The sperm is otherwise normal and current fertility techniques overcome this issue once diagnosed.
- Laterality, which determines how organs are laid down during embryogenesis, is dependent on ciliary motility. In the absence of a normal ciliary beat this process appears to occur by random chance and almost half of all patients with PCD have situs abnormalities. Most commonly this is situs inversus totalis, but occasionally there can be an ambiguous arrangement of the organs with associated major organ abnormalities such as heart defects. In this latter group the term heterotaxy is often applied and clinicians should screen for heart and abdominal organ abnormalities.
- Ventriculomegaly or hydrocephalus may occur due to immotility of the ependymal cilia. This is an uncommon but definite association in any condition where a motile ciliopathy exists, although it generally does not appear to disrupt the normal flow of cerebrospinal fluid in such patients.

Clinical features

PCD is a condition where many of the symptoms are non-specific and the typical clinical features mimic those of the common childhood illnesses. Classically, it is a disease of the upper and lower respiratory tracts, with associated mirror image arrangement of the organs in almost half of all patients. However, more complex disease associations can arise in individual patients with complex organ anomalies or other ciliopathic syndromes (Bush et al. 2007; Krawczynski et al. 2004). Extended kindreds where PCD and other ciliopathies co-exist, either in individuals or in different members of a family, have been described and it is becoming clear that PCD may not always be a single disease entity. This is best described in the X-linked *RPGR* mutations where some males are affected with both PCD and retinal degenerative disease (Moore et al. 2006; Zito et al. 2003).

Like many inherited conditions there is a wide spectrum of disease severity. This range exists between patients of the same family, patients with the same ultrastructural defect and prognostic assumptions cannot be made based on the biological variants determined on diagnostic testing. In patients at the milder end of the disease spectrum, in particular if situs is normal, the diagnosis may be very delayed (Coren et al. 2002). The result of a delayed diagnosis can lead to irreversible lung damage, as a result of recurrent infections in retained airway secretions, leading to collapse and consolidation and ultimately the onset of bronchiectasis. The burden of long term morbidity, and ultimately the mortality, in the patient with classical PCD (i.e. without cardiac or other organ disease, or a co-existing ciliopathy) lies in the lung pathology. Despite this lung function can be stabilised over many years (Ellerman & Bisgaard, 1997), although a recent report from a large Danish cohort suggests that up to a third of patients have progressive deterioration in spirometry (Marthin et al. 2010). Furthermore, end stage respiratory failure may lead to transplantation (Date et al. 2001; Schertler et al. 2007). Early diagnosis, with institution of therapeutic interventions, is therefore a paramount aim to prevent progression of lung disease and to improve long-term outcomes.

Although the lung disease often determines long term prognosis, the upper airway and middle ear disease requires aggressive investigation and management strategies to improve quality of life and optimise hearing in the formative years. Poor social and educational development with the possibility that key milestones are either delayed or missed altogether is a very real risk if the middle ear disease is poorly managed. In moderate-to-severe cases these missed milestones may lead to a lifetime of under-achievement at school, in further education and in the workplace. The socio-economic burden of a patient with poor health is theoretically exacerbated if such patients leave school unable to compete in the educational or economic arenas. Longitudinal studies are needed to assess the impact of poor hearing, speech and developmental delay in the setting of PCD, where improved educational attainment may positively influence understanding of good lung health and compliance with therapy, with improved long-term outcomes.

As a result the clinician needs to be vigilant to the cardinal signs that should direct further investigations to confirm or exclude PCD as the cause. The following list aims to outline both the clinical clues and to expand on the clinical features associated in each system (Bush & O'Callaghan, 2002; Bush et al. 2007):

- *Situs anomalies* of any kind, although most commonly there is situs inversus. In around 6% of patients there is situs ambiguous (heterotaxy). This sub-group of patients has a much higher incidence of cardiac anomalies, in particular isomerism, and an ambiguous arrangement of abdominal organs (Kennedy et al. 2007). The abdominal organs may have a predictable distribution and set of anomalies, in that left atrial isomerism is most commonly associated with poly-splenia and right atrial isomerism with asplenia. Odd placement of the liver and bowel are more common and mal-rotation of the gut can be the presenting feature leading to subsequent diagnosis of PCD. Dextrocardia may be detected on antenatal scanning. If co-existing ventriculomegaly is detected PCD becomes more likely. All patients with PCD should undergo echocardiogram, and where situs anomalies exist, an abdominal ultrasound.
- *Neonatal respiratory distress* in the term infant. Seventy-five per cent of PCD patients have a history of neonatal pneumonia of unknown cause. In the absence of situs inversus PCD as a cause is often not considered, but where a defined cause remains elusive PCD should be excluded as a routine.

- *Early onset rhinitis*: PCD patients will typically have a history of rhinitis and nasal congestion from the first week of life. It is persistent in nature, tends to continue between infections and some studies report it is present in 100% of patients.
- *Recurrent and persistent glue ear* with associated hearing deficit. Delayed speech development, poor enunciation and delayed educational milestones are a common consequence if this aspect of PCD is missed and inadequately monitored. Traditional surgical management of middle ear effusions usually results in persistent otorrhoea through the ventilation tubes. The discharge is socially disabling as it is both unsightly and malodorous, and the surgery does little to improve hearing. Specialist ENT input is essential to managing the hearing problems and transient hearing aid use may be employed to assist with development and education. By mid to late teens most patients have resolution of the hearing deficit and use of aids can be discontinued.
- *Persistent or daily wet cough* from the first weeks of life. Prior to diagnosis and institution of specific treatment regimens cough is persistent even when infection free. If the other features listed in this section are also present a high index of suspicion for PCD should exist and the patient referred for diagnostic testing.
- *Recurrent sinusitis* is a common feature in older children and adult patients and assessment of sinus pathology needs to consider the patients age. Sinus development occurs following dentition (maxillary sinuses) or during puberty (frontal and sphenoid sinuses). More recently sinus agenesis has been described as a possible feature suggestive of PCD and may be unilateral (Pifferi et al. 2011). Nasal polyps may be a feature, although rare in PCD.
- *Sub-fertility or infertility*: As both egg and sperm are functionally normal early referral to fertility specialists is recommended early in family planning.
- *Ciliopathies* now account for many rare and very rare conditions. Apart from PCD, retinitis pigmentosa and polycystic kidney disease are the most common examples, but you will read about many rare ciliopathies in this text including Bardet–Biedl syndrome, Alström syndrome and Jeune disease. When such disorders are suspected or diagnosed a wider consideration of co-existing ciliopathies is prudent (Fliegauf et al. 2007; Hildebrandt et al. 2011; Waters & Beales, 2011).

Finally, in patients with idiopathic bronchiectasis or 'atypical' asthma, where a productive cough is evident, then any form of chronic suppurative lung disease should be considered. Exclusion of cystic fibrosis, PCD and other causes of bronchiectasis should be pursued aggressively.

Range of the phenotype

In the setting of PCD the need for accurate clinical phenotyping poses problems for both clinicians and geneticists. The diagnosis of even classical PCD is often delayed due to the non-specific nature of many of the clinical features, which in some cases can be relatively mild albeit it typical of the disease (Coren et al. 2002).

Furthermore, like cystic fibrosis (CF) and cystic fibrosis transmembrane conductance regulator-related disease (CFTR), the definition of atypical PCD has not been worked out. From a geneticists point of view PCD causing mutations could involve genes involved in the actual ciliary structure itself, and these could be subdivided at least in theory into those coding for the purely structural proteins, those with metabolic functions and those involved in signalling to cilia (Duquesnoy et al. 2009; Loges et al. 2009; Omran et al. 2008). Therefore, clinical PCD could be caused by assembly genes whose products are not seen in cilia at all, or by mutations, which cause abnormal function of an apparently normally constructed cilium. This is borne out in the large cohorts, of well defined patients, where up to 15% of patients have a typical clinical phenotype, abnormal functional tests in terms of ciliary beat pattern and frequency, but a normal axonemal ultrastructure when examined by transmission electron microscopy (TEM) (Escudier et al. 2002).

From a purely clinical phenotype aspect, PCD is like many genetic diseases with no clear cut association between specific gene mutations and clinical phenotype. Furthermore, the functional tests that underpin the current diagnostic tools in PCD appear to have no association with phenotypic features or disease severity, which appear to be independent of both gene mutation and specific ciliary defect. Differences in ciliary beat pattern have been linked to ultrastructural defects (Chilvers et al. 2003), but neither of these findings can be used to predict clinical disease patterns or severity. In only one defined ultrastructural defect, that where the central microtubular pair of the axoneme is defective, known as a transposition defect, is any clinical association made. This is the finding that these patients appear to uniquely have no association with situs anomalies (Kennedy et al. 2007), although they have all the other phenotypic features typical of PCD.

Clinical phenotyping is essential to determining the true prevalence of PCD using genetic studies. Defining atypical PCD is essential for gene discovery in milder cases of PCD. Similar problems are faced in cystic fibrosis, where CF-like disease defies current diagnostic criteria; but it is well described that CFTR mutations are much commoner than expected in idiopathic bronchiectasis (Groman et al. 2002). Where does the border between pancreatic sufficient CF with bronchiectasis and idiopathic bronchiectasis with CFTR mutations lie, and in the latter cases are these causal mutations or modifiers? Like in CF, PCD may have similarly fuzzy boundaries as yet undefined, which could alter our current definitions of disease phenotypes.

Screening and diagnosis

Recognising the cardinal clues will lead to a presumptive clinical diagnosis of PCD. Because of the non-specific nature of many of the symptoms most patients will undergo more standard, and readily available, investigations before they are referred for ciliary function testing. These should include a sweat test and immune function.

Specific diagnostic testing for PCD is concentrated in a few specialist centres where experience, and appropriate equipment, is available and these investigations are offered at both tertiary and quaternary levels (O'Callaghan et al. 2007). Despite the development of specialist teams, diagnostic doubt remains in some cases, and repeated testing is often undertaken after aggressive treatment for upper and lower airway infection and inflammation. The majority of patients obtain a definitive diagnosis based on an ultrastructural anomaly or genetic mutation, but debate is ongoing as to what constitutes a diagnosis of PCD, and more importantly what level of diagnostic uncertainty is permissible within the current gold standards. This latter point is increasingly important as experience grows and more atypical cases are being identified using genetic studies, three-dimensional electron tomography (Burgoyne et al. 2012), mucociliary clearance techniques and immunofluorescence techniques (Omran & Loges, 2009).

Screening tests

Screening for PCD in the United Kingdom involves the measurement of nasal nitric oxide (nNO). This is available in three specialist diagnostic centres, and in some other tertiary level respiratory and ENT services. It is also routinely used in many European and North American specialist centres. Radioisotope tests and the saccharine tests are other options that can be utilised where nNO is not available.

It is long established that nNO is very low in PCD (Karadag et al. 1999), and studies show that nasal nitric oxide synthase-2 expression is reduced in PCD (Pifferi et al. 2011). A normal or high nNO of over 250 parts per billion (ppb) combined with a low clinical suspicion for PCD is a powerful argument that a patient's symptoms are not due to PCD (Narang et al. 2002). However, in patients with milder spectrum, or atypical, symptoms a nNO within the normal range cannot be used to exclude PCD and further diagnostic tests will be needed.

Nasal nitric oxide (nNO) measurements are conventionally undertaken during a breath-hold, and are most reliable in patients over 5 years of age. More recently, nNO measurements have been explored during tidal breathing in a bid to extend screening into infants and pre-school children (Mateos-Corral et al. 2011). There is good evidence this method can discriminate between PCD patients and controls at older ages, but less convincing data in infants where sinus development is rudimentary, and so its use as a screening tool in this group must be used with caution.

False positives also occur in patients with nasal obstruction from other causes, where a low nNO may be reversed following treatment of the nasal pathology. Polyps, septal deviation and adenoid hypertrophy may all present in this way, and may prompt referral for diagnostic tests from ENT services. Clinical history should elucidate if a typical PCD phenotype exists, and repeat testing after treatment usually yields a normal result. Other conditions such as CF (Balfour-Lynn et al. 1996) or panbronchiolitis (Nakano et al. 2000) are often associated with a nNO level at the lower end of normal (around 250 ppb), and overlap may exist with some cases of PCD. Confirmatory tests for PCD are essential as are further testing for other causes of lung disease.

Although still designated as a screening test, there is compelling evidence that reliable nNO measurements during a proper breath-hold should become part of the diagnostic panel of investigations. Audit of the UK diagnostic services indicates that nNO values are as consistently accurate as ciliary beat and frequency tests, and arguably more reliable than conventional TEM.

A nNO value of <200 ppb is highly indicative of PCD and all of these patients should proceed to diagnostic testing.

Radioisotope mucociliary clearance tests are emerging as a reliable way to discriminate between PCD patients and normals by demonstrating globally delayed clearance of labelled isotopes from the lungs (De Boeck et al. 2005; Marthin et al. 2007). However, it is expensive and time consuming and may be limited to atypical cases where further investigation is needed to make a diagnosis.

The saccharine test represents the oldest of the screening tests. A particle of saccharine is placed on the inferior turbinate and transit times to the naso-pharynx, where the taste indicates its arrival at the back of the nose, is measured. It requires patients to sit still for an hour, without sniffing, and so its use in children is very limited (Stanley et al. 1984).

Diagnostic tests

Accepted diagnostic tests involve analysis of ciliary motility, axonemal ultrastructure and genotyping. The majority of cases can be confirmed using the combined results of these tests, although the current 'gold standard' is by identification of an ultrastructural defect using TEM. When a diagnosis is clear cut no further investigation is required, but it has become clear that a combination of functional and molecular techniques offer the best chance of capturing the atypical cases that currently offer a diagnostic dilemma and this section will include some of these novel techniques.

The diagnostic tests that examine ciliary function require a nasal brush biopsy, nasal scrape or sampling of bronchial brushings taken via a bronchoscope. These samples are then processed through the following diagnostic steps:

1 *Analysis of ciliary beat pattern and ciliary beat frequency* (CBF) using light microscopy and fast video recording (Stannard et al. 2010). This functional test remains the most consistently abnormal test in patients with PCD. Certain beat patterns correlate with specific ultrastructural abnormalities later determined by transmission electron microscopy (TEM), although no link with clinical phenotype has been established (Chilvers et al. 2003). Both ciliary function and structure may be abnormal as a secondary feature, commonly due to recent viral infection, oxygen therapy or other noxious stimuli, and a repeat test should be performed to rule out such phenomena.

2 The sample is analysed for exact *ultrastructural defects* using TEM (Papon et al. 2010; Shoemark et al. 2012). TEM remains the 'gold standard' test, and no further investigation is required to confirm a diagnosis where a known PCD producing defect, such as absence of the outer dynein arms exists. However, undoubted cases of PCD with normal ultrastructure have been reported and several large centres have reported 10–15% of patients with PCD will have no identifiable defect (Escudier et al. 2002). Secondary defects are an additional cause for diagnostic confusion and if doubt remains after TEM

for any of the reasons noted then repeat testing or cell culture is essential (O'Callaghan et al. 2011).

3 *Ciliary cell culture* is only needed in cases where diagnostic doubt exists. When ciliated epithelium is regenerated in culture, secondary defects disappear (Jorissen & Willems, 2004; Pifferi et al. 2009). Primary defects persist, and can be examined by light and electron microscopy to confirm the diagnosis (Hirst et al. 2010). The technique is time consuming and expensive and presents significant technical challenges. It is available in a few specialist centres.

In cases where TEM cannot confirm the diagnosis an array of other options exist, but may not be available at all testing centres. These include:

1 *Genotyping*. The genetics of PCD are discussed later in this chapter, but it is likely that numerous genes can cause PCD. Around 20 have so far been identified although around 12 of these appear to confirm a diagnosis in about half of all patients with a clinical diagnosis of PCD (Knowles et al. 2012). Experience in PCD, and other less heterogeneous diseases, such as CF, has shown that in atypical cases, just when a genetic diagnosis would be most useful, the results may be inconclusive. The diagnostics of PCD therefore require ongoing expertise in the functional analysis of ciliary beat pattern and frequency alongside structural observations. Research and development to expand our knowledge of ciliary function, transport and signalling mechanisms is a rapid growth area in ciliopathy research.

2 *Specific monoclonal antibody immunofluorescence* staining of ciliary proteins is not widely available and is limited by the small number of proteins that can be stained for directly. There is evidence that some of these proteins do not reach the normal position in the cell (Fliegauf et al. 2005; Olbrich et al. 2006). Furthermore, alterations in staining may be the direct result of a gene mutation encoding for that protein, or may be secondary a mutation in another gene. Exclusion of secondary changes and determination of the sensitivity and specificity of these tests is required to adopt this technique into clinical practice, but with growing experience it is likely they will become part of mainstream diagnostics in the future.

3 *Dual-axis electron tomography* using computer averaging of electron microscopy images is emerging as a new research tool to visualise axonemal structures in three dimensions (Burgoyne et al. 2012). Sub-tomographic averaging can achieve a resolution of around 5 nm, sufficient to distinguish molecular complexes and provide a novel insight into conformation of structures and their interactions. Much work has been done in flagellated organisms, such as the highly conserved blue green algae *Chlamydomonas reinhardtii* (Nicastro et al. 2011), but more recently the focus has begun to shift to the examination of the human respiratory cilium. Recently, electron tomography, genetics and molecular biology, used in combination have identified the absence of hydin as another cause of PCD (Olbrich et al. 2012). Tomography showed that projections associated with the central pair apparatus (C2b and C2c) are absent in patients with a hydin defect, suggesting a location for hydin in the normal axoneme. Whilst the role

of electron tomography is assured as a research tool, in combination with molecular genetics, to unravel the small and highly complex structures of the ciliary axoneme, it may also have a significant role in specialist cases of atypical PCD diagnosis.

Even including all of these possible tests, around 10–15% of cases will not have a confirmed diagnosis, despite a typical clinical phenotype, low nasal NO and abnormal ciliary beat pattern. These cases should still be considered as PCD and treated as such.

Interesting features of the condition

It has become evident that motile cilia do more than merely drive mucociliary clearance, and many diseases are now attributed to ciliary dysfunction. Although most patients with PCD only have the classical manifestations of the disease, overlap with other ciliopathies has been described (Fliegauf et al. 2007; Waters and Beales, 2011). Furthermore, some clinical associations with PCD, such as atrial isomerism, may occur in patients without PCD as well, leading to speculation that ciliary dysfunction may present in different ways despite overlapping clinical presentations.

Currently it appears that defects in motile cilia cause PCD. Other sites where motile cilia exist include the oviduct and the cerebral ependymal. Sperm tails have a similar structure, with tail motility essential for normal fertility in males. But immotile sperm are not invariable in PCD, suggesting the sperm tail is under a different genetic control pathway or that there is redundancy at that site (Munro et al. 1994).

Nodal cilia are also motile, but with a circular, rotatory beat. They occur in the embryonic node and determine organ situs and are probably the only motile primary cilia (Nonaka et al. 2002). Unlike respiratory cilia, however, they do not have a central pair, and thus PCD mutations involving the central pair or radial spokes are not associated with situs abnormalities.

Finally, primary cilial abnormalities that are not the prime cause of PCD are usually non-motile. They have a 9 + 0 axonemal structure, no central pair and there is usually one per cell. They are found on many cell types and probably have chemo-, osmo- and photo-transduction sensors. In some locations, such as the inner ear, where there is a pair of central microtubules, they may be motile (Colantonio et al. 2009). Some patients with PCD have associated sensori-neural deafness, unrelated to the respiratory ciliary defect that leads to middle ear disease more typical of the classical phenotype.

Where does this leave the role of these different cilia in disease entities? Over compartmentalisation must be avoided as it is clear overlapping roles exist, implying that a single mutation may affect more than one cilia type in some but not all cases. Usher's syndrome, a so-called non-motile ciliopathy, is associated with altered nasal ciliary function, suggesting an overlap in the terms 'motile' and 'non-motile' ciliopathy. In PCD we have already discussed the role of nodal cilia which may, or may not, be involved depending on whether the central pair is affected. In summary, although there are discrete disease entities, of which PCD is one, there may also be overlapping conditions with, for example, PCD and retinitis pigmentosa manifesting in the same extended family.

Genetics

Cilia are extraordinarily complex and highly conserved structures, with a repeating array of interconnecting proteins from the basal body throughout the length of the axoneme (Ostrowski et al. 2011). There are several hundred candidate genes (Ishikawa et al. 2012; Kim et al. 2010; Konno et al. 2012; Mizuno et al. 2012), although not all may be relevant to human disease and given the pivotal role of cilia during embryogenesis many may in fact be embryo lethal. There may be many ciliopathies that therefore have no postnatal phenotype.

This chapter focuses on PCD and the detection of novel PCD genes has massively expanded over recent years. Unlike many ciliopathies, gene discovery in PCD can be linked in with direct visual assessment of the motile cilia, ultrastructural defects and functional impairments that has greatly enhanced research in this area.

Some of the problems arising from establishing a genetic cause for PCD, and other inherited diseases, have been discussed in the genotyping section of 'Diagnostic tests'. It is recognised that finding an alteration in a DNA sequence alone cannot confirm a diagnosis, and that the mutation must be shown to have functional consequences, usually as a result of absence or loss of stability in the gene product. In CF, where around 1800 gene mutations are described, fewer than 50 have been identified as disease causing (Castellani et al. 2008). The picture for PCD, a much more heterogeneous disease, is likely to be more complex still. Whilst CF mutations are almost exclusively limited to the CFTR complex (Groman et al. 2002), in PCD the mutations appear to affect a whole range of genes responsible for ciliary assembly, from purely structural proteins to those encoding for proteins involved in cell metabolism, transport and signalling (Duquesnoy et al. 2009; Loges et al. 2009; Omran et al. 2008; van Rooijen et al. 2008).

The commonest PCD causing genes known at present are in *DNAI1* and *DNAH5*, both of which encode proteins in the outer dynein arm (Hornef et al. 2006; Zietkiewicz et al. 2010). Other genes involved in dynein arm assembly include *DNAI2*, *DNAH11* and *DNAL1*(Loges et al. 2008; Mazor et al. 2011; Pifferi et al. 2010). Defects in *RSPH9* and *RSPH4A* result in disarrangement of the radial spokes (Castleman et al. 2009). Mutations in *RFX3*, *DNAAF3*, *KTU*, *DNAAF1* and *CCDC103* affect ciliary assembly proteins (El Zein et al. 2009; Mitchison et al. 2012; Panizzi et al. 2012), and in *CCDC39* and *CCDC40* lead to defects in the coiled-coil domain proteins responsible for the central pair microtubules (Becker-Heck et al. 2011; Merveille et al. 2011), defective inner arms and problems with the dynein regulatory complex (DRC), which may be attributed to nexin link assembly (Heuser et al. 2009). *TXND3*, a member of the thioredoxin family, is a novel gene that affects metabolic function is likely to be the first disease causing gene of many that result in the PCD clinical phenotype (Duriez et al. 2007).

Underlying ciliary defect

The explosion of research into ciliopathies has led to a greater understanding of how defects of human cilia lead to disease phenotypes. PCD results from a defect in the motile cilia lining the respiratory tract resulting in the classical features described above. Although our understanding of ciliary structural biology is increasing, especially within the motile respiratory cilia, there is still much that we do not understand about ciliary metabolic and signalling functions or their role in ciliary disorders such as PCD. Structural defects are well described and can be identified in 85–90% of all patients with the classical phenotype. In a small proportion of cases no structural abnormality is detected, although beat pattern and clinical phenotype are both abnormal and typical of PCD. Some of these cases have an identified gene mutation (*DNAH11*), but all should be considered as PCD and managed as such.

Physiological effect of the ciliary defect

Where PCD is concerned, we know that the structural defects result in movement abnormalities (dyskinesia), as a result of a genetic mutation (primary). Hence in PCD the beat pattern is the most important measurement of functional abnormality and is the most reliable test in making a positive diagnosis. There is a range of recognised abnormalities from static through stiff or chaotic beat patterns. The uniform consequence of this is that the muco-ciliary escalator is interrupted and the airways accumulate secretions, which results in the classical symptoms of PCD. But ciliary function encompasses more than ciliary beating and mucus clearance. Complex intraflagellar transport, including retrograde recycling of proteins, occurs via 'A' and 'B' rafts which have been shown to move to and from the cilium tip (Bisgrove & Yost, 2006). These transport mechanisms may at least be for nutritive functions, but some ciliopathies such as retinitis pigmentosa and Bardet–Biedl syndrome have clearly disordered intraflagellar transport systems resulting in the array of severe and profound phenotypic features including blindness (Armengot et al. 2012). Less is known about the role of intraflagellar transport mechanisms in PCD, although mutations in 'A' and 'B' raft genes is associated with short, stumpy cilia. Finally, motile cilia have been found to have a putative sensory function as chemoreceptors, possibly regulating ciliary motility in response to environmental chemicals. This is likely a primary defensive mechanism designed to clear the respiratory tract of potentially harmful insults.

Clinical management

Evidence-based medicine does not exist in PCD as no long-term, randomised, controlled trials have been carried out. Therefore, many aspects of managing the patient with PCD are empirically based and are similar to that of any chronic suppurative lung disease, in particular cystic fibrosis (Bush et al. 2007). Best practice guidelines, primarily based on experience from management practices of large cohorts in specialist centres (Barbato et al. 2009; Smyth et al. 2010) have recently been reviewed and are analogous in many respects to current guidelines for CF. Regular review in a specialist centre, with multi-system care, and an aggressive approach to airway clearance and management of infections provides the cornerstone of therapy. Although much of the management practices of PCD lung disease are similar to CF, PCD should not be managed alongside CF. The multi-system aspects of PCD differ from, and require different specialist input, to CF and these needs may get ignored in the CF setting. Dedicated PCD clinics should incorporate same day access to ENT and audiology, specialist physiotherapy, nursing and lung function testing. Ready access to cardiology, imaging and fertility services are also essential.

The lower respiratory tract

Physiotherapy

Secretion retention is treated with airway clearance techniques, with physiotherapy the mainstay of treatment supplemented by regular exercise. To date there have been no comparisons between physiotherapy techniques, which include postural drainage, autogenic drainage and positive pressure or huffing adjuncts such as the PEP or Acapella. Experience in different centres will often dictate what is prescribed and taught to patients, but involvement of experienced physiotherapists is essential and should be continued throughout a patient's life. Techniques can be modified to suit all ages and life-styles and this input greatly improves compliance. In addition to the commonly used adjuncts nebulised treatments with mucolytics, hypertonic saline or antibiotics may be employed. Cough-assist techniques, where mechanical augmentation of cough via the use of inspiratory or inspiratory and expiratory pressure support is adopted, can be delivered through specialist devices or a simple BiPAP machine. Although expensive, these latter 'adjuncts' have been useful in carefully selected patients. Randomised controlled studies, conducted on a multicentre basis, are needed to determine efficacy of commonly used techniques and therapies.

Antibiotic therapy

Key issues are:

1 Identification of microorganisms and targeted therapy.

2 Prolonged courses of antibiotics. A minimum of 2 weeks treatment is recommended to combat reduced muco-ciliary clearance. Continuous therapy is used where recurrent isolates of the same organism, or other pathogens is identified

3 Lower threshold for use of antibiotics for any respiratory tract infection. Despite the majority of infections being viral, antibiotics are recommended to reduce opportunistic infections.

4 Specific and aggressive treatment of pseudomonas infections, using CF eradication protocols with intravenous and nebulised antibiotics, plus regular monitoring of sputum samples.

Bronchoscopy to obtain samples for microbiology may be necessary to direct antibiotic choices and this procedure is becoming more commonplace in specialist centres to assist with therapy dilemmas.

Routine respiratory care is also essential. Immunisations, including annual influenza vaccination, avoidance of tobacco smoke, directly or via passive smoking, and reduced exposure to environmental pollution are all recommended. Although segregation of PCD patients is not essential within health setting scenarios, as it is in cystic fibrosis, a common sense approach to infection control measures is crucial. Patients with PCD should not be accommodated in shared bays on the ward with other bronchiectatic patients, whatever the underlying pathology. Physiotherapy should not be undertaken in a shared environment, especially where siblings are affected, and nebulisers must be appropriately vented where antibiotics are used. Careers advice for patients in terms of infection risks for specific jobs is an important part of transitional care and should be supported into young adult life. The use of inhaled steroids and bronchodilators is advised where co-existent asthma, atopy and airflow obstruction is evident. It should be noted that exercise is a powerful bronchodilator (Phillips et al. 1998), as well as an excellent form of physiotherapy.

Monitoring respiratory function

Spirometry continues to provide the first line in monitoring lung function, although it has recently been shown to be an insensitive marker of deterioration in respiratory health. In recent studies high resolution computed tomography scans have shown deterioration in respiratory status whilst spirometry has remained stable in both CF and PCD (Maglione et al. 2012). The search is on to establish an affordable, readily accessible and reproducible tool to assist in the monitoring of lung health in PCD patients. Multiple breath washout techniques, or lung clearance index, is likely to prove one such tool in specialist centres. It is more sensitive that spirometry and research has shown that distal airway involvement is common in PCD (Green et al. 2012). Early promise with such techniques now requires prospective studies to determine the optimal monitoring strategies in PCD.

The upper respiratory tract

In general, conservative management is best, alongside close review and non-surgical interventions where needed. The impact of chronic serous otitis media and hearing impairment, which is common and sometimes severe, should not be trivialised and furthermore benefits from specialist attention to ensure measures are implemented to optimise speech development and educational attainment. The majority respond to simple measures such as engaging teachers and placing the child at the front of class, but some may require transient use of hearing aids. Most patients' hearing improves with age and aids can be discarded in the teenage years (Majithia et al. 2005). Routine audiology screening and assessment should be adopted from the point of diagnosis.

In some patients nasal obstruction and chronic rhinitis can be so severe that quality of life is adversely affected (quality-of-life surveys) and in one cohort over 50% of patients have evidence of significant obstructive sleep apnoea (Oktem et al. 2012). Use of nasal sprays and drops has mixed results with many patients reporting reluctance or difficulty in their use. Anecdotal reports appear to show that saline douching may be most effective in upper airway clearance, although acceptance and compliance rates are low. Like most therapies employed in the management of PCD, there is no evidence for the use of topical steroid preparations in these patients.

Surgical interventions are avoided where possible. Insertion of tympanostomy tubes usually results in persistent otorrhoea, which can be profuse, malodorous and does little to improve hearing impairment (Hadfield et al. 1997). Where this occurs, infection with *Pseudomonas aeruginosa* is almost inevitable and the use of ciprofloxacin eardrops is needed. Functional endoscopic sinus surgery can lead to good results in selected cases, and is more common in adult patients.

Other manifestations

Like many complex diseases, PCD also requires a multisystem approach to management. The main considerations are:

- *Screening for organ situs* and associated abnormalities. Heterotaxy, in particular may have associated cardiac, liver and spleen anomalies and some of these patients are liable to suffer malrotation of the intestine.
- *Fertility*. All adult patients should have access to fertility clinics and counselling when family planning. Male patients may need assisted conception techniques such as intracytoplasmic sperm injection to overcome infertility due to immotile sperm (Matsumoto et al. 2010).
- The risk of *ectopic pregnancy* in the female patient is raised due to impaired oviduct transport (Blyth and Wellesley, 2008), and women should book early into obstetrics clinics to establish that the embryo has implanted appropriately in the uterus. These issues should be discussed from the point of diagnosis, and in particular during adolescence, with reassurance that both ovum and sperm are normal and fertility treatments allow for normal pregnancies to develop.

Finally, the aspects of management mentioned so far should be considered for cases with atypical PCD. As with CF, it is likely that PCD mutations may have a higher incidence in bronchiectasis and severe sinusitis, but the difficulties of securing a diagnosis are formidable. As in cases of atypical CF, it is advisable to remain vigilant, ensure pulmonary health measures, including physiotherapy, and treat any infections or other complications vigorously.

Prognosis and outcomes

The prognosis for the majority of patients is good with a potentially normal life span. There is, however, a wide spectrum of disease severity and some patients will develop severe lung disease and respiratory failure. Bronchiectasis is the main consequence of late diagnosis, poor compliance and recurrent infections, and its severity in the absence of other major organ abnormalities, is the main determinant of outcome. Age at diagnosis still varies widely but lung disease can stabilise after commencing appropriate treatment (Noone et al. 2004), and every effort to encourage compliance with treatment should be made to limit further lung damage. However, there is no room for complacency, since a recent report on a large Danish cohort suggests that up to one-third will have a progressive deterioration in spirometry (Marthin et al. 2010), and there are reports of lung transplantation for end stage respiratory failure (Date et al. 2001; Schertler et al. 2007). From a diagnostic perspective early detection, with institution of treatment, is crucial as early onset of lung disease, including bronchiectasis, has been described in infants with PCD (Brown et al. 2008).

Finally, attention to hearing impairment is crucial to improve educational attainment and improve career prospects. When badly managed many patients may miss educational milestones and be disadvantaged when entering the employment market. The socio-economic and healthcare burden of ignoring this aspect of care in PCD patients is a convincing argument why these complex patients should have access to specialist multidisciplinary care.

References

Afzelius, B. A. 1976. A human syndrome caused by immotile cilia. *Science* **193**, 317–319.

Armengot, M, Salom, D., Diaz-Llopis, M. Millan, J. M., Milara, J., Mata, M., et al. 2012. Nasal ciliary beat frequency and beat pattern in retinal ciliopathies. *Invest Opthalmol Vis Sci* **53**, 2076–2079.

Balfour-Lynn, I. M., Laverty, A. & Dinwiddie, R. 1996. Reduced upper airway nitric oxide in cystic fibrosis. *Arch Dis Child* **75**, 319–322.

Barbato, A., Frischer, T., Kuehni, C. E., Snijders, D., Azevedo, I., Baktai, G., et al. 2009. Primary ciliary dyskinesia: A consensus statement on diagnosis and treatment approaches in children. *Eur Respir J* **34**, 1264–1276.

Becker-Heck, A., Zohn, I. E., Okabe, N., Pollock, A., Lenhart, K. B., Sullivan-Brown, J., et al. 2011. The coiled-coil domain containing protein CCDC40 is essential for motile cilia function and left-right axis formation. *Nat Genet* **43**, 79–84.

Bisgrove, B. W. & Yost, H. J. 2006. The roles of cilia in developmental disorders and disease. *Development* **133**, 4131–4143.

Blyth, M. & Wellesley, D. O. 2008. Ectopic pregnancy in primary ciliary dyskinesia. *J Obstet Gynaecol* **28**, 358.

Brown, D. E., Pittman, J. E., Leigh, M. W., Fordham, L. & Davies, S. D. 2008. Early disease in young children with primary ciliary dyskinesia. *Pediatr Pulmonol* **43**, 514–516.

Burgoyne, T., Dixon, M., Luther, P., Hogg, C. & Shoemark, A. 2012. Generation of a three dimensional ultra-structural model of human respiratory cilia. *Am J Resp Cell Biol* **47**, 800–806.

Bush, A., Chodhari, R., Collins, N., Copeland, F., Hall, P., Harcourt, J., et al. 2007. Primary ciliary dyskinesia: Current state of the art. *Arch Dis Child* **92**, 1136–1140.

Bush, A. & O'Callaghan, C. 2002. Primary ciliary dyskinesia. *Arch Dis Child* **87**, 363–365.

Carlen, B., Lindberg, S. & Stenram, U. 2003. Absence of nexin links as a possible cause of primary ciliary dyskinesia. *Ultrastruct Pathol* **27**, 123–126.

Castellani, C., Cuppens, H., Macek Jr, M., Cassiman, J. J., Kerem, E., Durie, P., et al. 2008. Consensus on theuse and interpretation of cystic fibrosis mutation analysis in clinical practice. *J Cyst Fibros* **7**, 179–196.

Castleman, V. H., Romio, L., Chodhari, R., Hirst, R. A., de Castro, S. C., Parker, K. A., et al. 2009. Mutations in radial spoke head protein genes RSPH9 and RSPH4A cause primary ciliary dyskinesia with central-microtubular-pair abnormalities. *Am J Hum Genet* **84**, 197–209.

Chilvers, M. A. & O'Callaghan, C. 2000. Analysis of ciliary beatpattern and beat frequency using digital high speed imaging: Comparison with the photomultiplier and photodiode methods. *Thorax* **55**, 314–317.

Chilvers, M. A., Rutman, A. & O'Callaghan, C. 2003. Ciliary beat pattern is associated with specific ultra-structural defects in primary ciliary dyskinesia. *J Allergy Clin Immunol* **112**, 518–524.

Colantonio, J. R., Vermot, J., Wu, D., Langenbacher, A. D., Fraser, S., Chen, J. N., et al. 2009. The dynein regulatory complex is required for ciliary motility and otolith biogenesis in the inner ear. *Nature* **457**, 205–209.

Coren, M. E., Meeks, M., Morrison, I., Buchdahl, R. M.& Bush, A. 2002. Primary ciliary dyskinesia: Age at diagnosis and symptom history. *Acta Paediatr* **91**, 667–669.

Date, H., Yamashits, M., Nagahiro, I., Aoe, M., Andou, A. & Shimizu, N. 2001. Living-donor lobar lung transplantation for primary ciliary dyskinesia. *Ann Thorc. Surg* **71**, 2008–2009.

De Boeck, K., Proesmans, M., Mortlemans, L., Van Billoen, B., Willems, T. & Jorrisen, M. 2005. Mucociliary transport using ^{99m}Tc-albumin colloid: A reliable screening test for primary ciliary dyskinesia. *Thorax* **60**, 414–417.

Duquesnoy, P., Escudier, E., Vincensini, L., Freshour, J., Bridoux, A. M., Coste, A., et al. 2009. Loss-of-function mutations in the human ortholog of *Chlamydomonas reinhardtii* ODA7 disrupt dynein arm assembly and cause primary ciliary dyskinesia. *Am J Hum Genet* **85**, 890–896.

Duriez, B., Duquesnoy, P., Escudier, E., Bridoux, A. M., Escalier, D., Rayet, I., et al. 2007. A common variant in combination with a nonsense mutation in a member of the thioredoxin family causes primary ciliary dyskinesia. *Proc Natl Acad Sci U.S.A.* **104**, 3336–3341.

Ellerman, A. & Bisgaard, H. 1997. Longitudinal study of lung function in a cohort of primary ciliary dyskinesia. *Eur Respir J* **10**, 2376–2379.

El Zein, L., Ait-Lounis, A., Morle, L., et al. 2009. RFX3 governs growth and beating efficiency of motile cilia in mouse and controls the expression of genes involved in human ciliopathies. *J Cell Sci* **122**, 3180–3189.

Escudier, E., Couprie, M., Duriez, B., Roudot-Thoraval, F., Millepied, M. C., Prulière-Escabasse, V., et al. 2002. Computer assisted analysis helps detect inner dynein abnormalities. *Am J Respir Crit Care Med* **166**, 1257–1262.

Fliegauf, M., Benzing, T. & Omran, H. O. 2007. When cilia go bad: Cilia defects and ciliopathies. *Nat Rev Mol Cell Biol* **8**, 880–893.

Fliegauf, M., Olbrich, H., Horvath, J., Wildhaber, J. H., Zariwala, M. A., Kennedy, M., et al. 2005. Mislocalization of DNAH5 and DNAH9 in respiratory cells from patients with primary ciliary dyskinesia. *Am J Respir Crit Care Med* **171**, 1343–1349.

Green, K., Buchvald, F. F., Marthin, J. K., Hanel, B., Gustafsson, P. M. & Nielson, K. G. 2012. Ventilation inhomogeneity in children with primary ciliary dyskinesia. *Thorax* **67**, 49–53.

Groman, J. D., Meyer, M. E., Wilmott, R. W., Zeitlin, P. L. & Cutting, G. R. 2002. Variant cystic fibrosis phenotypes in the absence of CFTR mutations. *N Engl J Med* **347**, 401–407.

Hadfield, P. J., Rowe-Jones, J. M., Bush, A. & Mackay, I. S. 1997. Treatment of otitis media with effusion in children with primary ciliary dyskinesia. *Clin Otolaryngol Allied Sci* **22**, 302–306.

Heuser, T., Raytcher, M., Krell, J., Porter, M. E. & Nicastro, D. 2009. The dynein regulatory complex is the nexin link and a major regulatory node in cilia and flagella. *J Cell Biol* **187**, 921–933.

Hildebrandt, F., Benzing, T. & Katsanis, N. 2011. Ciliopathies. *N Engl J Med* **364**, 1533–1543.

Hirst, R. A., Rutman, A., Williams, G. & O'Callaghan, C. 2010. Ciliated air–liquid cultures as an aid to diagnostic testing of primary ciliary dyskinesia. *Chest* **138**, 1441–1447.

Hornef, N., Olbrich, H., Horvath, J., Zariwala, M. A., Fliegauf, M., Loges, N. T., et al. 2006. DNAH5 mutations are a common cause of primary ciliary dyskinesia with outer dynein arm defects. *Am J Respir Crit Care Med* **174**, 120–126.

Ishikawa, H., Thompson, J., Yates, J. R. 3rd & Marshall, W. F. 2012. Proteomin analysis of mammalian primary cilia. *Curr Biol* **22**, 414–419.

Jorissen, M. & Willems, T. 2004. The secondary nature of ciliary (dis)orientation in secondary and primary ciliary dyskinesia. *Acta Otolaryngol* **124**, 527–531.

Karadag, B., James, A. J. Gulrekin, E., Wilson, N. M. & Bush, A. 1999. Nasal and lower airway level of nitric oxide in children with primary ciliary dyskinesia. *Eur Respir J* **13**, 1402–1405.

Kartagener, M. 1933. Zur pathogenese der bronchiectasien. I Meitteilung:Bronchiectasien bei situs viscerum inversus. *Betr Klin Tuberk* **83**, 498–551.

Kennedy, M. P., Omran, H., Leigh, M. W., Dell, S., Morgan, L., Molina, P. L., **et al.** 2007. Congenital heart disease and other heterotaxic defects in a large cohort of patients with primary ciliary dyskinesia. *Circulation* **115**, 2814–2821.

Kim, J., Lee, J. E., Heynen-Genel, S., Suyama, E., Ono, K., Lee, K., et al. 2010. Functional genomic screen for modulators of ciliogenesis and cilium length. *Nature* **463**, 1048–1051.

Knowles, M. R., Leigh, M. W., Carson, J. L., Davis, S. D., Dell, S. D., Ferkol, T. W., et al. 2012. Mutations of DNAH11 in patients with primary ciliary dyskinesia with normal ciliary ultrastructure. *Thorax* (Epub ahead of print).

Konno, A., Setou, M. & Ikegami, K. 2012. Ciliary and flagella structure and function—their regulations by post-translational modifications of axonemal tubulin. *Int Rev Cell Mol Biol* **294**, 133–170.

Krawczynski, M. R., Dmenska, H. & Witt, M. 2004. Apparent X-linked primary ciliary dyskinesia associated with retinitis pigmentosa and a hearing loss. *J Appl Genet* **45** 107–110.

Loges, N. T., Olbrich, H., Becker-Heck, A., Häffner, K., Heer, A., Reinhard, C., et al. 2009. Deletions and point mutations at LRRC50 cause primary ciliary dyskinesia due to dynein arm defects. *Am J Hum Genet* **85**, 883–889.

Loges, N. T., Olbrich, H., Fenske, L., Mussaffi, H., Horvath, J., Fliegauf, M., et al. 2008. DNAI2 mutations cause primary ciliary dyskinesia with defects in the outer dynein arm. *Am J Hum Genet* **83**, 547–558.

Maglione, M., Bush, A., Montella, S., Mollica, C., Manna, A., Esposito, A., et al. 2012. Progression of lung disease in primary ciliary dyskinesia: Is spirometry less accurate than CT? *Pediatr Pulmonol* **47**, 498–504.

Majithia, A., Fong, J., Hariri, M. & Harcourt, J. 2005. Hearing outcomes in children with primary ciliary dyskinesia—a longitudinal study. *Int J Pediatr Otorhinolaryngol* **69**, 1061–1064.

Marthin, J. K., Mortenson, J., Pressler, T. & Nielsen, K. G. 2007. Pulmonary radio-aerosol mucociliary clearance in diagnosis of primary ciliary dyskinesia. *Chest* **132**, 966–967.

Marthin, J. K., Peterson, N., Skovgaard, L. T. & Nielson, K. G. 2010. Lung function in patients with primary ciliary dyskinesia: A cross sectional and 3 decade longitudinal study. *Am J Resp Crit Care Med* **181**, 1262–1268.

Mateos-Corral, D., Coombs, R., Grasemann, H., Ratjen, F. & Dell, S. D. 2011. Diagnostic value of nasal nitric oxide measured with non-velum closure techniques for children with primary ciliary dyskinesia. *J Pediatr* **159**, 420–424.

Matsumoto, Y., Goto, S., Hashimoto, H., Kokeguchi, S., Shiotani, M. & Okada, H. 2010. A healthy birth after intracytoplasmic sperm injection using ejaculated spermatozoa from a patient with Kartagener's syndrome. *Fertil Steril* **93**, 2074.e17–2074.e19.

Mazor, M., Alkrinawi, S., Chalifa-Caspi, V., et al. 2011. Primary ciliary dyskinesia caused by homozygous mutation in DNAL1 encoding dynein light chain 1. *Am J Hum Genet* **88**, 599–607.

Merveille, A. C., Davis, E. E., Becker-Heck, A., Legendre, M., Amirav, I., Bataille, G., et al. 2011. CCDC39 is required for assembly of inner dynein arms and the dynein regulatory complex and for normal ciliary motility in humans and dogs. *Nat Genet* **43**, 72–78.

Mitchison H. M., Schmidts, M., Loges, N. T., Freshour, J., Dritsoula, A., Hirst, R. A., et al. 2012. Mutations in axonemal dynein assembly factor DNAAF3 cause primary ciliary dyskinesia. *Nat Genet* **44**, 381–389, S1.

Mizuno, N., Taschner, M., Engel, B. D. & Lorentzen, E. 2012. Structural studies of ciliary components. *J Mol Biol* **422**, 163–180.

Moore, A., Escudier, E., Roger, G., Tamalet, A., Pelosse, B., Marlin, S., et al. 2006. RPGR is mutated in patients with a complex X-linked phenotype combining primary ciliary dyskinesia and retinitis pigementosa. *J Med Genet* **43**, 326–333.

Munro, N. C., Currie, D. C., Lindsay, K. S., Ryder, T. A., Rutman, A., Dewar, A., et al. 1994. Fertility in men with primary ciliary dyskinesia presenting with respiratory infection. *Thorax* **49**, 684–687.

Nakano, H., Ide, H., Imada, M., Osanai, S., Takahashi, T., Kikuchi, K., et al. 2000. Reduced nasal nitric oxide in diffuse pan-bronchiolitis. *Am J Respire Crit Care Med* **162**, 2218–2220.

Narang, I., Ersu, R., Wilson, N. M. & Bush, A. 2002. Nitric oxide in chronic airway inflammation in children: Diagnostic use and pathophysiological significance. *Thorax* **57**, 586–589.

Nicastro, D., Fu, X., Heuser, T., Tso, A., Porter. M. & Linck, R. 2011. Cryo-electron tomography reveals conserved features of doublet microtubules in flagella. *Proc Natl Acad Sci U.S.A.* **108**, 845–853.

Nonaka, S., Shiratori, H., Saijoh, Y. & Hamada, H. 2002. Determination of left–right patterning of the mouse embryo by artificial nodal flow. *Nature* **418**, 96–99.

Noone, P. G., Leigh, M. W., Sannuti, A., Minnix, S. L., Carson, J. L., Hazucha, M., et al. 2004. Primary ciliary dyskinesia: Diagnostic and phenotypic features. *Am J Respir Crit Care Med* **169**, 459–467.

O'Callaghan, C., Chilvers, M. A., Hogg, C., Bush, A. & Lucas, J. 2007. Diagnosing primary ciliary dyskinesia. *Thorax* **62**, 656–657.

O'Callaghan, C., Rutman, A., Williams, G. M. & Hirst, R. A. 2011. Inner dynein arm defects causing primary ciliary dyskinesia: Repeat testing required. *Eur Respir J* **38**, 603–607.

Oktem, S., Karadag, B., Erdem, E., Gokdemir, Y., Karakoc, F., Dagli, E., et al. 2012. Sleep disordered breathing in patients with primary ciliary dyskinesia. *Pediatr Pulmonol* Nov 20, (Epub ahead of print).

Olbrich, H., Horváth, J., Fekete, A., Loges, N. T., Storm van's Gravesande, K., Blum, A., et al. 2006. Axonemal localization of the dynein component DNAH5 is not altered in secondary ciliary dyskinesia. *Pediatr Res* **59**, 418–422.

Olbrich, H., Schmidts, M., Werner, C., Onoufriadis, A., Loges, N. T., Raidt, J., et al. 2012. Recessive HYDIN mutations cause primary ciliary dyskinesia without randomisation of left–right body asymmetry. *Am J Hum Genet* **91**, 672–684.

Omran, H., Kobayashi, D., Olbrich, H., Tsukahara, T., Loges, N. T., Hagiwara, H., et al. 2008. Ktu/PF13 is required for cytoplasmic pre-assembly of axonemal dyneins. *Nature* **456**, 611–616.

Omran, H. & Loges, N. T. 2009. Immunoflourescence staining of ciliated respiratory epithelial cells. *Methods Cell Biol* **91**, 123–133.

Ostrowski, L. E., Dutcher, S. K. & Lo, C. W. 2011. Cilia and models for studying structure and function. *Proc Am Thorac Soc* **8**, 423–429.

Panizzi, J. R., Becker-Heck, A., Castleman, V. H., Al-Mutairi, D. A., Liu, Y., Loges, N. T., et al. 2012. CCDC103 mutations cause primary ciliary dyskniesia by disrupting assembly of ciliary dynein arms. *Nat Genet* **44**, 714–719.

Papon, J. F., Coste, A., Roudot-Thoraval, F., Boucherat, M., Roger, G., Tamalet, A., et al. 2010. A twenty year experience of electron microscopy in primary ciliary dyskinesia. *Eur Resp J* **35**, 1057–1063.

Phillips, G. E., Thomas, S., Heather, S. & Bush, A. 1998. Airway response of children with primary ciliary dyskinesia to exercise and beta2-agonist challenge. *Eur Repir J* **11**, 1389–1391.

Pifferi, M., Bush, A., Caramella, D., Di Cicco, M., Zangani, M., Chinellato, I., et al. 2011. Agenesis of paranasal sinuses and nasal nitric oxide in primary ciliary dyskinesia. *Eur Respir J* **37**, 566–571.

Pifferi, M., Bush, A., Maggi, F., Michelucci, A., Ricci, V., Conidi, M. E., et al. 2011. Nasal nitric oxide and nitric oxide synthase expression in primary ciliary dyskinesia. *Eur Resp J* **37**, 572–577.

Pifferi, M., Michelucci, A., Conidi, M. E., et al. 2010. New DNAH11 mutations in primary ciliary dyskinesia with normal axonemal ultrastucture. *Eur Respir J* **35**, 1413–1416.

Pifferi, M., Montemurro, F., Cangiotti, A. M., et al. 2009. Simplified call culture method for the diagnosis of atypical primary ciliary dyskinesia. *Thorax* **64**, 1077–1081.

Plesec, T. P., Ruiz, A., McMahon, J. T. & Prayson, R. A. 2008. Ultrastructural abnormalities of respiratory cilia: A 25 year experience. *Arch Pathol Lab Med* **132**, 1786–1791.

Schertler, T., Lardinois, D., Boehm, T., Weder, W., Wildermuth, S. & Alkadhi, H. O. 2007. Lung transplantation in Kartagener syndrome and situs inversus: Potential of multi-detector row computed tomography and three dimensional postprocessing. *J Thorac Cardiovasc Surg* **134**, 814–815.

Shoemark, A., Dixon, M., Corrin, B. & Dewar, A. 2012. Twenty year review of quantitative transmission microscopy for the diagnosis of primary ciliary dyskinesia. *J Clin Pathol* **65**, 267–271.

Smyth, A. R., Barbato, A., Beydon, N., Bisgaard, H., de Boeck, K., Brand, P., et al. 2010. Respiratory medicines for children: Current evidence, unlicensed use and research priorities. *Eur Respir J* **35**, 247–265.

Stanley, P., MacWilliam, L., Greenstone, M., Mackay, I. & Cole, P. 1984. Efficacy of a saccharin test for screening to detect abnormal muco-ciliary clearance. *Br J Dis Chest* **78**, 62–65.

Stannard, W. A., Chilvers, M. A., Rutman, A. R., Williams, C. D. & O'Callaghan, C. 2010. Diagnostic testing of patients suspected of primary ciliary dyskinesia. *Am J Resp Crit Care Med* **181**, 307–314.

van Rooijen, E., Giles, R. H., Voest, E. E., van Rooijen, C., Schulte-Merker, S. & van Eeden, F. J. 2008. LRRC50, a conserved ciliary protein implicated in polycystic kidney disease. *J Am Soc Nephrol* **19**, 1128–1138.

Waters, A. E. & Beales, P. L. 2011. Ciliopathies: An expanding disease spectrum. *Paediatr Nephrol* **26**, 1039–1056.

Ziętkiewicz, E., Nitka, B., Voelkel, K., Skrzypczak, U., Bukowy, Z., Rutkiewicz, E., et al. 2010. Population specificity of the DNAI1 gene mutation spectrum in primary ciliary dyskinesia. *Respir Res* **103**, 109–130.

Zito, I., Downes, S. M., Patel, R. J., Cheetham, M. E., Ebenezer, N. D., Jenkins, S. A., et al. 2003. RPGR mutation associated with retinitis pigmentosa, impaired hearing, and sino-respiratory infections. *J Med Genet* **40**, 609–615.

Chapter 13

Usher syndrome

Maria Bitner-Glindzicz and Zubin Saihan

Introduction

The term Usher syndrome is given to a group of recessively inherited conditions encompassing the combination of deafness and retinitis pigmentosa (RP) resulting in dual sensory impairment. Intellect is normal. There are over 50 known human syndromes which include symptoms of combined visual and hearing loss (sometimes termed 'deafblindness') of which Usher syndrome is the most common, representing over 50% of deafblind adults. As both hearing loss and RP can occur in isolation, Usher syndrome is often referred to as a syndromic form of hearing impairment or a syndromic form of retinitis pigmentosa.

History of the eponym

Although the condition is named after the British ophthalmologist Charles Usher, it is widely acknowledged that the occurrence of a pigmentary retinopathy with deafness was observed in three out of five siblings by the German ophthalmologist Alfred von Graefe and reported by his cousin Albrecht von Graefe in 1858 (von Graefe, 1858). Three years later, in 1861, his student Richard Liebreich reported a survey of deaf inhabitants from Berlin, in whom he noted the frequent presence of retinal pigmentation in individuals with congenital deafness (Liebreich, 1861). The genetic nature of the disease was emphasised by commenting on the presence of these findings in Jewish consanguineous families or in families with several affected members. Following these reports of heritable pigmentary retinopathy and deafness, the disease finally received its eponymous name from the Scottish ophthalmologist Charles Usher, who examined a cohort of 69 cases of retinitis pigmentosa reporting 19 cases with additional hearing loss (Usher, 1914). He also emphasised the heritable nature of their disease. Although the name 'Usher syndrome' is still used to this day, it has also been referred to as Hallgren syndrome, particularly in the Scandinavian literature, Usher–Hallgren syndrome, RP–dysacusis syndrome, and dystrophia retinae dysacusis syndrome by others.

Epidemiology

The prevalence of Usher syndrome can be defined as the number of people affected with Usher syndrome in a given population at any given time and reports range from 1 in 23,000 in the United States (Boughman et al. 1983), 1 in 29,000 in Scandinavia (Grondahl, 1987) and 1 in 16,000 in Germany (Spandau & Rohrschneider, 2002) based upon case ascertainment. These are in broad agreement with our own figures from the United Kingdom based on a finding of four carriers of the common USH2A:p.Glu767SerfsX21 (c.2299delG) mutation in 846 control chromosomes (c.2299delG accounts for 33.7% of all mutations in *USH2A*, itself accounting for 80% of all USH2 cases,) (Le Quesne et al. 2012). More recently, however, it has been re-estimated to have a much higher frequency of 1 in 6000 based on molecular genetic analysis of children with hearing impairment; however, it is possible that not all those with mutations will develop RP, since some mutations may cause either Usher syndrome or non-syndromic deafness (Kimberling et al. 2010).

Founder effects have led to increased prevalence of particular alleles among some populations; in Finland the two most common alleles in *USH3A* (Finnmajor, p.Tyr176* and Finnminor, p.Met120Lys), which are both associated with a common ancestral haplotype, account for 40% of all Usher cases (Joensuu et al. 2001). Common alleles also segregate in the Ashkenazi Jewish population, with founder mutations in both *USH3A* as well as in *PCDH15* (USH1F) (Brownstein et al. 2004; Ness et al. 2003) and a recent molecular survey reported that four *USH2A* mutations accounted for 64% of all pathogenic alleles in Jewish families of non-Ashkenazi descent (Auslender et al. 2008). Perhaps the most well known is the Acadian population of Louisiana segregating a splice mutation c.216G>A in *USH1C* accounting for nearly all Acadian USH1 cases; the same *USH1C* mutation accounts for 40% of disease alleles among Quebecois Usher patients, due to shared ancestry with the Acadians (Ebermann et al. 2007, 2009). This common descent is mirrored by an *USH2A* mutation shared between these two populations, c.4338_4339delCT, accounting for 10 out of 18 disease alleles (55.6%) of USH2 cases from Quebec and New Brunswick (formerly Acadia) and also found in Acadian family from Louisiana.

Clinical features

The cardinal features are of course hearing loss and RP. Usher syndrome can be divided into three clinical subgroups, types 1, 2 and 3, which differ according to their audiovestibular features.

Audiovestibular features

Type 1 Usher syndrome

In type 1 Usher, which is the most severe form, deafness is congenital (present at birth) and profound, and as such infants will be diagnosed following Newborn Hearing Screening programmes. The profound nature of the hearing impairment implies that children do not benefit from hearing aids and therefore will almost never develop normal speech unless they receive cochlear implants. Before the advent of Newborn Hearing Screening therefore, the vast majority of children with type 1 Usher would become sign language users, but nowadays one would hope to implant these children, especially in view of the fact that they may lose the ability to lip read in later years as a result of their retinal dystrophy.

A typical audiogram of an older child with type 1 Usher is shown in Figure 13.1, with a normal audiogram for comparison in (Figure 13.2). In addition to profound hearing loss, children with type 1 Usher syndrome also have vestibular arreflexia and absent vestibular function. Over time children compensate well for their lack of balance, using proprioception and vision to help their balance but in infancy the vestibular problems are nearly always manifest by a delay in achievement of motor milestones. This means that they will be slow to gain head control, slow to sit unsupported and rarely walk independently before the age of 18 months. Although other conditions can cause the combination of profound hearing loss and absent vestibular function, a young child presenting in this way should raise a strong suspicion of Usher syndrome, especially if magnetic resonance imaging fails to show morphogenetic abnormalities of inner ear development (i.e. absence of dysplasia of the vestibular system), and the ECG is normal (one of the differential diagnoses in a

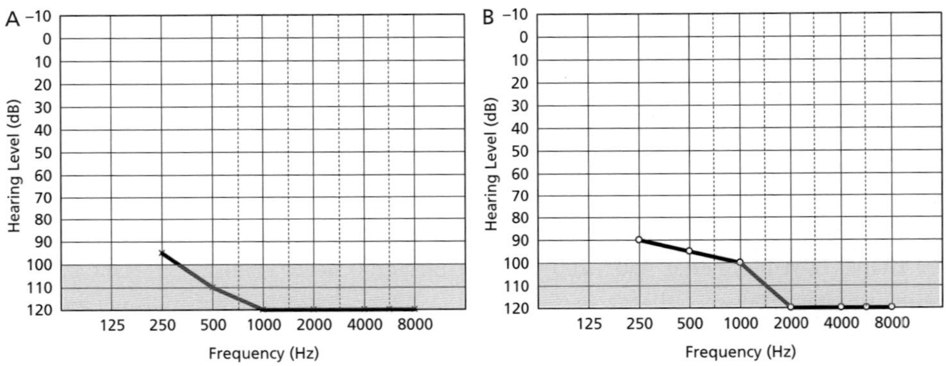

Figure 13.1 Typical audiogram of a person with Usher type 1. Hearing loss is profound with only a 'corner' preserved in the lower frequencies: (a) left ear; (b) right ear.

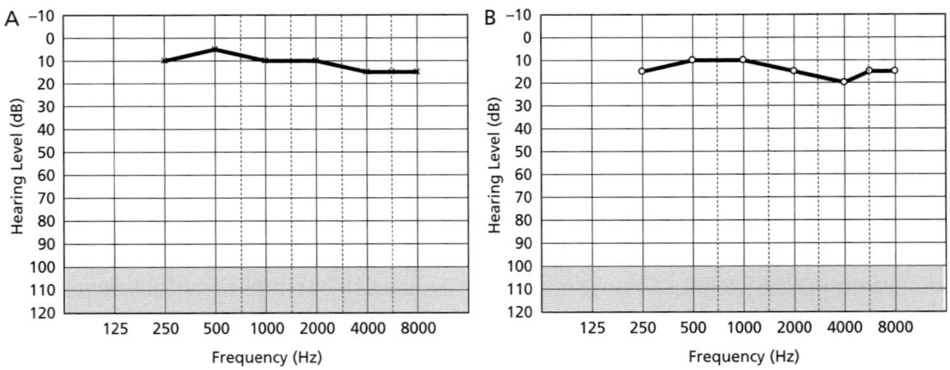

Figure 13.2 Normal audiogram. Frequencies in kHz are along the horizontal axis and sound intensity (dBHL) along the vertical axis. Circles represent thresholds in the right ear and crosses represent thresholds in the left ear: (a) left ear; (b) right ear.

child with profound hearing and balance loss is the cardio-auditory Jervell and Lange–Nielsen syndrome). Profound hearing loss leading to delayed language development combined with motor delay can be misinterpreted as global developmental delay, but careful clinical observation and questioning should distinguish between these.

In adults, signs of vestibular failure can be elicited by heel–toe walking, Romberg sign and objectively by posturography, electronystagmography and caloric testing.

Type 2 Usher syndrome

Type 2 Usher syndrome is clinically distinct from type 1. Hearing loss is also congenital but can be mild to profound with the higher frequencies being more severely affected. A typical audiogram in an older person is shown in Figure 13.3 and the audiometric pattern is described as 'gently sloping'. Hearing loss is also detectable by Newborn Hearing Screening but these children *do* gain benefit from conventional hearing aids and learn to speak normally. Vestibular function is normal, and because of the lack of distinguishing clinical features, and the overall moderate degree of hearing loss, children are initially diagnosed as having 'non-syndromic hearing loss'. The diagnosis of Usher is often not made until the symptoms of RP, night blindness and constricted visual fields, become apparent or the child undergoes routine ophthalmic examination, when RP is manifest at or around puberty.

Type 3 Usher syndrome

Type 3 Usher syndrome is the most variable of all the subtypes but is characterised by *progression* of the hearing loss. Onset may be either prelingual but is usually post-lingual and it is therefore not usually detectable by Newborn Hearing Screening. It is usually within the first decade in the majority of cases, but onset in adult life has also been reported (Sadeghi et al. 2005). Audiometric configuration is also variable, with most patients showing 'sloping' patterns similar to type 2 with moderate loss in the low frequencies and severe to profound loss at high frequencies (Plantinga et al. 2005; Ness et al. 2003;) but some

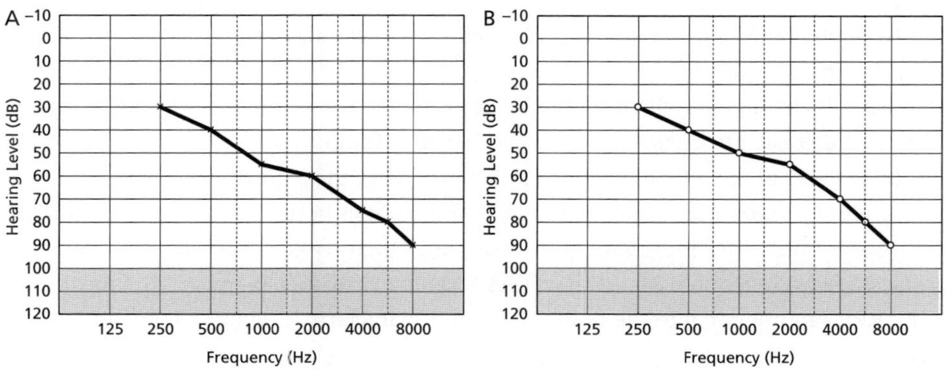

Figure 13.3 Typical audiogram of a person with Usher type 2. Hearing loss is mild to moderate in the lower frequencies but severe in the high frequencies: (a) left ear; (b) right ear.

patients have been reported with U-shaped audiograms. Excellent clinical audiovestibular studies on genotyped patients with type 3 Usher indicate significant progression is apparent in the first two decades of life with a period of stabilisation between 20 and 40 years, and then further progression between the fourth and fifth decades (Plantinga et al. 2005). Thus, early on, hearing loss in similar to that seen in patients with type 2 Usher, but with progression over several decades, severity may ultimately be more similar to that commonly associated with type 1. However, it was noted that in some subjects hearing loss was already profound within the first decade of life. Initially patients do benefit from hearing aids but with progression, which may be rapid, many have had successful cochlear implantation (Pietola et al. 2012).

Further variability in USH3 is observed in terms of the vestibular system; the majority have a normal age of walking (80–90%) although this is not universal. On caloric testing, 45% of patients had hypofunction or bilateral arreflexia whereas about half of people have normal vestibular function (Sadeghi et al. 2005), suggesting that vestibular dysfunction may be progressive. Variability in terms of both hearing and balance between siblings may exist.

Retinitis pigmentosa

The retinitis pigmentosa in Usher syndrome is referred to as a rod–cone dystrophy as the degeneration of the rod photoreceptors occurs early in the disease, whilst cone photoreceptor degeneration occurs later in the disease. Rod photoreceptors are located throughout the retina and are responsible for vision under low-light conditions, explaining the initial visual symptoms of poor vision in dim illumination (nyctalopia) and reduced visual fields. The nyctalopia gets progressively worse and the constriction in visual fields leads to 'tunnel vision' as shown in Figure 13.4. Later in the disease, cone system dysfunction and can manifest as difficulties with colour vision and blurred central vision. Visual field loss may go unnoticed until quite severe loss has occurred and is often mistaken for clumsiness, as individuals bump into and trip over objects. In time, loss of cones, spread throughout the retina, leads to the development of scotomas, and blindspots are apparent on testing. Central vision may be preserved for many years but may deteriorate later in the disease, either due to death of the cones, cystoid macular oedema or cataracts.

By the time symptoms have developed, clinical signs will be evident by retinal examination with dilated pupils, which should be performed by an ophthalmologist specialising in retinal disease. Other non-invasive tests such as fundus autofluorescence utilise the optical features of the retinal pigment epithelium (RPE) the pigmented layer between the neurosensory retina and vascular choroid. Abnormalities in the RPE can be seen early in the disorder and reveal more widespread pathology than is often observed by fundoscopy alone as shown in Figure 13.5. As the retina degenerates further, pigment from the underlying RPE can migrate into the retina giving the characteristic 'bone spicule' appearance which the term 'retinitis pigmentosa' describes. The retinal degeneration causes secondary attenuation of retinal blood vessels and optic disc pallor. Visual field tests can identify visual field defects which occur in the mid or extreme peripheral visual field initially.

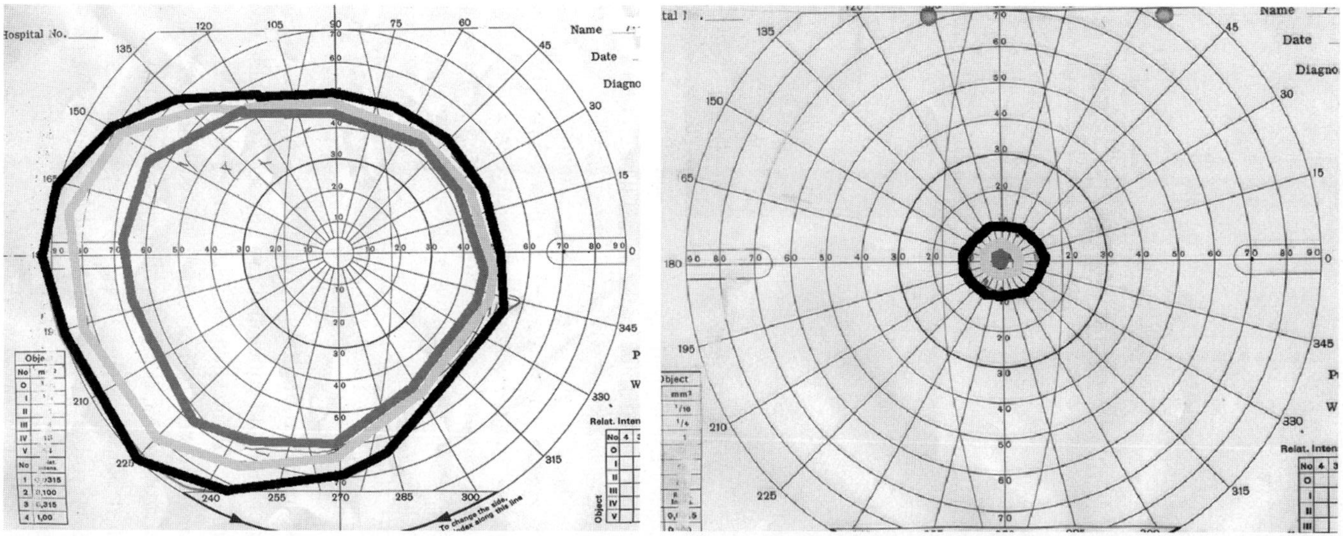

Figure 13.4 Visual fields taken in the left eye of two individuals with (left) a normal visual field; (right) an individual with USH2 due to disease in *USH2A* restricted to a central 10 degrees. The isopters represented are for the different target sizes used to map visual fields. Largest (V4e)—black; Intermediate (II4e)—light grey; Smallest (I4e)—dark grey.

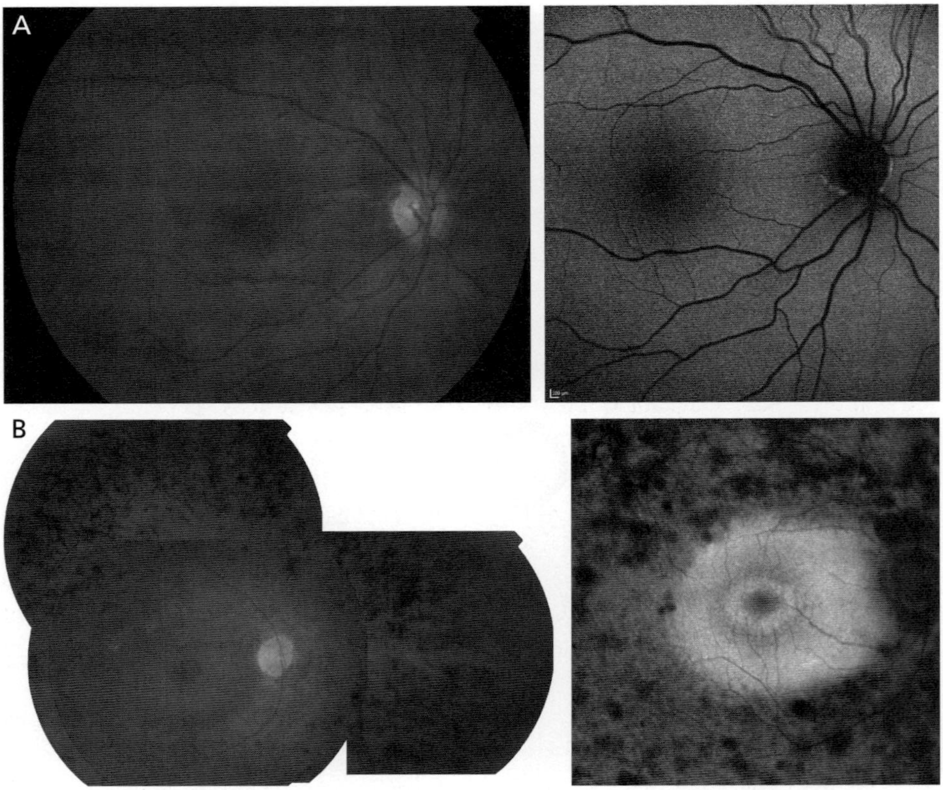

Figure 13.5 (Top row) Retinal photograph (left) as would be seen on dilated retinal examination of a normal right eye and retinal autofluorescence image (right) taken with a scanning laser ophthalmoscope in the same eye. (Bottom row) Retinal photograph (left) as would be seen on dilated retinal examination and retinal autofluorescence image (right) taken with a scanning laser ophthalmoscope in the right eye of a patient with USH2 due to disease in the *USH2A* gene. The retinal photograph on the left demonstrates attenuated retinal vessels, pale optic disc and the classic 'bone spicule' pigmentation, which can be seen in the superior retina, peripheral to the vascular arcades. The autofluorescence image on the right demonstrates patchy hypofluorescent spots (representing areas of RPE dysfunction) that are located within the vascular arcades in areas of non-pigmented retina. Additionally a hyperfluorescent ring can be seen around the fovea.

The most sensitive test of retinal function is the electro-retinogram (ERG) which can interrogate the function of rods and cones and can be used as an objective test of retinal dysfunction. ERGs record electrical responses of different retinal cell types, including rods and cones, bipolar and amacrine cells (inner retina), and ganglion cells. Usually electrodes are positioned on the cornea and the skin below the eye, although in children ERGs may be recorded from skin electrodes.

Responses primarily from rods are elicited by flash ERGs performed on a dark-adapted eye. Activity of cones is elicited by flash ERGs after light adaptation. Bright flashes result in an initial negative deflection (a-wave) the initial part of which is produced by the photoreceptors and then a positive deflection (b-wave) produced by photoreceptors, bipolar, amacrine, and Muller cells. Both amplitudes of the waves and implicit times (peak latencies) can be measured. In RP it is usually the rods that are affected first resulting in a loss of peripheral vision but sometimes vision is lost from the centre outwards. Degeneration is progressive. The ERG may be of use early in the disease, when clinical signs are less obvious; however, when retinal degeneration has been present for many years, the diagnosis of RP can be made on clinical findings alone.

Although the clinical subdivision of Usher syndrome in to its three clinical subtypes is made on audiovestibular features alone, the onset of retinal disease occurs earlier in USH1 compared to USH2. In a national Usher syndrome study performed in the UK we found the median age of onset of visual symptoms was 9 years of age for USH1 ($n = 57$: range 3–14 years) and 17 years of age for USH2 ($n = 129$: range 2–45 years) (Z. Saihan, PhD thesis, London). This highlights the importance of considering the diagnosis of USH2 in adults with congenital hearing loss, as visual symptoms may not occur until late in to adult life.

Observational data from the same study showed good central vision (defined as less than 0.22 logMAR units) was maintained by 50% of the cohort up to the ages of 35 years of age for USH1 and 55 years of age for USH2. Visual field constriction of less than 10 degrees defines the criteria for severe sight registration impairment in the UK and 50% of the USH1 cohort maintained 10 degree fields until their mid-forties whilst 50% of USH2 cohort did better and maintained 10 degree fields through until their mid-fifties.

The age of onset of visual symptoms also varies but is usually apparent by the second decade. Deterioration in vision in Usher type 3 patients is similar to that in other types but at a given age, visual fields in patients with Usher type 3 was fairly similar to that in USH1B patients but significantly poorer than in USH2A (Plantinga et al. 2006).

Range of phenotypes

The clinical subclassification of Usher syndrome particularly that of type 1 Usher, is generally very robust, so screening all genes is unnecessary for molecular diagnosis in most patients. This is borne out by molecular genetic testing; the overwhelming majority of patients with a type 1 Usher phenotype (profound hearing loss and absent vestibular function, together with RP) will have mutations in the associated genes and the same is true for types 2 and 3, although these are sometimes more difficult to distinguish clinically. However, families with atypical phenotypes have been reported at a consistent but low level.

We have previously described two siblings with sector RP, normal language development and moderate-to-severe hearing loss, with vestibular hypofunction (but not absent function) both of whom had compound heterozygous mutations in *USH1C*. Thus both audiovestibular and visual findings are atypical for Usher syndrome caused by this gene (Saihan et al. 2011). Garcia-Garcia and colleagues also found atypical phenotypes associated with mutations in *USH2A*; in one case the patient had profound sensorineural hearing loss, RP by the age of 6 years and vestibular dysfunction at the age of 19 years (more suggestive of USH1) and in a second case hearing loss was diagnosed aged 64 with onset of visual symptoms at 50 years (Garcia-Garcia et al. 2011). Jaijo also described a patient described as a 'double' homozygote for *USH2A* mutations p.Leu555Val and c.1841–2A>G, previously found in linkage disequilibrium. The patient had slowly progressive hearing loss with onset in childhood, left vestibular hypofunction and RP, thus an atypical audiovestibular phenotype; interestingly another patient in their cohort with the same mutations had a phenotype typical for that of USH2 (Jaijo et al. 2010).

Some of this phenotypic variability associated with USH1 and USH2 genes may reflect the nature of the particular mutations; certainly in type 1 Usher syndrome, it is well documented that different mutations may give rise to either typical Usher or non-syndromic deafness (without RP) (Ahmed et al. 2002, 2003; Bork et al. 2001; Liu et al. 1997b; Mburu et al. 2003; Ouyang et al. 2002). The assumed mechanism is that mutations with residual function probably result in the less severe phenotype of non-syndromic hearing loss. Recently, a large study of families ascertained through genotype with mutations in *CDH23* (USH1D), demonstrated an 'allelic hierarchy' which is to say that a DFNB12 mutation in trans with an USH1D mutation appeared to preserve vision and balance in deaf individuals, indicating 'phenotypic dominance' of the DFNB12 allele over the USH1D allele (Schultz et al. 2011). Since *USH2A* has also been reported to be responsible for about 20% of non-syndromic RP (McGee et al. 2010), it remains to be seen whether there is also an allelic hierarchy of mutations giving rise to different, possibly intermediate, phenotypes.

Interesting features of the condition

The classic description of Usher syndrome is one simply of deafness (with or without vestibular dysfunction) and RP. In fact Usher is unusual among the ciliopathies in that only the audiovestibular and visual systems appear to be affected. There are reports in the literature of additional clinical features but these are rare and it is unknown whether they are chance associations or true effects. The most common associated feature is that of psychiatric (psychotic) illness most notably reported in the early series by Hallgren who reported that 23% of 114 patients with Usher had psychotic illness which seems extraordinarily high but which is still occasionally cited (Hallgren, 1958, 1959).

A recent study of mental and behavioural disorders among children found that six of the 26 children were diagnosed with a variety of problems but only one had schizophrenia which was associated with mild learning difficulties; the others had a variety of problems which included atypical autism with or without learning disability, and conduct disorders (Dammeyer, 2012). Other studies have noted much lower rates of psychosis than did Hallgren, but at a significant prevalence (Grondahl & Mjoen, 1986; Nuutila, 1970).

Two explanations have been put forward for this: one is that genes which cause Usher may in some way predispose to behavioural and mental problems, but there is little evidence to support this; the second is that the mental problems are secondary to stress as a result of sensory deprivation. Mental problems appear to be more common in children with more severe symptoms (type 1 Usher) and many case reports associate the onset of mental symptoms co-incident with the worsening of vision on a background of hearing loss, consistent with sensory deprivation. Interestingly it is suggested that autism is over-represented among children with sensory impairment but that it may be difficult to distinguish features of autism or learning difficulty in children with communication disorders and dual sensory impairment. Such symptoms may resolve when appropriate visual, tactile or oral communication needs have been adequately addressed (Dammeyer, 2012).

There are occasional reports of associations more commonly associated with ciliopathies; bronchiectasis, sinusitis and reduced nasal mucociliary clearance was reported in two male siblings with type 1 Usher syndrome, reduced sperm motility (Hunter et al. 1986) and abnormal nasal cilia (Arden & Fox, 1979; Fox et al. 1980) but these do not appear to be consistent observations or must occur at a subclinical level in Usher patients.

There is one association which does have a proven biological basis and that is one of hyperinsulinaemia, enteropathy, renal tubulopathy and Usher syndrome caused by a contiguous gene deletion syndrome involving the *USH1C* gene, which led to the identification of this gene (Bitner-Glindzicz et al. 2000). The same deletion has since been found in other Arab kindreds and is likely to be a founder mutation (Al Mutair et al. 2012).

Diagnosis

The clinical features required to make a clinical diagnosis of Usher syndrome are those of deafness and of RP (or early retinal dystrophy in a child). Most patients can be classified into one of the three types, although as noted earlier in this chapter, atypical cases with mutations in the known genes do exist. However, the majority of people with an atypical presentation do *not* have mutations in the genes described for Usher to date (Le Quesne et al. 2012); they may therefore either have 'Usher-like' recessive disease, mitochondrial mutations (since deafness and RP are common manifestations of mitochondrial DNA mutations) or possibly mutations in more than one gene giving rise to separate phenotypes co-existing in a family (Ebermann et al. 2010; Eisenberger et al. 2012; Mansergh et al. 1999).

In infants presenting with profound congenital sensorineural hearing loss, the possibility of Usher syndrome type 1 should always be carefully considered and must be high on the list of differential diagnoses where there is co-existing motor delay. It is important to make the diagnosis of USH1 in a young child for management reasons; this will ensure that the child is offered timely cochlear implantation in view of the eventual dual sensory impairment, and that parents can be offered genetic counselling when it is still relevant for them.

The diagnosis of Usher syndrome type 2 is a much more difficult one to make clinically; the type of hearing loss is not uncommon and the age at which ERG shows abnormalities is not really known. Given the wide variation in severity of visual symptoms associated with USH2, it is also likely that the age at which abnormalities can be detected on ERG in a child with USH2 is potentially also wide (Pennings et al. 2004; Sandberg et al. 2008). For both USH2 and USH3, diagnosis tends to be made once visual symptoms of reduced visual fields and night blindness become apparent, which is in the teenage years, unless the hearing impaired patient is from an ethnic group in which Usher is common and diagnostic suspicion is high.

Until recently, the genetic heterogeneity underlying Usher syndrome and the large size of many Usher genes meant that molecular testing for diagnostic confirmation and genetic counselling was onerous, time-consuming and very expensive. Usher genotyping microarrays have proven useful as a 'first-pass' screen for Usher mutations but suffer from the limitation that only previously known variants are on the chip; as many Usher mutations are 'private', the sensitivity of the array is limited (Cremers et al. 2007). Molecular diagnostics will improve significantly with the advent of Next Generation sequencing in which multiple genes can be screened quickly and in parallel, reducing costs. This will allow earlier diagnosis in children at a time when genetic information may be helpful for counselling parents. It will also allow the presymptomatic diagnosis of USH2 and USH3 which may be beneficial in terms of preventive or restorative visual therapies which are being developed and trialled. However, since many of the Usher genes can also cause non-syndromic deafness (or non-syndromic RP in the case of *USH2A*) and genotype–phenotype correlations are imperfect, and interpreting whether a particular genotype will result in Usher syndrome may be problematic and possibly lead to overdiagnosis.

Furthermore, Next Generation sequencing is not yet sufficiently sensitive to detect single or multiple exon intragenic deletions which occur at a significant frequency in some of the Usher genes (Le Guedard et al. 2007). Currently, they still need to be specifically sought by techniques such as multiplex ligation-dependent probe amplification (**MLPA**) or comparative genomic hybridisation (CGH) microarray, and splicing mutations, which are often non-coding or whose effect may be difficult to predict with certainty, require fresh samples for RNA analysis (Vache et al. 2010).

Genetics

To date, mutations in ten genes have been shown to cause Usher syndrome in humans, with a further gene, PDZD7 having a potential role as a disease modifier (Table 13.1). For type 1 Usher syndrome, the most commonly mutated gene is *MYO7A* which accounts for about 50% of cases in most populations studied; *CDH23* and *PCDH15* are frequently the next most commonly mutated accounting for around 10% each. *USH1C* also accounts for a smaller proportion of cases, although in some populations, such as French Acadians or Quebecois, it may be the major USH1 gene. Mutations in *SANS* and *CIB2* appear to account for very few families with USH1.

For cases with type 2 Usher, about 80% of people have disease caused by *USH2A*, with around 10% or less caused by *GPR98*, and only a handful of reported cases to date caused by mutations in *WHRN/DFNB31*. Molecularly proven cases of USH3 appear to be rare in many populations with the exception of Finland where this is the most common type and among Ashkenazi Jews in whom it also appears to be a significant cause.

In most molecularly screened cohorts, there still remain a significant number of families with heterozygous mutations despite full sequencing of all known coding exons. This,

Table 13.1 Usher genes and their encoded proteins

Usher	Approved gene symbol	Protein	Protein type
USH1B	*MYO7A*	Myosin VIIA	Motor protein
USH1C	*USH1C*	Harmonin	Scaffold protein
USH1D	*CDH23*	Cadherin 23	Cell–cell adhesion protein
USH1E	Unknown		
USH1F	*PCDH15*	Protocadherin 15	Cell–cell adhesion protein
USH1G	*USH1G/SANS*	SANS	Scaffold protein
USH1H	Unknown		
USH1J	*CIB2*	CIB2	Ca^{2+}–integrin binding protein
USH2A	*USH2A*	Usherin	Transmembrane protein
USH2B	UNKNOWN		
USH2C	*GPR98*	VLGR1B (GPR98)	G protein coupled receptor
USH2D	*DFNB31*	Whirlin	Elongation protein
USH3A	*USH3A*	Clarin1	Transmembrane protein

together with the polymorphic nature of the genes, has led some to speculate that there may be 'digenic' inheritance in Usher syndrome, as has been proposed in other retinal diseases and ciliopathies (Bonnet et al. 2011; Vozzi et al. 2011). However, stringent analysis of novel variants and assignment of pathogenic status reveals very few, if any convincing cases with bi-allelic truncating mutations in more than one Usher gene in humans (Le Quesne et al. 2012).

Underlying ciliary defect and its physiological effects

All of the Usher proteins interact with at least one other in a network which has become known as the Usher 'interactome': in the assembly of this network, harmonin, SANS and whirlin are essential scaffold proteins, with harmonin being able to bind all of the other USH1 proteins by its PDZ domains. Interactions occur at several locations and functional points in both the sensory cells of the cochlea (hair cells) and in the photoreceptors of the retina (Adato et al. 2005a,b; Boeda et al. 2002; Reiners et al. 2005a,b; Siemens et al. 2002; Weil et al. 2003). For example, in the cochlea, the USH1 protein interactions are crucial in both the developing hair bundles as well as in the function of the process of mechanotransduction (MET) in mature hair cells, and in the ribbon synapses (reviewed in Kremer et al. 2006 and Pan & Zhang, 2012).

In order to understand the effects of mutations which disrupt the Usher 'interactome' a little should be outlined about the structure of the cochlea and particularly the sensory hair cells within it. In brief, when sound waves enter the ear canal they cause the eardrum to vibrate, movement which is amplified by the ossicles, and transmitted through the oval window into the fluid filled cochlea. The cochlea is divided lengthways by the basilar membrane, which vibrates as a result of the travelling sound wave. The sensory hair cells sit on the basilar membrane. They are so-called because of the hair-like projections on their apical surfaces, stereocilia, filled with actin, and arranged in step-like rows according to height. These are not true cilia but specialised microvilli filled with actin. The tallest of these is embedded in the overlying, relatively immobile, tectorial membrane. Adjacent stereocilia from neighbouring rows are tethered together as a bundle by a series of extracellular links, top connectors, shaft connectors and ankle links. Tip links connect the tip of each stereocilium to the side of the taller stereocilium next to it, and are thought to be indirectly coupled with the MET channels at the tip of the smaller stereocilium. When the basilar membrane moves, the hair bundle is deflected. This results in a change of tension on the tip links which physically opens the cation MET channels, allowing flow of potassium from the K^+-rich endolymph into the hair cells leading to depolarisation and neurotransmitter release from ribbon synapses at the base of the cell.

In developing hair bundles, cadherin 23 and protocadherin 15 are thought to interact to form transient lateral links between adjacent stereocilia and between the stereocilia and the transient structure, the kinocilium, a true ciliary structure (Goodyear et al. 2005; Lagziel et al. 2005; Michel et al. 2005; Webb et al. 2011). These lateral links are themselves anchored to the actin cytoskeleton, likely through harmonin b isoforms and myosin VIIA, and are necessary for bundle cohesion and orientation (Boeda et al. 2002). In the absence of USH1 proteins the developing hair bundles appear disoriented and disorganised (Lefevre et al. 2008). Similarly, USH2 proteins, usherin and VLGR1/GPR98, together with whirlin, comprise transient ankle links between the stereocilia (Adato et al. 2005a; Lefevre et al. 2008; McGee et al. 2006; Michalski et al. 2007).

Regarding MET function, cadherin 23 and protocadherin 15 interact to form the tip links of the stereocilia which physically open the MET channels in response to sound

Figure 13.6 Schematic representation of the protein interactions involved in the ciliary/periciliary protein network of photoreceptor cells. (A) The membranes of the connecting cilium and the inner segment are connected by ectodomains of USH2A isoform b and VLGR1b; they are anchored via whirlin to the USH protein network in the cytoplasm of the adjacent domains. Known direct protein–protein interactions are indicated by solid lines, and dotted lines indicate putative interactions. The asterisk indicates the inner segment collar. (B) Longitudinal section and (C) cross-section of the ciliary/periciliary USH protein network in a mammalian rod photoreceptor cell. Cytoplasmic dynein mediates vesicle transport along microtubules to the apical inner segment collar (asterisk). Docking and fusion membrane sites (indicated in red) are predefined by the USH protein network arrangement.

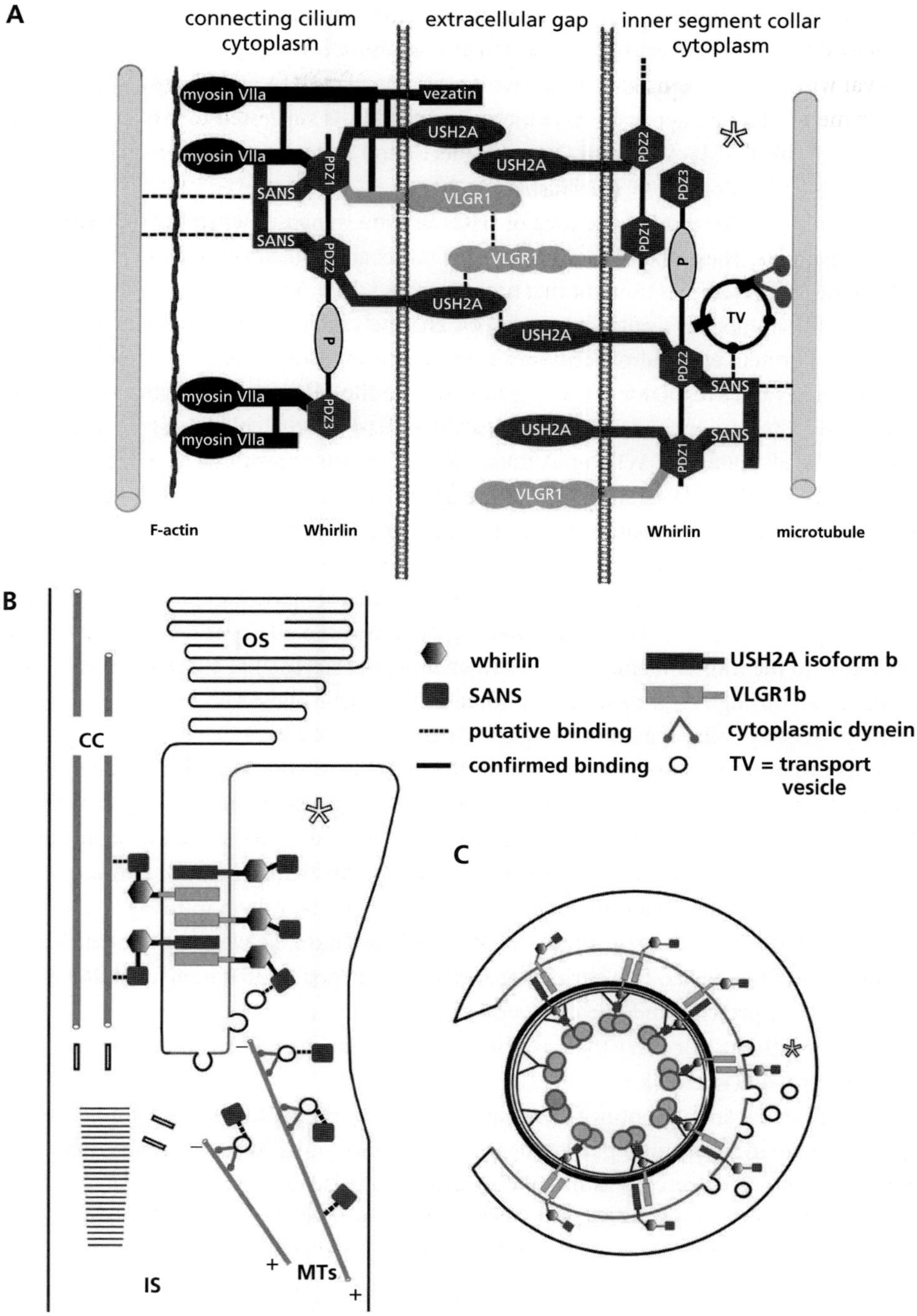

Reprinted from Tina Maerker et al., A novel Usher protein network at the periciliary reloading point between molecular transport machineries in vertebrate photoreceptor cells, *Human Molecular Genetics*, Volume 17, Issue 1, pp. 71–86, Copyright © The Author 2007, by permission of Oxford University.

stimulation (Alagramam et al. 2011; Kazmierczak et al. 2007; Siemens et al. 2004); the intracellular part of cadherin 23 interacts with harmonin b which in turn is likely to connect the tip link to the central actin filaments. Also myosin VIIA and SANS are part of the mechanotransduction apparatus in which myosin VIIA is suggested to influence tip link tension. SANS also plays a role in actin polymerisation at the tip of the stereocilia. (Boeda et al. 2002; Grillet et al. 2009; Michalski et al. 2009).

Harmonin also localises at the base of IHCs in some synapses where it associates with $Ca_v1.3$ channels. These channels mediate Ca^{2+} influx and exocytosis and are required for normal hearing. Here it is thought that harmonin tags $Ca_v1.3$ channels for ubiquitination, reducing the number of available presynaptic channels in IHCs. So, as well as acting as a scaffolding protein elsewhere, harmonin may also be an integral component of ion channel complexes and a regulator of Ca^{2+} signalling in auditory hair cells (Gregory et al. 2011).

Similarly the Usher protein network is critical to normal photoreceptor function (Maerker et al. 2008; van Wijk et al. 2006). Usher proteins form macromolecular complexes in the periciliary collar region of the apical inner segment and the adjacent connecting cilium of mammalian photoreceptors (Figure 13.6). Here, whirlin and SANS are thought to act as the major scaffold proteins. It is suggested that the ectodomains of VLGR1b and Usherin isoform b, form homomeric or heteromeric complexes between the periciliary membrane of the apical inner segment and the adjacent connecting cilium, analogous to the transient ankle links between inner ear stereocilia. These complexes may be anchored through their cytoplasmic domains by whirlin and SANS.

The periciliary collar region is thought to be similar to the periciliary ridge complex of amphibian photoreceptors. The grooves of the periciliary ridge complex are specialised membrane subdomains which act as docking sites for inner segment transport vesicles, transporting cargo by cytoplasmic dynein along microtubules to the photoreceptor outer segments from the trans-Golgi network of the inner segments. Thus, in common with other 'ciliopathies', the Usher interactome may fulfil a critical role in intraflagellar/microtubule transport of proteins through the connecting cilium, as well as a structural role in cilia, or modified cilia. It is perhaps therefore not surprising that a physical link exists between Usher proteins and one involved in a classical ciliary disorder, Leber's congenital amaurosis, with the discovery that usherin and lebercilin, are linked via a ninein-like centrosomal protein (van Wijk et al. 2009).

USH1 proteins are present in the calyceal processes, microvilli-like structures that are known to extend from apical part of the inner segments to the basal half of the outer segments of photoreceptor cells in human, macaque and also *Xenopus*. These calyceal processes are thought to form an adhesion belt essential for photoreceptor maintenance and a role in the formation of outer segments has been proposed (Sahly et al. 2012).

The Usher proteins also localise in the ribbon synapses of the photoreceptor cells, where they may play a role in trafficking of synaptic vesicles (Reiners et al. 2003, 2005a). A third site of expression of myosin VIIA in particular is in the retinal pigment epithelium. Here, absence of the protein causes a decrease in phagocytosis of photoreceptor outer segment discs and intracellular transport of melanosomes, as well as reduced light-dependent translocation of RPE65, an enzyme required in the retinoid pathway (Gibbs et al. 2003; Liu et al. 1997a, 1998; Lopes et al. 2011).

Clinical management

Management of Usher syndrome currently involves general 'rehabilitative strategies' for both hearing loss and visual loss, as well as supportive counselling and practical advice to cope with the dual sensory handicap.

Audiology

For profound hearing losses such as those seen in USH1, the most effective management strategy in a person who will develop RP, is cochlear implantation, whereas those with USH2 and USH3 benefit from hearing aids. Hearing aids cannot provide sufficient amplification for those with USH1 and, in the past, the overwhelming majority of these children without cochlear implants have learned to communicate using sign language. As vision decreases, patients may need to develop tactile communication. Cochlear implantation may also be useful in older patients with USH3 in whom deterioration in hearing means that they no longer gain benefit from conventional amplification (Pietola et al. 2012).

The norm is now to offer simultaneous bilateral implantation in young children, preferably before the age of 2 years, and to offer a second implant in those who received a unilateral implant initially. In USH1 in the absence of any other clinical or intellectual problems outcomes are good, with good open set speech recognition scores (understanding speech without lip reading), especially if children are implanted early (Blanchet et al. 2007; Liu et al. 2008; Pennings et al. 2006) similar to children with deafness due to other causes. Given the very different potential outcomes with different modes of rehabilitation and age at implantation, it is clear that early diagnosis in children with type 1 Usher is of paramount importance and that an index of suspicion among physicians looking after deaf infants is key. As previously mentioned, profound deafness together with delay in motor development should prompt automatic consideration of an ERG. Whilst ERGs in young children require expertise and time and in many centres entail general anaesthesia and the use of corneal electrodes, it is possible to perform ERGs in children using skin electrodes (Fulton et al. 2006; Gore et al. 2010). Suspiciously low amplitudes for laboratory norms, combined with serial reduction in amplitude several months later is strongly suspicious of Usher in this context. However, as this is not universal practice it is fortunate that currently many children with profound deafness, often diagnosed by Newborn Hearing Screening, are assessed for cochlear implants regardless of the aetiology of their deafness.

Patients should be advised that the combination of hearing loss, tunnel vision, defective dark adaptation and in some cases poor balance, may increase their risk of accidents. Patients with USH1 should be advised not to swim and especially dive under water unaccompanied as their lack of balance can lead to disorientation when swimming underwater. Many patients with Usher can learn to drive (even those with USH1) but progressive loss of peripheral vision means that this becomes unsafe as their symptoms worsen.

Vision

Whilst RP caused by Usher is currently untreatable, routine ophthalmic review is advised in order to monitor for potentially treatable complications such as cataract or cystoid macula oedema. Cataracts (lens opacities) may be visually significant and can be managed surgically. Central retinal dysfunction occurs later in the disease resulting in reduced visual acuity. Cystic spaces at the macular, termed cystoid macular oedema can be seen and demonstrated using non-invasive retinal imaging using optical coherence tomography see Figure 13.7. In some cases this can be treated with topical or oral carbonic anhydrase inhibitors, which may have a beneficial effect on visual function and retinal anatomy (Genead & Fishman, 2010).

Sequential visual acuity and visual field measurements provide useful indicators of visual function, whilst non-invasive imaging using fundus autofluorescence imaging (Figure 13.5) enables documentation of progression. Features such as hyperfluorescent rings have been identified in patients with RP (Figure 13.3), which are not visible on routine fundoscopy. These rings have been shown to constrict over time and encircle areas of photoreceptor integrity and photopic retinal function. These hyperfluorescent rings may reach a critical minimum before disappearing, at which stage central visual loss occurs (Robson et al. 2011, 2012). Clearly it will be of great importance to have objective

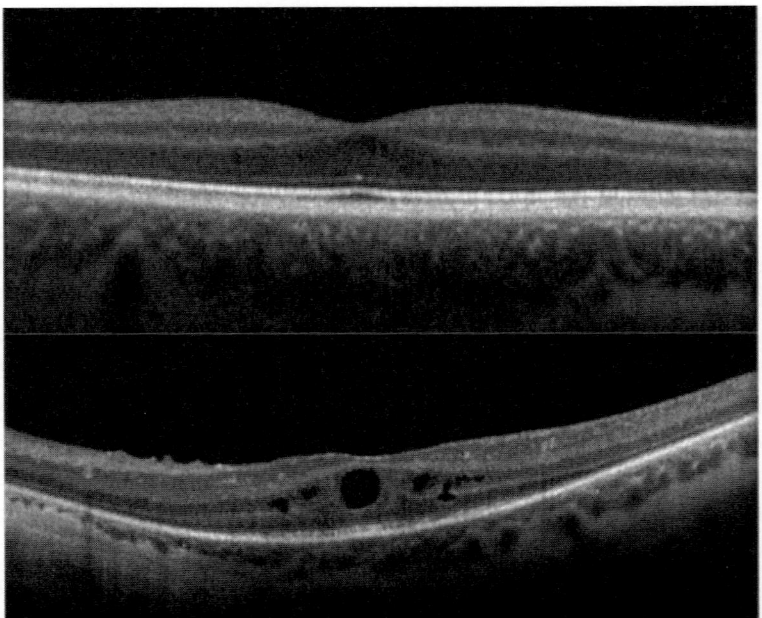

Figure 13.7 Optical coherence tomography (OCT) scan through a normal macula (above) demonstrating normal retinal layers and a central foveal depression; and a 19 year old with USH2 due to disease in the *USH2A* gene (below) showing cystic spaces within the retina centrally and retinal atrophy, seen as thinning of the retinal layers either side of the fovea.

measures of disease progression or quiescence to assess the response to future therapies, some of which have reached clinical trial stage.

Genetic counselling

Usher syndrome is inherited as an autosomal recessive condition, which means that risk of recurrence is high (25%) for parents who have an affected child. Because diagnosis is frequently made late in USH2 and USH3 patients, genetic counselling is of limited value. However, in cases of USH1 where it is possible to make early diagnoses, parents may benefit from genetic counselling and prenatal diagnosis, if the mutation in the family can be identified, and if this is culturally acceptable to the couple.

Future treatments

A number of avenues are being investigated for the treatment of Usher syndrome and other retinal dystrophies, including retinal implants, stem cell therapy, gene therapy, and pharmacological agents. One of the drawbacks to treatment of this condition is the lack of suitable animal models for Usher syndrome; although many knockout and naturally occurring mouse models exist, they do not manifest retinal degeneration as seen in humans under normal laboratory conditions.

However, in the wake of successful gene therapy trials for the treatment of Leber congenital amaurosis (Bainbridge et al. 2008; Maguire et al. 2008), the first clinical trial for gene therapy in patients with Usher syndrome is currently under way using a lentiviral vector to deliver *MYO7A*. The advantage of lentiviral vectors is that they can deliver large genes such as those causing Usher syndrome, which previously hindered investigation. Results in the MYO7A-null mouse have shown that defects in phagosome digestion and melanosome motility, used as measures of gene delivery in primary cultures of RPE cells, were corrected, and *in vivo,* melanosomes in RPE cells were correctly located, and opsin was correctly transported through the connecting cilium of photoreceptor cells (Hashimoto et al. 2007). Nanoparticles are a novel method of gene delivery, which have been used to deliver very large genes in mouse models of Stargardts disease and Leber's congenital amaurosis (Han et al. 2012; Koirala et al. 2013) and are being investigated for the delivery of USH2A. More 'mutation-specific' approaches using synthetic aminoglycosides to suppress nonsense mutations in *USH1C* (Goldmann et al. 2012) and *PCDH15* (Rebibo-Sabbah et al. 2007) as well as antisense oligonucleotides to suppress the effect of splice mutations, are separate foci of treatment-based research. More generalisable therapies, such as stem cell transplantation, have shown promise in a variety of animal models of retinal dystrophy (Barber et al. 2013; Pearson et al. 2012) and retinal implants, potentially applicable to a wide range of retinal disorders, have enabled patients with inherited retinal dystrophies to be able to locate and recognise bright objects on a dark table, identify geometric patterns, and to read large letters (Zrenner et al. 2011).

In conclusion, the diagnosis of Usher syndrome remains a devastating one for individuals with an initial diagnosis of non-syndromic hearing loss, who had hoped to lead a near-normal life. The knowledge that they or their children will become severely visually impaired or even legally blind is difficult and often life-changing. Early diagnosis, especially in children with type 1 Usher, is particularly important whilst there is a window for successful cochlear implantation and will become increasingly important in types 2 and 3 when clinical trials for therapies to conserve vision are realised.

Acknowledgements

We would like to thank Dr Hannie Kremer for reading the section on 'Underlying Ciliary Defect' and for her helpful comments.

References

Adato, A., Lefevre, G., Delprat, B., Michel, V., Michalski, N., Chardenoux, S., et al. 2005a. Usherin, the defective protein in Usher syndrome type IIA, is likely to be a component of interstereocilia ankle links in the inner ear sensory cells. *Hum Mol Genet* **14**, 3921–3932.

Adato, A., Michel, V., Kikkawa, Y., Reiners, J., Alagramam, K. N., Weil, D., et al. 2005b. Interactions in the network of Usher syndrome type 1 proteins. *Hum Mol Genet* **14**, 347–356.

Ahmed, Z. M., Riazuddin, S., Ahmad, J., Bernstein, S. L., Guo, Y., Sabar, M. F., et al. 2003. PCDH15 is expressed in the neurosensory epithelium of the eye and ear and mutant alleles are responsible for both USH1F and DFNB23. *Hum Mol Genet* **12**, 3215–3223.

Ahmed, Z. M., Smith, T. N., Riazuddin, S., Makishima, T., Ghosh, M., Bokhari, S., et al. 2002. Nonsyndromic recessive deafness DFNB18 and Usher syndrome type IC are allelic mutations of USHIC. *Hum Genet* **110**, 527–531.

Al Mutair, A. N., Brusgaard, K., Bin-Abbas, B., Hussain, K., Felimban, N., Al, S. A., et al. 2013. Heterogeneity in phenotype of Usher–congenital hyperinsulinism syndrome: hearing loss, retinitis pigmentosa, and hyperinsulinemic hypoglycemia ranging from severe to mild with conversion to diabetes. *Diabetes Care*. **36**, 557–561.

Alagramam, K. N., Goodyear, R. J., Geng, R., Furness, D. N., van Aken, A. F., Marcotti, W., et al. 2011. Mutations in protocadherin 15 and cadherin 23 affect tip links and mechanotransduction in mammalian sensory hair cells. *PLoS One* **6**, e19183.

Arden, G. B. & Fox, B. 1979. Increased incidence of abnormal nasal cilia in patients with retinitis pigmentosa. *Nature* **279**, 534–536.

Auslender, N., Bandah, D., Rizel, L., Behar, D.M., Shohat, M., Banin, E., et al. 2008. Four USH2A founder mutations underlie the majority of Usher syndrome type 2 cases among non-Ashkenazi Jews. *Genet Test* **12**, 289–294.

Bainbridge, J. W., Smith, A. J., Barker, S. S., Robbie, S., Henderson, R., Balaggan, K., et al. 2008. Effect of gene therapy on visual function in Leber's congenital amaurosis. *N Engl J Med* **358**, 2231–2239.

Barber, A. C., Hippert, C., Duran, Y., West, E. L., Bainbridge, J. W., Warre-Cornish, K., et al. 2013. Repair of the degenerate retina by photoreceptor transplantation. *Proc Natl Acad Sci U.S.A.* **110**, 354–359.

Bitner-Glindzicz, M., Lindley, K. J., Rutland, P., Blaydon, D., Smith, V. V., Milla, P. J., et al. 2000. A recessive contiguous gene deletion causing infantile hyperinsulinism, enteropathy and deafness identifies the Usher type 1C gene. *Nat Genet* **26**, 56–60.

Blanchet, C., Roux, A. F., Hamel, C., Ben, S. S., Artieres, F., Faugere, V., et al. 2007. [Usher type I syndrome in children: Genotype/phenotype correlation and cochlear implant benefits). *Rev Laryngol Otol Rhinol (Bord.)* **128**, 137–143.

Boeda, B., El-Amraoui, A., Bahloul, A., Goodyear, R., Daviet, L., Blanchard, S., et al. 2002. Myosin VIIa, harmonin and cadherin 23, three Usher I gene products that cooperate to shape the sensory hair cell bundle. *EMBO J* **21**, 6689–6699.

Bonnet, C., Grati, M., Marlin, S., Levilliers, J., Hardelin, J. P., Parodi, M., et al. 2011. Complete exon sequencing of all known Usher syndrome genes greatly improves molecular diagnosis. *Orphanet J Rare Dis* **6**, 21.

Bork, J. M., Peters, L. M., Riazuddin, S., Bernstein, S. L., Ahmed, Z. M., Ness, S. L., et al. 2001. Usher syndrome 1D and nonsyndromic autosomal recessive deafness DFNB12 are caused by allelic mutations of the novel cadherin-like gene CDH23. *Am.J Hum Genet* **68**, 26–37.

Boughman, J. A., Vernon, M. & Shaver, K. A. 1983. Usher syndrome: Definition and estimate of prevalence from two high-risk populations. *J Chronic Dis* **36**, 595–603.

Brownstein, Z., Ben-Yosef, T., Dagan, O., Frydman, M., Abeliovich, D., Sagi, M., et al. 2004. The R245X mutation of PCDH15 in Ashkenazi Jewish children diagnosed with nonsyndromic hearing loss foreshadows retinitis pigmentosa. *Pediatr Res* **55**, 995–1000.

Cremers, F. P., Kimberling, W. J., Kulm, M., de Brouwer, A. P., van, W. E., Te, B. H., et al. 2007. Development of a genotyping microarray for Usher syndrome. *J Med Genet* **44**, 153–160.

Dammeyer, J. 2012. Children with Usher syndrome: Mental and behavioral disorders. *Behav Brain Funct* **8**, 16.

Ebermann, I., Koenekoop, R. K., Lopez, I., Bou-Khzam, L., Pigeon, R. & Bolz, H. J. 2009. An USH2A founder mutation is the major cause of Usher syndrome type 2 in Canadians of French origin and confirms common roots of Quebecois and Acadians. *Eur J Hum Genet* **17**, 80–84.

Ebermann, I., Lopez, I., Bitner-Glindzicz, M., Brown, C., Koenekoop, R. K. & Bolz, H. J. 2007. Deafblindness in French Canadians from Quebec: A predominant founder mutation in the USH1C gene provides the first genetic link with the Acadian population. *Genome Biol* **8**, R47.

Ebermann, I., Phillips, J. B., Liebau, M. C., Koenekoop, R. K., Schermer, B., Lopez, I., et al. 2010. PDZD7 is a modifier of retinal disease and a contributor to digenic Usher syndrome. *J Clin Invest* **120**, 1812–1823.

Eisenberger, T., Slim, R., Mansour, A., Nauck, M., Nurnberg, G., Nurnberg, P., et al. 2012. Targeted next-generation sequencing identifies a homozygous nonsense mutation in ABHD12, the gene underlying PHARC, in a family clinically diagnosed with Usher syndrome type 3. *Orphanet J Rare Dis* **7**, 59.

Fox, B., Bull, T. B. & Arden, G. B. 1980. Variations in the ultrastructure of human nasal cilia including abnormalities found in retinitis pigmentosa. *J Clin Pathol* **33**, 327–335.

Fulton, A. B., Brecelj, J., Lorenz, B., Moskowitz, A., Thompson, D. & Westall, C. A. 2006. Pediatric clinical visual electrophysiology: A survey of actual practice. *Doc Ophthalmol* **113**, 193–204.

Garcia-Garcia, G., Aparisi, M. J., Jaijo, T., Rodrigo, R., Leon, A. M., Avila-Fernandez, A., et al. 2011. Mutational screening of the USH2A gene in Spanish USH patients reveals 23 novel pathogenic mutations. *Orphanet J Rare Dis* **6**, 65.

Genead, M. A. & Fishman, G. A. 2010. Efficacy of sustained topical dorzolamide therapy for cystic macular lesions in patients with retinitis pigmentosa and usher syndrome. *Arch Ophthalmol* **128**, 1146–1150.

Gibbs, D., Kitamoto, J. & Williams, D. S. 2003. Abnormal phagocytosis by retinal pigmented epithelium that lacks myosin VIIa, the Usher syndrome 1B protein. *Proc Natl Acad Sci U.S.A.* **100**, 6481–6486.

Goldmann, T., Overlack, N., Moller, F., Belakhov, V., van Wyk, M., Baasov, T., et al. 2012. A comparative evaluation of NB30, NB54 and PTC124 in translational read-through efficacy for treatment of an USH1C nonsense mutation. *EMBO Mol Med* **4**, 1186–1199.

Goodyear, R. J., Marcotti, W., Kros, C. J. & Richardson, G. P. 2005. Development and properties of stereociliary link types in hair cells of the mouse cochlea. *J.Comp Neurol.* **485**, 75–85.

Gore, S., Carr, L., Moore, A. & Thompson, D. 2010. Hereditary primary lateral sclerosis with cone dysfunction. *Ophthalmic Genet* **31**, 221–226.

Gregory, F. D., Bryan, K. E., Pangrsic, T., Calin-Jageman, I. E., Moser, T. & Lee, A. 2011. Harmonin inhibits presynaptic Cav1.3 Ca(2)(+) channels in mouse inner hair cells. *Nat Neurosci* **14**, 1109–1111.

Grillet, N., Xiong, W., Reynolds, A., Kazmierczak, P., Sato, T., Lillo, C., et al. 2009. Harmonin mutations cause mechanotransduction defects in cochlear hair cells. *Neuron*, **62**, 375–387.

Grondahl, J. 1987. Estimation of prognosis and prevalence of retinitis pigmentosa and Usher syndrome in Norway. *Clin Genet* **31**, 255–264.

Grondahl, J. & Mjoen, S. 1986. Usher syndrome in four Norwegian counties. *Clin Genet* **30**, 14–28.

Hallgren, B. 1958. Retinitis pigmentosa in combination with congenital deafness and vestibulocerebellar ataxia; with psychiatric abnormality in some cases; a clinical and genetic study. *Acta Genet Stat Med* **8**, 97–104.

Hallgren, B. 1959. Retinitis pigmentosa combined with congenital deafness; with vestibulo-cerebellar ataxia and mental abnormality in a proportion of cases: A clinical and genetico-statistical study. *Acta Psychiatr Scand Suppl* **34**, 1–101.

Han, Z., Conley, S. M., Makkia, R. S., Cooper, M. J. & Naash, M. I. 2012. DNA nanoparticle-mediated ABCA4 delivery rescues Stargardt dystrophy in mice. *J Clin Invest* **122**, 3221–3226.

Hashimoto, T., Gibbs, D., Lillo, C., Azarian, S. M., Legacki, E., Zhang, X. M., et al. 2007. Lentiviral gene replacement therapy of retinas in a mouse model for Usher syndrome type 1B. *Gene Ther* **14**, 584–594.

Hunter, D. G., Fishman, G. A., Mehta, R. S. & Kretzer, F. L. 1986. Abnormal sperm and photoreceptor axonemes in Usher's syndrome. *Arch Ophthalmol* **104**, 385–389.

Jaijo, T., Aller, E., Garcia-Garcia, G., Aparisi, M. J., Bernal, S., vila-Fernandez, A., et al. 2010. Microarray-based mutation analysis of 183 Spanish families with Usher syndrome. *Invest Ophthalmol Vis Sci* **51**, 1311–1317.

Joensuu, T., Hamalainen, R., Yuan, B., Johnson, C., Tegelberg, S., Gasparini, P., et al. 2001. Mutations in a novel gene with transmembrane domains underlie Usher syndrome type 3. *Am J Hum Genet* **69**, 673–684.

Kazmierczak, P., Sakaguchi, H., Tokita, J., Wilson-Kubalek, E. M., Milligan, R. A., Muller, U., et al. 2007. Cadherin 23 and protocadherin 15 interact to form tip-link filaments in sensory hair cells. *Nature* **449**, 87–91.

Kimberling, W. J., Hildebrand M. S., Shearer A. E., Jensen M. L., Halder J. A., Trzupek K., et al. 2010. Frequency of Usher syndrome in two pediatric populations: Implications for genetic screening of deaf and hard of hearing children. *Genet Med* **12**, 512–516.

Koirala, A., Makkia, R. S., Conley, S. M., Cooper, M. J. & Naash, M. I. 2013. S/MAR-containing DNA nanoparticles promote persistent RPE gene expression and improvement in RPE65-associated LCA. *Hum Mol Genet.* **22**, 1632–1642.

Kremer, H., van Wijk, E., Marker, T., Wolfrum, U. & Roepman, R. 2006. Usher syndrome: Molecular links of pathogenesis, proteins and pathways. *Hum Mol Genet* **15**(Spec No 2), R262–270.

Lagziel, A., Ahmed, Z. M., Schultz, J. M., Morell, R. J., Belyantseva, I. A. & Friedman, T. B. 2005. Spatiotemporal pattern and isoforms of cadherin 23 in wild type and waltzer mice during inner ear hair cell development. *Dev Biol* **280**, 295–306.

Le Guedard, S., Faugere, V., Malcolm, S., Claustres, M. & Roux, A. F. 2007. Large genomic rearrangements within the PCDH15 gene are a significant cause of USH1F syndrome. *Mol Vis* **13**, 102–107.

Le Quesne, S. P., Saihan, Z., Rangesh, N., Steele-Stallard, H. B., Ambrose, J., Coffey, A., et al. 2012. Comprehensive sequence analysis of nine Usher syndrome genes in the UK National Collaborative Usher Study. *J Med Genet* **49**, 27–36.

Lefevre, G., Michel, V., Weil, D., Lepelletier, L., Bizard, E., Wolfrum, U., et al. 2008. A core cochlear phenotype in USH1 mouse mutants implicates fibrous links of the hair bundle in its cohesion, orientation and differential growth. *Development* **135**, 1427–1437.

Liebreich, R. 1861. Abkunft aus Ehen unter Blutsverwandten als Grund von Retinitis pigmentosa. *Dtsch Klin* **13**, 53.

Liu, X., Ondek, B. & Williams, D. S. 1998. Mutant myosin VIIa causes defective melanosome distribution in the RPE of shaker-1 mice. *Nat Genet* **19**, 117–118.

Liu, X., Vansant, G., Udovichenko, I. P., Wolfrum, U. & Williams, D. S. 1997a. Myosin VIIa, the product of the Usher 1B syndrome gene, is concentrated in the connecting cilia of photoreceptor cells. *Cell Motil Cytoskeleton* **37**, 240–252.

Liu, X. Z., Angeli, S. I., Rajput, K., Yan, D., Hodges, A. V., Eshraghi, A., et al. 2008. Cochlear implantation in individuals with Usher type 1 syndrome. *Int J Pediatr Otorhinolaryngol* **72**, 841–847.

Liu, X. Z., Walsh, J., Mburu, P., Kendrick-Jones, J., Cope, M. J., Steel, K. P., et al. 1997b. Mutations in the myosin VIIA gene cause non-syndromic recessive deafness. *Nat Genet* **16**, 188–190.

Lopes, V. S., Gibbs, D., Libby, R. T., Aleman, T. S., Welch, D. L., Lillo, C., et al. 2011. The Usher 1B protein, MYO7A, is required for normal localization and function of the visual retinoid cycle enzyme, RPE65. *Hum Mol Genet* **20**, 2560–2570.

Maerker, T., van Wijk, E., Overlack, N., Kersten, F. F., McGee, J., Goldmann, T., et al. 2008. A novel Usher protein network at the periciliary reloading point between molecular transport machineries in vertebrate photoreceptor cells. *Hum Mol Genet* **17**, 71–86.

Maguire, A. M., Simonelli, F., Pierce, E. A., Pugh Jr, E. N., Mingozzi, F., Bennicelli, J., et al. 2008. Safety and efficacy of gene transfer for Leber's congenital amaurosis. *N Engl J Med* **358**, 2240–2248.

Mansergh, F. C., Millington-Ward, S., Kennan, A., Kiang, A. S., Humphries, M., Farrar, G. J., et al. 1999. Retinitis pigmentosa and progressive sensorineural hearing loss caused by a C12258A mutation in the mitochondrial MTTS2 gene. *Am J Hum Genet* **64**, 971–985.

Mburu, P., Mustapha, M., Varela, A., Weil, D., El-Amraoui, A., Holme, R. H., et al. 2003. Defects in whirlin, a PDZ domain molecule involved in stereocilia elongation, cause deafness in the whirler mouse and families with DFNB31. *Nat Genet* **34**, 421–428.

McGee, J., Goodyear, R. J., McMillan, D. R., Stauffer, E. A., Holt, J. R., Locke, K. G., et al. 2006. The very large G-protein-coupled receptor VLGR1: A component of the ankle link complex required for the normal development of auditory hair bundles. *J Neurosci* **26**, 6543–6553.

McGee, T. L., Seyedahmadi, B. J., Sweeney, M. O., Dryja, T. P. & Berson, E. L. 2010. Novel mutations in the long isoform of the USH2A gene in patients with Usher syndrome type II or non-syndromic retinitis pigmentosa. *J Med Genet* **47**, 499–506.

Michalski, N., Michel, V., Bahloul, A., Lefevre, G., Barral, J., Yagi, H., et al. 2007. Molecular characterization of the ankle-link complex in cochlear hair cells and its role in the hair bundle functioning. *J Neurosci* **27**, 6478–6488.

Michalski, N., Michel, V., Caberlotto, E., Lefevre, G. M., van Aken, A. F., Tinevez, J. Y., et al. 2009. Harmonin-b, an actin-binding scaffold protein, is involved in the adaptation of mechanoelectrical transduction by sensory hair cells. *Pflugers Arch* **459**, 115–130.

Michel, V., Goodyear, R. J., Weil, D., Marcotti, W., Perfettini, I., Wolfrum, U., et al. 2005. Cadherin 23 is a component of the transient lateral links in the developing hair bundles of cochlear sensory cells. *Dev Biol* **280**, 281–294.

Ness, S. L., Ben-Yosef, T., Bar-Lev, A., Madeo, A. C., Brewer, C. C., Avraham, K. B., et al. 2003. Genetic homogeneity and phenotypic variability among Ashkenazi Jews with Usher syndrome type III. *J Med Genet* **40**, 767–772.

Nuutila, A. 1970. Dystrophia retinae pigmentosa–dysacusis syndrome (DRD): A study of the Usher- or Hallgren syndrome. *J Genet Hum* **18**, 57–88.

Ouyang, X. M., Xia, X. J., Verpy, E., Du, L. L., Pandya, A., Petit, C., et al. 2002. Mutations in the alternatively spliced exons of USH1C cause non-syndromic recessive deafness. *Hum Genet* **111**, 26–30.

Pan, L. & Zhang, M. 2012. Structures of usher syndrome 1 proteins and their complexes. *Physiology (Bethesda)* **27**, 25–42.

Pearson, R. A., Barber, A. C., Rizzi, M., Hippert, C., Xue, T., West, E. L., et al. 2012. Restoration of vision after transplantation of photoreceptors. *Nature* **485**, 99–103.

Pennings, R. J., Damen, G. W., Snik, A. F., Hoefsloot, L., Cremers, C. W. & Mylanus, E. A. 2006. Audiologic performance and benefit of cochlear implantation in Usher syndrome type I. *Laryngoscope* **116**, 717–722.

Pennings, R. J., Huygen, P. L., Orten, D. J., Wagenaar, M., van, A. A., Kremer, H., et al. 2004. Evaluation of visual impairment in Usher syndrome 1b and Usher syndrome 2a. *Acta Ophthalmol Scand* **82**, 131–139.

Pietola, L., Aarnisalo, A. A., Abdel-Rahman, A., Vastinsalo, H., Isosomppi, J., Lopponen, H., et al. 2012. Speech recognition and communication outcomes with cochlear implantation in Usher syndrome type 3. *Otol Neurotol* **33**, 38–41.

Plantinga, R. F., Kleemola, L., Huygen, P. L., Joensuu, T., Sankila, E. M., Pennings, R. J., et al. 2005. Serial audiometry and speech recognition findings in Finnish Usher syndrome type III patients. *Audiol Neurootol* **10**, 79–89.

Plantinga, R. F., Pennings, R. J., Huygen, P. L., Sankila, E. M., Tuppurainen, K., Kleemola, L., et al. 2006. Visual impairment in Finnish Usher syndrome type III. *Acta Ophthalmol Scand* **84**, 36–41.

Rebibo-Sabbah, A., Nudelman, I., Ahmed, Z. M., Baasov, T. & Ben-Yosef, T. 2007. *In vitro* and *ex vivo* suppression by aminoglycosides of PCDH15 nonsense mutations underlying type 1 Usher syndrome. *Hum Genet* **122**, 373–381.

Reiners, J., Marker, T., Jurgens, K., Reidel, B. & Wolfrum, U. 2005a. Photoreceptor expression of the Usher syndrome type 1 protein protocadherin 15 (USH1F) and its interaction with the scaffold protein harmonin (USH1C). *Mol Vis* **11**, 347–355.

Reiners, J., Reidel, B., El-Amraoui, A., Boeda, B., Huber, I., Petit, C., et al. 2003. Differential distribution of harmonin isoforms and their possible role in Usher-1 protein complexes in mammalian photoreceptor cells. *Invest Ophthalmol Vis Sci* **44**, 5006–5015.

Reiners, J., van Wijk, E., Marker, T., Zimmermann, U., Jurgens, K., Te, B. H., et al. 2005b. Scaffold protein harmonin (USH1C) provides molecular links between Usher syndrome type 1 and type 2. *Hum Mol Genet* **14**, 3933–3943.

Robson, A. G., Lenassi, E., Saihan, Z., Luong, V. A., Fitzke, F. W., Holder, G. E., et al. 2012. Comparison of fundus autofluorescence with photopic and scotopic fine matrix mapping in patients with retinitis pigmentosa: 4- to 8-year follow-up. *Invest Ophthalmol Vis Sci* **53**, 6187–6195.

Robson, A. G., Tufail, A., Fitzke, F., Bird, A. C., Moore, A. T., Holder, G. E., et al. 2011. Serial imaging and structure-function correlates of high-density rings of fundus autofluorescence in retinitis pigmentosa. *Retina* **31**, 1670–1679.

Sadeghi, M., Cohn, E. S., Kimberling, W. J., Tranebjaerg, L. & Moller, C. 2005. Audiological and vestibular features in affected subjects with USH3: A genotype/phenotype correlation. *Int J Audiol* **44**, 307–316.

Sahly, I., Dufour, E., Schietroma, C., Michel, V., Bahloul, A., Perfettini, I., et al. 2012. Localization of Usher 1 proteins to the photoreceptor calyceal processes, which are absent from mice. *J Cell Biol* **199**, 381–399.

Saihan, Z., Stabej, P. Q., Robson, A. G., Rangesh, N., Holder, G. E., Moore Frcophth, A. T., et al. 2011. Mutations in the ush1c gene associated with sector retinitis pigmentosa and hearing loss. *Retina* **31**, 1708–1716.

Sandberg, M. A., Rosner, B., Weigel-DiFranco, C., McGee, T. L., Dryja, T. P. & Berson, E. L. 2008. Disease course in patients with autosomal recessive retinitis pigmentosa due to the USH2A gene. *Invest Ophthalmol Vis Sci* **49**, 5532–5539.

Schultz, J. M., Bhatti, R., Madeo, A. C., Turriff, A., Muskett, J. A., Zalewski, C. K., et al. 2011. Allelic hierarchy of CDH23 mutations causing non-syndromic deafness DFNB12 or Usher syndrome USH1D in compound heterozygotes. *J Med Genet* **48**, 767–775.

Siemens, J., Kazmierczak, P., Reynolds, A., Sticker, M., Littlewood-Evans, A., et al. 2002. The Usher syndrome proteins cadherin 23 and harmonin form a complex by means of PDZ-domain interactions. *Proc Natl Acad Sci U.S.A.* **99**, 14946–14951.

Siemens, J., Lillo, C., Dumont, R. A., Reynolds, A., Williams, D. S., Gillespie, P. G., et al. 2004. Cadherin 23 is a component of the tip link in hair-cell stereocilia. *Nature* **428**, 950–955.

Spandau, U. H. & Rohrschneider, K. 2002. Prevalence and geographical distribution of Usher syndrome in Germany. *Graefes Arch Clin Exp Ophthalmol* **240**, 495–498.

Usher, C. 1914. On the inheritance of retinitis pigmentosa with notes of cases. *Roy Lond Ophthalmol Hosp Rep* **19**, 130–236.

Vache, C., Besnard, T., Blanchet, C., Baux, D., Larrieu, L., Faugere, V., et al. 2010. Nasal epithelial cells are a reliable source to study splicing variants in Usher syndrome. *Hum Mutat* **31**, 734–741.

van Wijk, E., Kersten, F. F., Kartono, A., Mans, D. A., Brandwijk, K., Letteboer, S. J., et al. 2009. Usher syndrome and Leber congenital amaurosis are molecularly linked via a novel isoform of the centrosomal ninein-like protein. *Hum Mol Genet* **18**, 51–64.

van Wijk, E., van der Zwaag B., Peters, T., Zimmermann, U., Te, B. H., Kersten, F. F., et al. 2006. The DFNB31 gene product whirlin connects to the Usher protein network in the cochlea and retina by direct association with USH2A and VLGR1. *Hum Mol Genet* **15**, 751–765.

von Graefe, A. 1858. Exceptionelles verhalten des gesichtsfeldes bei pigmentenartung des netzhaut. *Archiv für Ophthalmologie* **4**, 250–253.

Vozzi, D., Aaspollu, A., Athanasakis, E., Berto, A., Fabretto, A., Licastro, D., et al. 2011. Molecular epidemiology of Usher syndrome in Italy. *Mol Vis* **17**, 1662–1668.

Webb, S. W., Grillet, N., Andrade, L. R., Xiong, W., Swarthout, L., Della Santina, C. C., et al. 2011. Regulation of PCDH15 function in mechanosensory hair cells by alternative splicing of the cytoplasmic domain. *Development* **138**, 1607–1617.

Weil, D., El-Amraoui, A., Masmoudi, S., Mustapha, M., Kikkawa, Y., Laine, S., et al. 2003. Usher syndrome type I G (USH1G) is caused by mutations in the gene encoding SANS, a protein that associates with the USH1C protein, harmonin. *Hum Mol Genet* **12**, 463–471.

Zrenner, E., Bartz-Schmidt, K. U., Benav, H., Besch, D., Bruckmann, A., Gabel, V. P., et al. 2011. Subretinal electronic chips allow blind patients to read letters and combine them to words. *Proc Biol Sci* **278**, 1489–1497.

Chapter 14

Syndromes not yet proven to be ciliopathies

Kate Baker

Introduction

When is a ciliopathy not a ciliopathy? In principle, three criteria should be fulfilled for a condition to be considered ciliopathic: (1) the condition is characterised by abnormalities (single organ or multi-system) of the types associated with ciliopathy; (2) the condition is caused by mutations in genes whose gene products are localised to the cilium and have a ciliary-specific function; and (3) abnormalities in the structure or function of the cilium have been demonstrated, ideally in patient-derived tissue or alternatively in an animal model that recapitulates both the genotype and phenotype of the human disorder under consideration. Conditions that meet one, two but not three of these criteria are suspected but not proven ciliopathies.

Given the breadth and diversity of ciliopathic clinical features, suspicion of ciliopathy on clinical grounds has been raised for a very large number of disorders. A systematic search of the Online Mendelian Inheritance in Man database (www.OMIM.org) applied a pairwise strategy to identify syndromes associated with at least two of nine core ciliopathic features (retinitis pigmentosa, renal cystic disease, polydactyly, mental retardation/developmental delay, situs inversus/isomerism, hypoplasia of the corpus callosum, posterior encephalocoele, Dandy–Walker malformation, hepatic disease). This search for potential ciliopathies yielded a list of 193 unique OMIM entries, of which only 14 were considered to be confirmed ciliopathies at the time of publication (Baker & Beales, 2009).

In this chapter, we will review progress in resolving whether the numerous additional conditions identified by Baker and Beales (2009) are ciliopathic. The number of suspected-but-not-confirmed ciliopathies has declined since 2009 for reasons that will be discussed in the section 'Syndromes suspected to be cilopathies now resolved'. However, of syndromes identified as possible or likely ciliopathies, many have yet to be fully resolved for reasons discussed in the section 'Syndromes suspected to be ciliopathies still unresolved'.

Diagnostic uncertainties remain for many families, either where a clinical syndrome has been defined but no definite cause identified, or a genotype discovered but with limited prognostic information available. The question 'What is the problem?' has to be tackled from multiple perspectives. It is hoped that the answer 'It's a ciliopathy' might be increasingly helpful for both clinicians and families, by highlighting potential medical problems requiring surveillance, and facilitating therapeutic options for progressive aspects of disease such as renal and visual impairment. The psychosocial importance of diagnostic identity and allegiance to a patient community (such as the Ciliopathy Alliance) should not be underestimated in reducing isolation for individuals and families affected by these rare, severe, lifelong disorders.

Syndromes suspected to be ciliopathies now resolved

Genetic confirmation of ciliopathy

The process of gene discovery for rare syndromes, and for families or singleton cases with apparently unique disorders, is becoming increasingly efficient. Identification of potentially pathogenic variants in candidate genes and at novel loci is achievable rapidly and with high yields via Next Generation sequencing strategies and multi-gene panel testing, although mining of genome-wide datasets is complex and assigning pathogenicity of novel variants is challenging. For many patients, there will be on-going uncertainty over the significance of genetic results. However, for several suspected ciliopathic conditions, genetic evidence supporting ciliopathy has recently emerged.

Mutations in at least four different ciliary genes implicated in intraflagellary transport have been identified in patients with Sensenbrenner syndrome (cranioectodermal dysplasia OMIM phenotype #218,330) via a variety of genetic strategies (Arts et al. 2011; Bredrup et al. 2011; Gilissen et al. 2010; Walczak-Sztulpa et al. 2010). Mainzer–Saldino syndrome (conorenal syndrome OMIM phenotype #266,920) has recently been associated with mutations in an IFT-A component in six independent families (Perrault et al. 2012). It is noteworthy that the majority of these recent genetic findings were accompanied by investigation of ciliary protein constituents or ciliary ultrastructure in patient-derived fibroblasts. Demonstration of ciliary disruption is increasingly important after identification of candidate genes and novel mutations, given the challenge of interpreting rare or unique variants as being pathogenic or benign.

A particularly diverse group of phenotypes has recently been associated with mutations in *KIF7*. Putoux et al. (2011) carried out genome-wide homozygosity mapping in a consanguineous family with four cases of severe structural brain abnormality, post-axial polydactyly and cleft palate (a phenotype consistent with hydrolethalus syndrome) and linkage analysis in a cohort of families with acrocallosal syndrome in whom *Gli3* mutations had not been identified. The smallest common interval of homozygosity and linkage at 15q26 contained 32 annotated genes including *KIF7*, a strong candidate gene for ciliopathy. A *KIF7* homologue in *Drosophila* (*Costal 2*) is cilia-associated and is required for ciliogenesis. *Costal2* belongs to a protein complex involved in activation of sonic hedgehog targets via the Ci/Gli regulatory pathway. Moreover *Kif7* knockout mice have polydactyly and severe brain abnormality (exencephaly). Sequencing of *KIF7* revealed a homozygous splice site deletion in the hydrolethalus syndrome family, and truncating mutations in all six acrocallosal syndrome families showing linkage to 15q26. Structural analysis of cilia in cultured fibroblasts from two patients with acrocallosal syndrome showed that cilia were present but were longer than cilia from control fibroblasts, suggesting impaired regulation of ciliogenesis, in addition to likely disruption of downstream signalling. Independently, Dafinger et al. (2011) mapped the 15q26 region as a locus underlying Joubert syndrome, identifying truncating mutations in three families, in combination with *TMEM67* hypomorphic mutations in one case. Putoux et al. (2011) went on to explore whether mutations in *KIF7* might contribute to other ciliopathic syndromes, and sequenced the gene in 130 cases with diverse syndrome diagnoses including Bardet–Biedl syndrome (BBS), Joubert syndrome and oral–facial–digital syndrome type VI. Eight non-synonymous changes were identified, and pathogenicity assessed *in vivo* using a zebrafish experimental system, confirming that these hypomorphic mutations disrupted embryological patterning (somite formation). Since patients with BBS carrying these apparently deleterious mutations also carried mutations in *BBS1*, *BBS7*, *BBS9* and *BBS10*, the proposal is that *KIF7* (and potentially other ciliary genes) can participate in oligogenic inheritance or in phenotype modification, via mutational load and combinatorial genotype–phenotype relationships.

Genetic confirmation of non-ciliopathy

For some conditions previously suspected to be ciliopathies, determination of genetic cause has disproven this suspicion. For example Kabuki syndrome (OMIM phenotype #147,920) is a neurodevelopmental disorder sometimes associated with renal and hepatic abnormalities and with under-development of the corpus callosum, phenotypic manifestations raising the suspicion of ciliopathy. Kabuki syndrome is now known to be caused by mutations in *MLL2* (Banka et al. 2012; Ng et al. 2010). *MLL2* is a histone methyltransferase involved in chromatin modification and epigenetic regulation of transcription, with no known ciliary-specific functions discovered to date. Johanson–Blizzard syndrome (#243,800) is characterised by growth failure, abnormal skin, hair and teeth, genital anomalies and malabsorption, and identified as a possibly ciliopathy due to the presence of mental retardation and an infrequent association with situs inversus. Zenker et al. (2005) identified mutations in *UBR1* in twelve unrelated families with Johanson–Blizzard syndrome. *UBR1* is now known to be an E3 ubiquitin protein ligase, involved in regulating protein degradation via a mechanism known as the Arg/N-end rule pathway (Hwang et al. 2011) which is not ciliary-specific.

A large number of craniofacial conditions were identified as possibly ciliopathic in the search by Baker and Beales (2009). One such condition, a form of acrofacial dysostosis known as Nager syndrome (OMIM #154,400) characterised by micrognathia and anterior limb defects and sometimes associated with agenesis of the corpus callosum, Dandy–Walker malformation and developmental delay, has recently been attributed to mutations in *SF3B4*, a component of the spliceosome complex involved in mRNA processing (Bernier et al. 2012). Interestingly, microcephalic osteodysplastic primordial dwarfism type I (MOPD I), another potential ciliopathy, has been attributed to mutations in a small nuclear RNA (snRNA) spliceosome component (He et al. 2011). These findings appear to reduce the likelihood that these conditions (and other similar craniofacial and skeletal syndromes) are caused with ciliopathic mechanisms. However, there remains the possibility that post-transcriptional processing of ciliary components could be preferentially disrupted, or that these regulatory factors could have additional functions not limited to splice regulation. Moreover, although clinical characteristics were indistinguishable between the 60% of cases in whom *SF3B4* mutations were identified and the remainder, there may be significant heterogeneity in genetic cause and pathological mechanism.

Elucidation of ciliary functions of causative genes

A key challenge in understanding ciliopathies is elucidation of the mechanisms by which ciliary signalling influences development of different organ systems. For several syndromes with clinical features suggestive of ciliopathy, causative genes have been known for several years but a link to ciliary structure or function has been elusive. Knowledge of the relationships between ciliary activation and downstream signalling cascades is enabling these links to be more firmly established.

Grieg cephalopolysyndactyly syndrome (GCPS) and acrocallosal syndrome (ACS) are caused (in most cases) by mutations or deletions of the transcription factor *Gli3*. Both conditions are characterised by polydactyly (usually preaxial in the feet and post-axial in the hands), hypertelorism, macrocephaly and variable degrees of developmental delay. These phenotypes reflect both a patterning defect (affecting the limbs and facial structures), and a proliferation defect (causing macrocephaly). Both of these regulatory aspects have been attributed to interactions between *Gli3* and *Shh* (sonic hedgehog), downstream of ciliary processing (Biesecker, 2006). Moreover, mutations in a specific exonic region at the C-terminal of *Gli3* cause a different disorder (Pallister–Hall syndrome, PHS) involving central and post-axial polydactyly and hypothalamic hamatorma, again emphasising a highly specific role for *Gli3* in transcriptional regulation of patterning and proliferation. However, the direct mechanisms of ciliary-*Gli3* interaction are only now becoming clear. In the presence of *Shh,* secreted in gradients from organising cells into the extra-cellular space, *Gli-3* translocates to the ciliary tip and is phosphorylated by a cascade of interactions to switch from transcriptional repressor mode to transcriptional activator (Goetz & Anderson, 2010). However, this translocation does not occur for PHS-mutant Gli3 protein (Zeng et al. 2010). Therefore in *Gli3*-associated disorders, although the structure of the cilium is intact, localisation of key embryological signalling molecules to the cilium and transmission of extra-cellular signals to the interior of the differentiating cell is impaired. At the cellular and morphological level, the balance between Gli3 activator and repressor forms, itself dependent on intact primary cilium, appears to regulate cell cycle length in neural progenitors, with a consequential impact on brain size explaining the presence of macrocephaly in Grieg syndrome (Wilson et al. 2011).

Chudley–McCullough syndrome, defined by sensorineural deafness and structural brain abnormalities including agenesis of the corpus callosum, polymicrogyria, ventriculomegaly and arachnoid cysts, has recently been associated with mutations in *GPSM2*

(Doherty et al. 2012), a gene previously associated with non-syndromic hearing loss. *GPSM2* is required for planar orientation of the mitotic spindle in apical neural progenitor cells (radial glia). Whilst *GPSM2* is not itself known to be localised to the primary cilium, a role for the cilium in this same process of neuronal progenitor polarity and regulation of proliferation and migration is strongly suspected (Wilsch-Brauninger et al. 2012). Hence, understanding the molecular biology of rare syndrome-associated genotypes is starting to explain similarities and differences in phenotypic expression between ciliopathy syndromes.

Elucidation of non-ciliary functions of causative genes

Cohen syndrome (OMIM #216,550) is a particularly interesting condition with some features that are suggestive of ciliopathy (obesity, developmental delay, progressive choroidoretinal degeneration) but additional non-ciliopathy features (joint hyperextensibility, intermittent neutropenia). Around 50% of patients with a clinical diagnosis of Cohen syndrome have mutations in *VPS13B*(El Chehadeh (El et al. 2010) with additional dosage abnormalities detectable by MLPA in another 40% of cases (Parri et al. 2010). *VPS13B* is a transmembrane protein with putative membrane recycling functions that has been localised to the Golgi matrix (Seifert et al. 2011), the key cellular compartment for post-translational protein modification and packaging of proteins to determine future cellular localisation. Given the specificity of phenotypes in Cohen syndrome, it is interesting to speculate whether processing of ciliary components within the Golgi apparatus could be selectively disrupted in Cohen syndrome, resulting in partial phenocopy. Indeed regulation of intracellular vesicle trafficking may underlie a number of conditions with ciliopathy-like single organ features, including choroideremia (Strunnikova et al. 2009).

Nosological reorganisation

Some suspected ciliopathy syndromes have now been resolved by being subsumed into broader diagnostic categories, because of both phenotypic and genotypic similarity. For example, COACH (**c**erebellar vermis hypoplasia, **o**ligophrenia, **a**taxia, **o**cular coloboma, and **c**ongenital **h**epatic fibrosis; OMIM phenotype #216,360) and arima syndrome (cerebellar vermis hypoplasia, ocular abnormalities, cystic kidney disease; OMIM %243,910) have been 'lumped' into the 'Joubert syndrome and related disorders' (JSRD) group (Sattar & Gleeson, 2011). These syndromes share partially overlapping phenotypic features and heterogeneous genetic aetiology with both Joubert syndrome (JS) and Meckel syndrome (MKS), with limited evidence for genotype–phenotype correlations. COACH has been associated with mutations in several JSRD genes, including *TMEM67*, which in one clinical series has been found to be mutated in 7% of MKS cases and 57% of COACH patients (Brancati et al. 2009). In another series, *TMEM67* was mutated in only 1% of classical JS families without hepatic fibrosis, and 83% (19/23) of families meeting criteria for COACH (Doherty et al. 2010). More rarely, the COACH phenotype is associated with *CC2D2A* or *RPGRIP1L* mutations, and the frequency of mutations in these genes appears fairly constant across the JSRD spectrum (Doherty et al. 2010).

Syndromes suspected to be ciliopathies still unresolved

Historical syndromes: the case for further lumping

One of the limitations of the OMIM catalogue as a tool for syndrome identification is the collation within the database of extremely rare entities, based on only a handful of case reports, prior to the availability of molecular genetic testing. Some of these conditions should now be considered variants of more frequently occurring conditions. For example, Biemond syndrome (%601,794) was identified as a possible ciliopathy via the search terms 'retinitis pigmentosa' and 'mental retardation'. Other clinical descriptors for this condition include obesity, hypogonadism and hydrocephalus. There have been no published case reports of this condition since 1997, and later authors have considered it to be a variant of BBS with prominent ocular features (Heon et al. 2005). Other conditions listed as distinct entities may be better considered as severe variants at the extreme of a phenotypic spectrum. For example, renal–hepatic–pancreatic dysplasia (OMIM #208,540) is associated with highly deleterious mutations in *NPHP3*, whereas hypomorphic mutations cause milder nephronophthisis. A large number of craniofacial syndromes with partially overlapping phenotypes have been described, and genetic investigation of these unresolved syndromes may benefit from pooling cases with similar features.

One approach to gene discovery across syndrome boundaries is to collect a large number of cases defined by a single phenotypic feature (irrespective of the presence or absence of other features that may point toward a particular syndrome diagnosis) and conduct genome-wide copy number variant analysis. This approach was successfully applied by Fakhro et al. (2011), who conducted single-nucleotide polymorphism genotyping in 262 patients with heterotaxy (abnormalities of the left–right axis of the heart) to identify rare deletion and duplication events. Rare dosage variants affecting 14 ciliary components were identified, providing a list of candidate genes for sequencing in further cases. There may be similar benefit to analysing cases with shared neurological phenotypes such as Dandy–Walker malformation and cystic brain lesions that occur across a number of rare, unresolved syndromes suspected to be ciliopathic.

Heterogeneity: the case for further splitting

Opitz GBBB syndrome involves a diverse spectrum of clinical features relating to the development of midline structures including hypertelorism, cleft lip or palate, and laryngotracheoesophageal abnormalities resulting in swallowing difficulties and respiratory dysfunction. Genitourinary abnormalities (in particular hypospadias), developmental delay, and congenital heart defects can also be present. Sporadic, X-linked and autosomal dominant forms are reported. The syndrome has been described in association with a number of different chromosomal aberrations including (most commonly) the recurrent 22q11.2 microdeletion. Ciliopathy-typical structural abnormalities (of the brain in particular) are seen in some but not all cases of Opitz GBBB, hence it seems possible that within this heterogeneous syndrome there may be ciliopathic and non-ciliopathic forms. Mutations in *MID1* are responsible for fewer than half of cases of X-linked Opitz GBBB, with milder manifestations of the condition including isolated hypertelorism observed in female carriers (So et al. 2005). Interestingly, *MID1* encodes an E3 ubiquitin ligase that is required for degradation of a protein phosphatase *PP.2CA* that targets microtubule-associated proteins and may be necessary for microtubule-associated protein translation (Aranda-Org et al. 2008), providing a potential link to ciliary structure. Other causes of Opitz GBBB may therefore converge on microtubule functions.

Conditions that remain of unknown genetic origin

The large number of craniofacial syndromes identified as possible ciliopathies are ideal targets for investigation by Next Generation sequencing, given likely genetic heterogeneity within syndromes and commonality in genes across syndromes. For example acromelic frontonasal dysplasia (%603,671) is a variant of frontonasal dysplasia with additional features suggestive of ciliopathy including polydactyly and agenesis of the corpus callosum, and no known genotypes as yet. Other unresolved syndromes with partially overlapping craniofacial features include cerebro-facio-thoracic syndrome, cerebro-oculo-nasal syndrome and oculo-auricular-fronto-nasal syndromes. Moreover, there is increasing evidence from animal models that the midline embryological abnormalities which characterise the frontonasal dysplasia group of conditions can result from ciliary dysfunction and hyperactivation of hedgehog function (Brugmann et al. 2010).

Various forms of oral–facial–digital syndrome (OFD) have yet to be linked to any genetic locus. Mohr syndrome (%252,100), also termed OFD type II, is characterised by lobate tongue, cleft palate and other craniofacial features, but with the addition of digital abnormalities and conductive hearing loss not typical of other forms of OFD. Similarly, Varadi–Papp (subtype of OFD VI), associated with central polydactyly and cerebellar abnormalities, has yet to be resolved and may provide additional insights into limb patterning abnormalities.

Neu–Laxova syndrome (%256,520) is a lethal multiple malformation syndrome associated with ciliopathy-typical structural brain abnormalities including cerebellar hypoplasia and agenesis of the corpus callosum. The unique diagnostic feature of Neu–Laxova syndrome is severe restrictive icthyosis of the skin. Genetic resolution of this condition would provide significant evidence for or against a critical function of the primary cilium for skin development. Investigation of ciliary biology in other predominantly dermatological conditions identified as potentially ciliopathic (basal cell nevus syndrome and Gorlin syndrome) has not yet been undertaken. Mutations have recently been identified in *DOCK6* (an actin cytoskeleton component) in two families affected by autosomal recessive Adams–Oliver syndrome, highlighting a molecular pathway of potential relevance (Shaheen et al. 2011).

Conditions of known genetic origin, but the link to cilia is uncertain

Endocrine dysfunction is not considered a core ciliopathic feature; however, hypogonadism and obesity (thought to be of hypothalamic origin) are key diagnostic features in BBS and Alström syndrome, and are occasional features in other confirmed ciliopathies. A number of syndromes with prominent endocrine features have phenotypic overlap with ciliopathy. For example endocrine–cerebro-osteodystrophy is an ultra-rare condition that may be confined to the Old Amish population, characterised by hypoplastic adrenal and pituitary glands plus midline brain and facial anomalies and cystic renal disease highly suggestive of ciliopathy. The causative mutation identified in the known cases of this condition is in *ICK*, an intestinal kinase without known ciliary function (Lahiry et al. 2009). The X-linked condition Kallman syndrome (hypogonadotropic hypogonadism, anosmia) is sometimes associated with developmental delay and agenesis of the corpus callosum, prompting suggestion of possible ciliopathy. *Anosmin-1* is a chemotactic molecule required for correct migration of GnRH-secreting cells and for olfactory axonal development (Hu & Bouloux, 2011). One possibility could be that *anosmin-1* signals via receptors located at the primary cilium, explaining partially overlapping phenotypes. Another endocrine condition with ciliopathy-like associations (in this case renal cystic disease and hepatic fibrosis) is neonatal diabetes and congenital hypothyroidism (NDH, OMIM #610,199), caused by mutations in a Gli-similar (*GLIS3*) transcription factor (Dimitri et al. 2011; Senee et al. 2006). Although predominantly nuclear, *GLIS3* has been shown to localise to the cilium and translocate to the nucleus, and is essential for pancreatic cell development and regulation of insulin production (Kang et al. 2010). Interestingly, polymorphic variation in *GLIS3* has been associated with risk of diabetes (types 1 and 2) in the general population (Dupuis et al. 2010).

Proteus syndrome (OMIM #176,920) is characterised by severe, disabling soft tissue overgrowth, but was identified as being possibly ciliopathic because of occasional associations with renal disease, brain malformations and mental retardation. Lindhurst et al. (2011) recently identified a recurrent activating somatic mutation in *AKT1* (a protein kinase regulated via the PI3K pathway) in biopsies of overgrown tissue but not unaffected tissue of all patients with proteus syndrome. This confirms the long-held hypothesis that somatic mosaicism is responsible for overgrowth in this disorder, and suggests that this same mutation may be responsible for other associated features depending on lineage of cells acquiring the mutation. *AKT1* has predominantly been investigated as a proto-oncogene because acquired mutations are identified in several different malignancies, and its role in normal growth and development is less clear. Sequence variation in *AKT1* has also been associated with risk of schizophrenia (Schwab et al. 2005), suggesting that disruption to the same pathway may have subtle neurodevelopmental effects. As yet no link between *AKT1* and cilia has been reported. Somatic mosaic mutations in three genes within the same pathway (*AKT3*, *PIK3R2* and *PIK3CA*) have been reported recently in patients with megalencephaly syndromes, including megalencephaly–polymicrogyria–polydactyly–hydrocephalus (MPPH, OMIM #603,387), a very rare disorder with severe ciliopathy-like clinical features (Riviere et al. 2012). In these patients it was possible to identify somatic mutations by 'deep-sequencing' of DNA in blood or saliva to identify mutations present at very low levels of mosaicism (down to 1% of gene copies in some patients). Lee et al. (2012) were able to identify brain-specific somatic mutations in the same group of genes, because the abnormal tissue had been surgically removed to treat intractable epilepsy secondary to hemimegalencephaly. The implications of these landmark studies are two-fold. First, somatic mutation may be responsible for ciliopathy-associated clinical presentations in some patients where germline mutations have not been detected. Second, the PIK3-AKT growth-regulatory pathway may be implicated in proliferation abnormalities seen in some ciliopathy syndromes.

Conclusion

The accelerated pace of genetic discovery will continue and will drive further progress in elucidating the diverse mechanisms (ciliogenesis, ciliary structural integrity, intraflagellary transport, ciliary signalling and downstream cellular regulation) underlying ciliopathic disorders. The diagnostic landscape remains complex for patients and health professionals, with ongoing tensions between syndrome-specific and spectrum-associated classifications, and evolving debate about genotype-first versus phenotype-first diagnostic approaches. Further ciliopathic syndromes may yet be identified, although these are likely to be ultra-rare or unique to individual patients and families. There is also potential for ciliopathic mechanisms to contribute to the pathogenesis of non-syndromic (single organ) congenital abnormalities of the types observed in ciliopathy syndromes.

Finally, one can speculate that inherited or acquired ciliary dysfunction could contribute to risk of common, multifactorial clinical problems such as obesity, developmental delay and psychiatric illness. For example, *DISC1* (disrupted-in-schizophrenia 1) is a neurodevelopmental gene identified at the breakpoint of an apparently balanced translocation in a family with multiple individuals affected by psychiatric illness, including schizophrenia (Millar et al. 2000). There has been controversy as to whether sequence variation within *DISC-1* is associated with risk of mental illness in the general population, unsurprising given the genetic and neurodevelopmental heterogeneity of schizophrenia (Schumacher et al. 2009). *DISC1* encodes a scaffolding protein and has recently been shown to regulate translocation of BBS1 to the centrosome, influencing the balance between neuronal proliferation and migration (Ishizuka et al. 2011). Patients with BBS demonstrate elevated rates of diverse mental health difficulties in addition to variable developmental delay and cognitive impairment (Bennouna-Greene et al. 2011). This provides early evidence for convergence of syndromic, familial and population genetic risk factors for common mental illness on ciliary biology. It is plausible that a proportion of apparently idiopathic psychiatric illness could be attributable to ciliopathy-associated risk genes and ciliopathic mechanisms, but evidence for this is lacking and methods for *in vivo* investigation of this hypothesis are required. Translating this knowledge to clinical interventions and improved outcomes for diverse patient populations is a realistic long-term prospect.

References

Aranda-Org, Trockenbacher, A., Winter, J., Aigner, J., Kohler, A., Jastrzebska, E., et al. 2008. The Opitz syndrome gene product MID1 assembles a microtubule-associated ribonucleoprotein complex. *Hum Genet* **123**, 163–176.

Arts, H. H., Bongers, E. M., Mans, D. A., van Beersum, S. E., Oud, M. M., Bolat, E., et al. 2011. C14ORF179 encoding IFT43 is mutated in Sensenbrenner syndrome. *J Med Genet* **48**, 390–395.

Baker, K. & Beales, P. L. 2009. Making sense of cilia in disease: The human ciliopathies. *Am J Med Genet, C Semin Med Genet* **151C**, 281–295.

Banka, S., Veeramachaneni, R., Reardon, W., Howard, E., Bunstone, S., Ragge, N., et al. 2012. How genetically heterogeneous is Kabuki syndrome?: MLL2 testing in 116 patients, review and analyses of mutation and phenotypic spectrum. *Eur J Hum Genet* **20**, 381–388.

Bennouna-Greene, V., Kremer, S., Stoetzel, C., Christmann, D., Schuster, C., Durand, M., et al. 2011. Hippocampal dysgenesis and variable neuropsychiatric phenotypes in patients with Bardet–Biedl syndrome underline complex CNS impact of primary cilia. *Clin Genet* **80**, 523–531.

Bernier, F. P., Caluseriu, O., Ng, S., Schwartzentruber, J., Buckingham, K. J., Innes, A. M., et al. 2012. Haploinsufficiency of SF3B4, a component of the pre-mRNA spliceosomal complex, causes Nager syndrome. *Am J Hum Genet* **90**, 925–933.

Biesecker, L. G. 2006. What you can learn from one gene: GLI3. *J Med Genet* **43**, 465–469

Brancati, F., Iannicelli, M., Travaglini, L., Mazzotta, A., Bertini, E., Boltshauser, E., et al. 2009. MKS3/TMEM67 mutations are a major cause of COACH syndrome, a Joubert syndrome related disorder with liver involvement. *Hum Mutat* **30**, E432–E442.

Bredrup, C., Saunier, S., Oud, M. M., Fiskerstrand, T., Hoischen, A., Brackman, D., et al. 2011. Ciliopathies with skeletal anomalies and renal insufficiency due to mutations in the IFT-A gene WDR19. *Am J Hum Genet* **89**, 634–643.

Brugmann, S. A., Allen, N. C., James, A. W., Mekonnen, Z., Madan, E. & Helms, J. A. 2010. A primary cilia-dependent etiology for midline facial disorders. *Hum Mol Genet* **19**, 1577–1592.

Dafinger, C., Liebau, M.C., Elsayed, S. M., Hellenbroich, Y., Boltshauser, E., Korenke, G. C., et al. 2011. Mutations in KIF7 link Joubert syndrome with Sonic Hedgehog signaling and microtubule dynamics. *J Clin Invest* **121**, 2662–2667.

Dimitri, P., Warner, J. T., Minton, J. A., Patch, A. M., Ellard, S., Hattersley, A. T., et al. 2011. Novel GLIS3 mutations demonstrate an extended multisystem phenotype. *Eur J Endocrinol* **164**, 437–443.

Doherty, D., Chudley, A. E., Coghlan, G., Ishak, G. E., Innes, A. M., Lemire, E. G., et al. 2012. GPSM2 mutations cause the brain malformations and hearing loss in Chudley–McCullough syndrome. *Am J Hum Genet* **90**, 1088–1093.

Doherty, D., Parisi, M. A., Finn, L. S., Gunay-Aygun, M., Al-Mateen, M., Bates, D., et al. 2010. Mutations in 3 genes (MKS3, CC2D2A and RPGRIP1L. cause COACH syndrome (Joubert syndrome with congenital hepatic fibrosis). *J Med Genet* **47**, 8–21.

Dupuis, J., Langenberg, C., Prokopenko, I., Saxena, R., Soranzo, N., Jackson, A. U., et al. 2010. New genetic loci implicated in fasting glucose homeostasis and their impact on type 2 diabetes risk. *Nat Genet* **42**, 105–116.

El, C. S., Aral, B., Gigot, N., Thauvin-Robinet, C., Donzel, A., Delrue, M. A., et al. 2010. Search for the best indicators for the presence of a VPS13B gene mutation and confirmation of diagnostic criteria in a series of 34 patients genotyped for suspected Cohen syndrome. *J Med Genet* **47**, 549–553.

Fakhro, K. A., Choi, M., Ware, S. M., Belmont, J. W., Towbin, J. A., Lifton, R. P., et al. 2011. Rare copy number variations in congenital heart disease patients identify unique genes in left-right patterning. *Proc Natl Acad Sci U.S.A.* **108**, 2915–2920.

Gilissen, C., Arts, H. H., Hoischen, A., Spruijt, L., Mans, D. A., Arts, P., et al. 2010. Exome sequencing identifies WDR35 variants involved in Sensenbrenner syndrome. *Am J Hum Genet* **87**, 418–423.

Goetz, S. C. & Anderson, K. V. 2010. The primary cilium: A signalling centre during vertebrate development. *Nat Rev Genet* **11**, 331–344.

He, H., Liyanarachchi, S., Akagi, K., Nagy, R., Li, J., Dietrich, R. C., et al. 2011. Mutations in U4atac snRNA, a component of the minor spliceosome, in the developmental disorder MOPD I. *Science* **332**, 238–240.

Heon, E., Westall, C., Carmi, R., Elbedour, K., Panton, C., Mackeen, L., et al. 2005. Ocular phenotypes of three genetic variants of Bardet–Biedl syndrome. *Am J Med Genet A* **132A**, 283–287.

Hu, Y. & Bouloux, P. M. 2011. X-linked GnRH deficiency: Role of KAL-1 mutations in GnRH deficiency. *Mol Cell Endocrinol* **346**, 13–20.

Hwang, C. S., Sukalo, M., Batygin, O., Addor, M. C., Brunner, H., Aytes, A. P., et al. 2011. Ubiquitin ligases of the N-end rule pathway: Assessment of mutations in UBR1 that cause the Johanson–Blizzard syndrome. *PLoS One* **6**, e24925.

Ishizuka, K., Kamiya, A., Oh, E. C., Kanki, H., Seshadri, S., Robinson, J. F., et al. 2011. DISC1-dependent switch from progenitor proliferation to migration in the developing cortex. *Nature* **473**, 92–96.

Kang, H. S., ZeRuth, G., Lichti-Kaiser, K., Vasanth, S., Yin, Z., Kim, Y. S., et al. 2010. Gli-similar (Glis) Kruppel-like zinc finger proteins: Insights into their physiological functions and critical roles in neonatal diabetes and cystic renal disease. *Histol Histopathol* **25**, 1481–1496.

Lahiry. P., Wang, J., Robinson, J. F., Turowec, J. P., Litchfield, D. W., Lanktree, M. B., et al. 2009. A multiplex human syndrome implicates a key role for intestinal cell kinase in development of central nervous, skeletal, and endocrine systems. *Am J Hum Genet* **84**, 134–147.

Lee, J. H., Huynh, M., Silhavy, J. L., Kim, S., Dixon-Salazar, T., Heiberg, A., et al. 2012. De novo somatic mutations in components of the PI3K-AKT3-mTOR pathway cause hemimegalencephaly. *Nat Genet* **44**, 941–945.

Lindhurst, M. J., Sapp, J. C., Teer, J. K., Johnston, J. J., Finn, E. M., Peters, K., et al. 2011. A mosaic activating mutation in AKT1 associated with the Proteus syndrome. *N Engl J Med* **365**, 611–619.

Millar, J. K., Wilson-Annan, J. C., Anderson, S., Christie, S., Taylor, M. S., Semple, C. A., et al. 2000. Disruption of two novel genes by a translocation co-segregating with schizophrenia. *Hum Mol Genet* **9**, 1415–1423.

Ng, S. B., Bigham, A. W., Buckingham, K. J., Hannibal, M. C., McMillin, M. J., Gildersleeve, H. I., et al. 2010. Exome sequencing identifies MLL2 mutations as a cause of Kabuki syndrome. *Nat Genet* **42**, 790–793.

Parri, V., Katzaki, E., Uliana, V., Scionti, F., Tita, R., Artuso, R., et al. 2010. High frequency of COH1 intragenic deletions and duplications detected by MLPA in patients with Cohen syndrome. *Eur J Hum Genet* **18**, 1133–1140.

Perrault, I., Saunier, S., Hanein, S., Filhol, E., Bizet, A. A., Collins, F., et al. 2012. Mainzer–Saldino syndrome is a ciliopathy caused by IFT140 mutations. *Am J Hum Genet* **90**, 864–870.

Putoux, A, Thomas, S., Coene, K. L., Davis, E. E., Alanay, Y., Ogur, G., et al. 2011. KIF7 mutations cause fetal hydrolethalus and acrocallosal syndromes. *Nat Genet* **43**, 601–606.

Riviere. J. B., Mirzaa, G. M., O'Roak, B. J., Beddaoui, M., Alcantara, D., Conway, R. L., et al. 2012. De novo germline and postzygotic mutations in AKT3, PIK3R2 and PIK3CA cause a spectrum of related megalencephaly syndromes. *Nat Genet* **44**, 934–940.

Sattar, S. & Gleeson, J. G. 2011. The ciliopathies in neuronal development: A clinical approach to investigation of Joubert syndrome and Joubert syndrome-related disorders. *Dev Med Child Neurol* **53**, 793–798.

Schumacher. J., Laje, G., Abou, J. R., Becker, T., Muhleisen, T. W., Vasilescu, C., et al. 2009. The DISC locus and schizophrenia: Evidence from an association study in a central European sample and from a meta-analysis across different European populations. *Hum Mol Genet* **18**, 2719–2727.

Schwab, S. G., Hoefgen, B., Hanses, C., Hassenbach, M. B., Albus, M., Lerer, B., et al. 2005. Further evidence for association of variants in the AKT1 gene with schizophrenia in a sample of European sib-pair families. *Biol Psychiatry* **58**, 446–450.

Seifert, W., Kuhnisch, J., Maritzen, T., Horn, D., Haucke, V. & Hennies, H. C. 2011. Cohen syndrome-associated protein, COH1, is a novel, giant Golgi matrix protein required for Golgi integrity. *J Biol Chem* **286**, 37665–37675.

Senee, V., Chelala, C., Duchatelet, S., Feng, D., Blanc, H., Cossec, J. C., et al. 2006. Mutations in GLIS3 are responsible for a rare syndrome with neonatal diabetes mellitus and congenital hypothyroidism. *Nat Genet* **38**, 682–687.

Shaheen, R., Faqeih, E., Sunker, A., Morsy, H., Al-Sheddi, T., Shamseldin, H. E., et al. 2011. Recessive mutations in DOCK6, encoding the guanidine nucleotide exchange factor DOCK6, lead to abnormal actin cytoskeleton organization and Adams–Oliver syndrome. *Am J Hum Genet* **89**, 328–333.

So, J., Suckow, V., Kijas, Z., Kalscheuer, V., Moser. B., Winter, J., et al. 2005. Mild phenotypes in a series of patients with Opitz GBBB syndrome with MID1 mutations. *Am J Med Genet A* **132A**, 1–7.

Strunnikova, N. V., Barb, J., Sergeev, Y. V., Thiagarajasubramanian, A., Silvin, C., Munson, P. J., et al. 2009. Loss-of-function mutations in Rab escort protein 1 (REP-1) affect intracellular transport in fibroblasts and monocytes of choroideremia patients. *PLoS One* **4**, e8402.

Walczak-Sztulpa, J., Eggenschwiler, J., Osborn, D., Brown, D. A., Emma, F., Klingenberg, C., et al. 2010. Cranioectodermal dysplasia, Sensenbrenner syndrome, is a ciliopathy caused by mutations in the IFT122 gene. *Am J Hum Genet* **86**, 949–956.

Wilsch-Brauninger, M., Peters, J., Paridaen, J. T. & Huttner, W. B. 2012. Basolateral rather than apical primary cilia on neuroepithelial cells committed to delamination. *Development* **139**, 95–105.

Wilson, B. J., Sundaram, S. K., Huq, A. H., Jeong, J. W., Halverson, S. R., Behen, M. E., et al. 2011. Abnormal language pathway in children with Angelman syndrome. *Pediatr Neurol* **44**, 350–356.

Zeng, H., Jia, J. & Liu, A. 2010. Coordinated translocation of mammalian Gli proteins and suppressor of fused to the primary cilium. *PLoS One* **5**, e15900.

Zenker, M., Mayerle, J., Lerch, M. M., Tagariello, A., Zerres, K., Durie, P. R., et al. 2005. Deficiency of UBR1, a ubiquitin ligase of the N-end rule pathway, causes pancreatic dysfunction, malformations and mental retardation (Johanson–Blizzard syndrome). *Nat Genet* **37**, 1345–1350.

Index

A

acrocallosal syndrome (ACS) 5, 265
acrofacial dysostosis 31, 38
acromelic frontonasal dysplasia 268
adherens junctions 114
adult polycystic kidney disease *see* autosomal dominant polycystic kidney disease (ADPKD)
AKT1 269
algorithm (diagnostic)
  Bardet—Biedl syndrome *98*
  Cilary chondrodysplasias *45*
  clinical diagnosis of ciliopathies *6*
  cystic kidney disease *195*
  nephronophthisis *154*
algorithm (testing)
  Alström syndrome (ALMS1 gene) *18*
ALMS1 *18*, 19, 114
alphafetoprotein (AFP), Meckel—Gruber syndrome (MGS) 139
Alström syndrome 8–29, 152
  annual review 14
  clinical features 9–10, 17
  coronary artery disease 11
  diagnosis 16–17
  dyslipidaemia 12–13
  epidemiology 8–9
  eponym (history) 8
  genetics 19–20
    ALMS1 18, 19, 114
    future studies 19–20
    prenatal diagnosis 20
    variation in genotype 19
  glucose intolerance 12–13
  growth hormone secretion 13
  heart 11
  history 10–11, *15*, 16–17
  infant cardiomyopathy 8–9, 10, 17, 18
  insulin resistance 12–13, 23
  kidney 11
  kyphoscoliosis 16, 19
  liver 12
  major organ fibrosis 15–16
  metabolic disorders 16
  musculoskeletal problems 13–14
  peripheral nervous system 19
  physiology 20–1
  predictable consequences 17–18
  puberty 13
  respiratory system 14
  reversible infant cardiomyopathy 8–9, 10, 17, 18
  sensory impairment 14–15
  smooth muscle dysfunction 12, 16
    caecal volvulus 12
    gastro-oesophageal reflux disease 12
    urological problems 12
  thyroid failure 13
  treatment 21–3
    atherosclerosis 23
    cardiomyopathy 22–3
    cirrhosis 24
    deafness 22
    fibrosis 24
    gastro-oesophageal disorders 24
    gene therapy 25
    hepatic steatosis 24
    infection 24–5
    musculoskeletal system 25
    psychosocial needs 21–2
    renal failure 24
    stem cell therapy 25
    urethral—detrusor dyssynergia 24
    visual impairment 22
  urological problems 12
  variability 18
Arima syndrome 152
aryl hydrocarbon receptor protein-like 1 (AIPL1) 112
asphyxiating thoracic dystrophy *see* Jeune asphyxiating thoracic dystrophy (JATD)
asthma, nasal nitric oxide 224–5
atherosclerosis, Alstrom syndrome 23
atrial isomerism 227
autosomal dominant polycystic kidney disease (ADPKD) 174–93
  additional resources 189
  characteristics *198–9*
  ciliary defect 184
    physiological effect 184–5
  clinical features 175–8, *175*, 182–3
    cystic/non-cystic 175–8
  clinical management 185–9
    caffeinated drinks 188
    fluid intake 188
    therapies 186
  complications 186–9
    end-stage renal disease 188
    hypertension 187
    loin pain 187–8
    pregnancy 188
  diagnosis 179–82
    ultrasound criteria for exclusion *181*
  differential diagnosis *180*
  epidemiology 174–5
  eponym history 174
  genetic counselling 188–9
  genetics 183–4
  phenotype range 178–9
  renal and vascular expression *183*
  ultrasound, CT and MRI *176*

autosomal recessive polycystic kidney disease (ARPKD) 194–217
characteristics *198–9*
ciliary defect, physiological effects 212
classification 194
clinical features 199–207
chronic renal failure 203–4
clinical spectrum 199–203
cysts 203
hepatobiliary complications 204–5
hypertension 205–6
intracerebral aneurysm 206–7
renal-hepatobiliary morbidity 205
clinical studies *201–2*
diagnosis 208–12
diagnostic algorithm *195*
differential diagnosis 194
epidemiology 195–9
eponym 194–5
genetics 208–12
infant, renal ultrasound *197*
management 199–207
microscopic and macroscopic appearance *200*
morphology 195–6
phenotypic variability, affected siblings 207–8
treatment prospects 212–13
axoneme 218

B
Baraitser–Burn syndrome 30
Bardet—Biedl syndrome (BBS) 92–107
biology 101–3
cilium ultrastructure *102*
clinical features 92–5
cognitive deficit 94
developmental delay 94
hypogonadism 93–4
obesity 93
overlap, other ciliopathies 95
post-axial polydactyly 93
renal abnormalities 94
rod—cone dystrophy 93
speech deficit 94
diagnosis 95–6, *97*
algorithm 95–6, *98*
dysmorphic features *96*
eponym history 92
genes *100*
genetic counselling 103
genotype—phenotype correlations 100–1
inheritance 92
management 96–9
prevalence 92, *97*
screening 95–6
basal cell nevus syndrome 268
BBS1 99, 100, 264
BBS4 95, 100, 114
BBS4 116
BBS10 99, 100, 264
Beemer—Langer type SRPS 30, 35
clinical, radiological and genetic findings 33
Best disease 119
Biemond syndrome 267
BiPAP machine 230
breath washout techniques 231
bronchiectasis 222

C
CaBP4 114
caecal volvulus 12
cardiac anomalies 221
cardiac defects
Alstrom syndrome 11
atrial isomerism 227
ciliary chondrodysplasias 54
heterotaxy 267
Joubert syndrome 70, 85
cardiomyopathy, Alstrom syndrome 8–9, 17, 18, 22–3
CCDC103 228
CEP290 114
mutation 111
CEP290 centrosomal and basal body protein 115
cerebellar vermis hypoplasia, COACH syndrome 266
cerebellar-ocular-renal syndrome (CORS) 70
cerebello-oculo-renal syndrome 5
cerebro-facio-thoracic syndrome 268
cerebro-oculo-nasal syndrome 268
CFTR 223, 228
cGMP-gated cation channel (CNGA3) 113
Chlamydomonas reinhardtii 226
chondrodysplasias *see* ciliary chondrodysplasias
chondroectodermal dysplasia *see* Ellis—van Creveld syndrome
chronic renal failure 203–4
chronic suppurative lung disease 222
Chudley–McCullough syndrome 265
ciliary beat frequency (CBF) test 225
ciliary cell culture 226
ciliary chondrodysplasias 30–63
ciliary defect 47–8
eye 51
kidney 49–51
skeleton, chondrogenic proliferation 49
ciliary hedgehog signalling 49, *50*
clinical, radiological and genetic findings 33–4
clinical features 32–8, *33–4*
clinical management 51–5
dysplastic teeth 54
gastroenterology 54
genetic counselling 54–5
heart 54
joints 52
liver, kidney, pancreas 53
neurodevelopment 54
nutrition 54
polydactyly 39, 52
psychosocial impact 55
retina 53
skeleton 51–3
stature and growth hormone treatment 52–3
thorax 51–2

cranial-ectodermal dysplasia 31, 38
diagnosis 41–3
clinical observation 41–2
echocardiography 43
electro-retinogram 43
fundoscopy 43
laboratory findings 43
patient history 41–2
pulmonary function evaluation 42
radiology 42
sonography 42–3
differential diagnosis *45*
Ellis—van Creveld syndrome 31–2, 38
epidemiology 31–2
eponym history 30–1
extraskeletal manifestation 40–1
brain 41
ectoderm (nails, hair and teeth) 41
eye 40
hearing impairment 41
heart 40
kidney 40
laterality defects 41
liver 40
pancreas 40
tongue 41
genetic findings *46*
genetic investigations 44–7
causative genes 45–7
electron microscopy (EM) 44
immunofluorescence in patient fibroblasts 44
translational diagnostics 44
JBTS and JSRD, clinical subtypes *75*
Jeune asphyxiating thoracic dystrophy (JATD) 30–1, 36
Mainzer—Saldino syndrome (MSS) 31, 37
oral-facial-digital syndrome (IV) 30, 35
proteins affected by genetic defects *48*
radiological, genetic and clinical findings *33–4*
radiological and clinical findings *32*
short-rib-polydactyly syndromes (SRPS) 30
I (Saldino—Noonan type) 30, 35–6
II (Majewski type) 30, 32
III (Verma—Naumoff type) 30, 36
IV (Beemer—Langer type) 30, 35
V 30, 37
skeletal manifestation 39–40, 49
chondrogenic proliferation 49
clefting 39
epiphyses 39
other craniofacial features 40
pelvis 39
polydactyly 39
thorax phenotype 38–9
tibia dysplasia 39
Weyers' acrofacial dysostosis 31, 38
ciliary proteins 5–6, 19–20, 50–1, 78, 80–1, 117, 119, 219
ciliary proteome ('ciliome') 80
ciliopathic features 262
Ciliopathy Alliance 158
ciliopathy diagnosis
algorithm *6*
classifying ciliopathies 3–4
clinical features 4–5
common associations *4*
intraflagellar transport 2
molecular 5–6
underlying genes *5*
cilium
absence of the dynein arms 219
axoneme 218
core ciliopathic features 262
diseases arising from cilia dysfunction 3
respiratory, normal ultrastructure *219*
structure and function 1–2
ultrastructure *2*
Bardet—Biedl syndrome *102*
underlying ciliary defect 47–8
see also primary ciliary dyskinesia
cirrhosis, Alstrom syndrome 24
classification of ciliopathies 3–4
clefting, ciliary chondrodysplasias 39
COACH syndrome, ocular coloboma, congenital ataxia, oligophrenia, cerebellar vermis hypo-/aplasia 70, 152, 266
cochlear implantation 253
Cogan's oculomotor apraxia 152
Cohen syndrome 266
cone–rod dystrophy (CRD) 118
see also rod—cone dystrophy
congenital hepatic fibrosis (CHF) *200*
cono-renal syndrome 31
coronary artery disease, Alstrom syndrome 11
corticomedullary cysts, nephronophthisis *151*
cough 222
cranial—ectodermal dysplasia (CED) 31, 38, 41
clinical, radiological and genetic findings 34
clinical follow-up recommendations *56*
craniocervical spine, Joubert syndrome 70
craniofacial features, ciliary chondrodysplasias 40
creatinine clearance 176
Curry–Hall syndrome 31
cystic fibrosis 222
cystic fibrosis transmembrane conductance regulator-related disease (CFTR) 223, 228
cysts
multiple renal and hepatic cysts 178
nephronophthisis *151*
renal cysts and diabetes (RCAD) 180

D

Dandy–Walker malformation 38, 73, 76, 134–5, 138, 143, 262, 264, 267
Danish National Registry on Regular Dialysis and Transplantation (NRDT) 175
deafness, Alstrom syndrome 22
Dekaban—Arima syndrome (DKA) 70
DISC1 270
DNA testing 139
DNAH11 228, 229
DNAI1 228

dual-axis electron tomography 226
duodenal atresia 70
dyskinesia, *see also* primary ciliary dyskinesia
dyslipidaemia, Alstrom syndrome 12–13

E
ectopic pregnancy, primary ciliary dyskinesia 220, 232
electron tomography 226
Ellis—van Creveld syndrome 31–2, 38, 152
  clinical, radiological and genetic findings 34
  clinical follow-up recommendations *56*
end-stage renal disease (ESRD) 174
  ADPKD 188
endocrine–cerebro-osteodystrophy 269
ependymal cilia 220
epiphyses, ciliary chondrodysplasias 39
European Directory of DNA Laboratories (EDDNAL) 167
eye defects
  ciliary chondrodysplasias 40, 51, 53
  Joubert syndrome 68
  Meckel—Gruber syndrome 136
  retinal degeneration 242–3
  retinal dystrophy *18*

F
Fallopian tube, dysfunctional cilia 220
fertility 220, 222, 232
fibrosis of major organs, Alstrom syndrome 15–16, 24
Franceschetti, oculodigital sign 110

G
gastro-oesophageal disorders, Alstrom syndrome 12, 24
gastro-oesophageal reflux disease 12, 24
Gene Tests Laboratory Directory 167
gene therapy
  Alstrom syndrome 25
  LCA 122
  Usher syndrome 255
genetic counselling 188–9
  (ADPKD) 188–9
  Bardet—Biedl syndrome 103
  ciliary chondrodysplasias 54–5
  oral-facial-digital type I syndrome 169–70
  Usher syndrome 255
Gli-similar (GLIS3) transcription factor 269
Gli3 265
glomerular hyperfiltration 176
glomerulocystic kidney disease (GCKD) 194
glue ear 222
Golgi, Cohen syndrome 266
Gorlin syndrome 268
GPSM2 265
Grieg cephalopolysyndactyly syndrome (GCPS) 265
growth hormone treatment, ciliary chondrodysplasias 52–3
GUCY2D 112

H
Hallgren syndrome *see* Usher syndrome
HALT (Halt Progression of Polycystic Kidney Disease) studies 177
harmonin 249
hearing impairment
  ciliary chondrodysplasias 41
  and glue ear 222
  see also Usher syndrome
hedgehog signalling 49–51, *50*, 81, 102–3, 142, 168, 212, 265, 268
hepatic steatosis, Alstrom syndrome 24
hepatobiliary complications, ARPKD 204–5
Hirschsprung's disease 70, 95
homeobox transcription factors 113
hydrocephalus 220
  Joubert syndrome 85
  MPPH 269
hydrolethalus syndrome 5, 73, 74, 263
hypertension
  Alström syndrome (ALTITUDE trial) 23
  ARPKD 187, 205–6
hypogonadotropic hypogonadism 269
hypothyroidism, Joubert syndrome 70

I
IMPDH1 117
infants
  respiratory distress at term 220, 221
  reversible cardiomyopathy 8–9, 17, 18, 22–3
infertility 220, 222, 232
insulin resistance, Alstrom syndrome 12–13
'interactome', ciliary defect 249
intracranial aneurysm (ICA)
  ADPKD 178
  ARPKD 206–7
intraflagellar transport (IFT) particles 44
ion channels (*KCNJ13, CaBP4*) 114
IQCB1 114
IQCB1/NPHP5 protein 116

J
Jeune asphyxiating thoracic dystrophy (JATD) 30–1, 36, 152
  ciliary defects 47–8, 51
  clefting 39
  clinical, radiological and genetic findings 33
  clinical follow-up recommendations *56*
  eye defects 40, 51
  genetic counselling 54–5
Johanson–Blizzard syndrome 264
Joubert syndrome 64–91, 152
  ciliary defect 80
    physiological effect 81
  clinical features 66–70
    behaviour 66–7
    breathing 67–8
    congenital hypothyroidism 70
    craniocervical spine involvement 70
    duodenal atresia 70
    dysmorphology 69

endocrine abnormalities 70
eyes 68
heart defects 70
Hirschsprung's disease 70
intellectual disability 66–7
kidneys 68–9
liver 69
neurology 66
polydactyly 69–70
vocal cord paralysis 70
clinical management 81–6
congenital cardiac defects 70, 85
hydrocephalus 85
liver 85
neonatal period onwards 84–5
older children 85
prenatal 86
diagnosis 74–6
epidemiology 65–6
eponym history 64–5
features 76–7
genetics 77–80
JBTS and JSRD
clinical checklist *82–3*
clinical subtypes *75*
phenotype range 70–4
cerebellar vermis hypo-/aplasia, oligophrenia, congenital ataxia, ocular coloboma, and hepatic fibrosis (COACH) syndrome 70
cerebellar-ocular-renal syndrome (CORS) 70
Dekaban—Arima syndrome (DKA) 70
oral-facial-digital (OFD) syndromes 70
Senior—Loken syndrome (SLS) 70
related disorders (JSRD) 266
brain MRI sections *65*
clinical features *72*
genes involved *73–4*

K

Kabuki syndrome 264
Kallman syndrome (hypogonadotropic hypogonadism) 269
Kartagener syndrome *see* primary ciliary dyskinesia
kidney defects *see* renal defects
KIF7 263–4
kyphoscoliosis, Alstrom syndrome 16, 19

L

laterality defects
ciliary chondrodysplasias 41
PCD 220
Laurence–Moon syndrome *see* Bardet—Biedl syndrome
LCA5 116
Leber congenital amaurosis (LCA) 108–31
associated genes and proteins 115–17
clinical features 108–9
clinical management 122–3
gene therapy 122–3
genetic manipulation 123
pharmacological therapies 123
epidemiology 108
eponym history 108
genetics 111–22
common mutations 119
cone—rod dystrophy 118
genetic heterogeneity 119
genotype—phenotype correlations 119–22
ion channels, transport across photoreceptor 114
photoreceptor development and structure (*CRX, OTX2, CRB1*) 113–14
phototransduction (*AIPL1, GUCY2D, CNGA3*) 112–13
retinoid cycle (*RDH12, LRAT, RPE65*) 113
non-syndromic retinal ciliopathies 118
cone dystrophy 118
genes and proteins *120–2*
macular dystrophy 118–19
molecular defects 112–13
retinitis pigmentosa 118
phenotype range 109–11
phenotypic and genetic overlap among non-syndromic retinal ciliopathies *110*
lebercilin 114–16
Levin syndrome 31
liver
Alström syndrome 12
ciliary chondrodysplasias 40
ductal plate malformations 183–4, 186, 196, 198–9, 200
Joubert syndrome 69, 85
Meckel—Gruber syndrome (MKS) 135
polycystic liver disease (PLD) 178, 180
liver, kidney, pancreas, ciliary chondrodysplasias 53
LRAT protein 113

M

McKusick–Kauffman syndrome 95
macula, optical coherence tomography scan *254*
macular dystrophy 118–19
Mainzer—Saldino syndrome (MSS) 31, 37, 263
clinical, radiological and genetic findings 34
clinical follow-up recommendations *56*
Majewski type SRPS 30, 32
clinical, radiological and genetic findings 33
Meckel—Gruber syndrome (MKS) 132–49, 152
clinical features 133–6
central nervous system (CNS) defects 134–5
craniofacial abnormalities 135
eye defects 136
genitourinary tract defects 136
kidney defects 133–4
liver defects 135
skeletal defects 135–6
visceral organ defects 136
diagnosis 137–40
ciliary defect 141–2
physiological effect 142
clinical management 142–4
differential diagnoses 139–40
DNA testing 139
embryofetoscopy 139

Meckel—Gruber syndrome (MKS) (*continued*)
features 140
genetics 140–1
magnetic resonance imaging (MRI) 138
testing for alphafetoprotein (AFP) 139
ultrasonography 137–8
epidemiology 132–3
eponym history 132
external and histopathological feature *134*
phenotype range 136–7
medullary cystic disease 2, UMOD nephropathy 180
medullary cystic disease of the kidney, *see also* nephronophthisis
megalencephaly syndromes 269
MERTK 114
mutations 117
mesoectodermal dysplasia 31
microcephalic osteodysplastic primordial dwarfism type I (MOPD I) 264
microphthalmia 136
midline embryological abnormalities 267, 268
mitochondrial DNA mutations 247
mitral valve prolapse, ADPKD 178
MLL2 264
Mohr syndrome *see* oral-facial-digital type II syndrome
Mohr—Majewski syndrome (OFD IV) 30, 167
clinical, radiological and genetic findings 33
'molar tooth sign' (MTS) 64, 70–4, 76, 82, 86
molecular testing, ADPKD diagnosis 181
monoclonal antibody immunofluorescence staining 226
MORM syndrome (mental retardation) 5, 73, 74
mosaicism 269
mucociliary clearance tests 225
MYO7A 114
MYO7A 116

N
Nager syndrome 264
nasal nitric oxide (nNO) 224
neonatal diabetes and congenital hypothyroidism (NDH) 269
neonatal respiratory distress 220, 221
nephrocystins 150
nephronophthisis 150–61
associated extra-renal manifestations, mutated genes *156–7*
associated syndromes *152–3*
ciliary defect 157
clinical features 150–1
clinical management 157–8
corticomedullary cysts, renal ultrasound scan *151*
diagnosis 153–5
diagnostic algorithm *154*
epidemiology 150
eponym history 150
genetics 154–7
and other disorders, diagnostic algorithm *154*
phenotype range 151–3
neural leucine-zipper (NRL) 113
Neu–Laxova syndrome 268
nicotinamide mononucleotide adenyltransferase NMNAT1 117
NMNAT1 117
NMNAT1 protein 117
NMNAT6 155
NPHP1 155
nyctalopia 242
nystagmus 8, 17

O
Oak Ridge mouse model, polycystic liver disease (PLD) 3
obesity
Alström syndrome 10
Bardet—Biedl syndrome 93
oculo-auricular-fronto-nasal syndromes 268
oculodigital sign of Franceschetti 110
OFD1 166
Opitz GBBB syndrome 267
oral-facial-digital (OFD) syndromes 70, 268
clinical follow-up recommendations *56*
oral-facial-digital type I syndrome 162–73
ciliary defect 168
physiological effect 168–9
clinical features 163–6
central nervous system (CNS) malformations 164
craniofacial abnormalities 164
oral abnormalities 163
renal abnormalities 164–6
skeletal abnormalities 164
clinical management 169–70
genetic counselling 169–70
surveilance 169
treatment 169
clinical spectrum *165*
diagnosis 166–7
differential diagnosis 180
epidemiology 162
eponym history 162
examples *163*
features 167
genetics 167–8
phenotype range 166
oral-facial-digital type II syndrome 162, 268
oral-facial-digital type IV syndrome 30, 35, 268

P
Pallister–Hall syndrome 265
pancreas, ciliary chondrodysplasias 40, 53
pelvis, ciliary chondrodysplasias 39
photophobia 8, 17
photoreceptor
development and structure (LCA) 113–14
retinal ciliopathy proteins *120*
phototransduction 112–13
PIK3-AKT growth-regulatory pathway 270
PKD1 177
PKD1 and *PKD2* mutations *182*, 183
polycstin-1 183, 185

polycystic liver disease (PLD) 178, 180
Oak Ridge mouse model 3
polycystin complex 185
polydactyly
Bardet—Biedl syndrome 93
ciliary chondrodysplasias 39, 52
see also short-rib—polydactyly syndromes (SRPS)
Potter syndrome type I *see* autosomal dominant polycystic kidney disease (ADPKD)
Potter syndrome types II-IV 194
premyelocytic leukaemia gene product (PML) bodies 117
primary ciliary dyskinesia (PCD) 218–37
ciliary biology 218–20, 228–9
clinical features 220–2
clinical management 229–32
antibiotic therapy 230
monitoring respiratory function 231
physiotherapy 230
diagnosis/tests 223–7
ectopic pregnancy risk 220, 232
epidemiology 218–20
eponym history 218
features 227
fertility 220, 222, 232
genetics 228
phenotype range 222–3
prognosis 232–3
screening 223–7
for organ situs 232
tests 224–5
Proteus syndrome 269

R
radioisotope mucociliary clearance tests 225
RD3 117
renal cAMP 185
renal cysts 196
and diabetes (RCAD) 180
renal defects
Bardet—Biedl syndrome 94
ciliary chondrodysplasias 49–51
Joubert syndrome 68–9
Meckel—Gruber syndrome (MGS) 133–4
renal failure
Alstrom syndrome 24
end-stage renal disease (ESRD) 174, 188
renal-hepatobiliary morbidity, ARPKD 205
renal–hepatic–pancreatic dysplasia 267
renin–angiotensin–aldosterone system 187
respiratory cilia, ultrastructure *219*
respiratory distress, neonatal 220, 221
retinal ciliopathy genes *121*
retinal ciliopathy proteins *120*
retinal degenerative disease 220
retinal dystrophy 8
gene augmentation 123
testing *18*
retinal pigment epithelium (RPE) 113
retinitis pigmentosa 118, 242–3
disordered intraflagellar transport systems 219
see also Usher syndrome
retinoid cycle (*RDH12, LRAT, RPE65*) 113
rhinitis, early onset 222
RHYNS syndrome 153
rod—cone dystrophy 93, 118, 242
Bardet—Biedl syndrome 93
see also retinitis pigmentosa
RP GTPase regulator interacting protein 1 (RPGRIP1) 5, 74, 75, 78–80, 109, 115, 119–23, 140, 141, 156, 266
RP GTPase regulator (RPGR) 115

S
saccharine test 225
Saldino—Noonan type SRPS 30, 35–6
clinical, radiological and genetic findings 33
scaffold proteins 249
schizophrenia 270
Senior–Løken syndrome (SLS) 5, 70, 116, 153
Sensenbrenner syndrome 5, 31, 38, 153, 263
see also see cranial—ectodermal dysplasia (CED)
sensory impairment, Alstrom syndrome 14–15
short-rib—polydactyly syndromes (SRPS) 30
clinical follow-up recommendations *56*
I (Saldino—Noonan type) 30, 35–6
II (Majewski type) 5, 32, 35
III (Verma—Naumoff type) 5, 30, 36
IV (Beemer—Langer type) 30, 35
V (OMIM 614091) 30, 37
Simpson–Golabi–Behmel syndrome type 2166
sinusitis, recurrent 222
situs anomalies 220
skeletal defects
ciliary chondrodysplasias 32, 39–40, 49, 51–3
Meckel—Gruber syndrome (MKS) 135, 139
nephronophthisis 152
smooth muscle disorders 12, 16
Alstrom syndrome 12, 16
SPATA7 114
SPATA7 117
SPATA7 protein 117
sperm motility 220
spirometry 231
Stargardt disease 119
stem cell therapy, Alstrom syndrome 25
syndromes (not yet proven as ciliopathies) 262–74
ciliary functions of causative genes 265–6
core ciliopathic features 262
genetic confirmation of ciliopathy/non-ciliopathy 263–4
heterogeneity, case for further splitting 267–8
historical syndromes, case for further lumping 267
known/unknown genetic orgin, link to cilia uncertain 268–70
non-ciliary functions of causative genes 266
nosological reorganisation 266–7
suspected, now resolved 263–6
suspected, still unresolved 267–70

T
teeth, ciliary chondrodysplasias 54
tetratricopeptide repeat (TPR) domains 112
thorax phenotype, JATD 38–9
tibia dysplasia, ciliary chondrodysplasias 39
TMEM67 264
TMEM216 167
tongue, ciliary chondrodysplasias 41
TRPP1 and TRPP2 (proteins) 183
Tubby-like protein (TULP) family 114
tuberous sclerosis complex (TSC) 180
TULP1 114
TXND3 228

U
UMOD nephropathy (medullary cystic disease 2) 180
urethral—detrusor dyssynergia, Alstrom syndrome 24
urological problems, Alstrom syndrome 12
Usher syndrome 238–61
  audiograms *240*
    Usher type 2 *241*
  ciliary defect 249–53
    'interactome' 249
  clinical features 239–45, 246–7
    audiovestibular features 239–42
    mental and behavioural disorders 246–7
    retinitis pigmentosa 242–5
    type 1 239–41
    type 2 241, 247–8
    type 3 241–2, 247–8
  clinical management 253–6
    audiology 253
    gene therapy 255
    genetic counselling 255
    vision 254–5
  diagnosis 247–8
  epidemiology 238–9
  eponym history 238
  features 246–7
  genes and encoded proteins *249*
  genetics 248–9
  macula, optical coherence tomography scan *254*
  mechanotransduction (MET) 249
  phenotype range 245–6
  photoreceptor cells, protein interactions *250–1*
  retinal photographs *244*
  visual fields *243*

V
Varadi–Papp syndrome 268
ventriculomegaly 220
Verma—Naumofftype SRPS 5, 30, 36
  clinical, radiological and genetic findings 33
visual impairment, Alstrom syndrome 22
vocal cord paralysis, Joubert syndrome 70
von Hippel–Lindau disease (VHL) 180, 194
VPS13B 266

W
Weyers' acrofacial dysostosis 31, 38
  clinical, radiological and genetic findings 34
whirlin 249–51